# Sampling

# Sampling

## Second Edition

STEVEN K. THOMPSON

Pennsylvania State University

A Wiley-Interscience Publication

JOHN WILEY & SONS, INC.

Copyright © 2002 by John Wiley & Sons, Inc., New York. All rights reserved.

Published simultaneously in Canada.

*Library of Congress Cataloging-in-Publication Data:*

Thompson, Steven K., 1945–
    Sampling / Steven K. Thompson.—2nd ed.
        p.   cm.—(Wiley series in probability and statistics)
    "A Wiley-Interscience publication."
    Includes bibliographical references and index.
    ISBN 0-471-29116-1 (acid-free paper)
    1. Sampling (Statistics)   I. Title. II. Series.

  QA276.6.T58 2002
  519.5′2—dc21                                2001046957

Printed in the United States of America.

10 9 8 7 6 5

# Contents

# Preface to the Second Edition

The Second Edition retains the general organization of the first, but incorporates new material interspersed throughout the text. For example, model-based ideas and alternatives are included from the earliest chapters, including those on simple random sampling and stratified sampling, rather than suddenly appearing along with ratio and regression estimation methods as has been traditional. Estimation methods deriving from a combination of design and model considerations receive added attention in this edition. Some useful ideas from the ever-developing theory of sampling are briefly described in the chapters on making the most of survey data.

Among the added sections is an expanded description of methods for adjusting for nonsampling errors. A wider discussion of link-tracing designs for sampling hidden human populations—or the Internet—has been added to the chapter on network sampling. New developments in the rapidly expanding field of adaptive sampling are briefly summarized.

Additional numerical examples, as well as exercises, have been added. A number of additional derivations of results have been tucked into the latter parts of chapters.

A brief history of sampling has been added to the introduction.

I would like to express my thanks and appreciation to the many people who have so generously shared with me their views on sampling theory and methods in discussions, collaborations, and visits to field sites. They include my colleagues at The Pennsylvania State University and those in the wider research community of sampling and statistics, as well as researchers in other fields such as ecology, biology, environmental science, computer science, sociology, anthropology, ethnography, and the health sciences. I would like to thank my editor Steve Quigley and editorial program coordinator Heather Haselkorn at John Wiley & Sons for their encouragement and assistance with this project. Research support for my work has been provided by grants from the National Science Foundation (DMS-9626102) and the National Institutes of Health (R01 DA09872).

*University Park, Pennsylvania*                                    STEVEN K. THOMPSON

# Preface to the First Edition

This book covers the basic and standard sampling design and estimation methods and, in addition, gives special attention to methods for populations that are inherently difficult to sample, elusive, rare, clustered, or hard to detect. It is intended as a reference for scientific researchers and others who use sampling and as a textbook for a graduate or upper-level undergraduate course in sampling.

The twenty-six chapters of the book are organized into six parts. Part I covers basic sampling from simple random sampling to unequal probability sampling. Part II treats the use of auxiliary data with ratio and regression estimation and looks at the ideas of sufficient data and of model and design in practical sampling. Part III covers major useful designs including stratified, cluster, systematic, multistage, double, and network sampling. Part IV examines detectability methods for elusive populations: Basic problems in detectability, visibility, and catchability are discussed and specific methods of line transects, variable circular plots, capture-recapture, and line intercept sampling are covered. Part V concerns spatial sampling, with the prediction or "kriging" methods of geostatistics, considerations of efficient spatial designs, and comparisons of different observational methods including plot shapes and detection aspects. Part VI introduces adaptive sampling designs, in which the sampling procedure depends on what is observed during the survey; for example, sampling effort may be increased in the vicinity of high observed abundance. The adaptive cluster sampling designs described can be remarkably effective for sampling rare, clustered populations, which by conventional methods are notoriously difficult to sample.

Researchers faced with such problems as estimating the abundance of an animal population or an elusive human population, predicting the amount of mineral or fossil-fuel resource at a new site, or estimating the prevalence of a rare disease must be aware that the most effective methods go beyond the material traditionally found in sampling books. At the same time, such researchers may not be aware of the potential usefulness of some of the relatively recent developments in sampling theory and methods—such as network sampling, adaptive sampling designs, and generalized ratio and regression estimation with unequal probability designs. For these reasons, the selection of topics covered in this book is wider than has been traditional for sampling texts.

Some important sampling methodologies have developed largely in particular fields—such as ecology, geology, or health sciences—seemingly in isolation from the mainstream of statistical sampling theory. In the chapters on such methods, I have endeavored to bring out the connections with and the advantages to be gained from basic sampling design, estimation, and prediction results. Thus, for instance, in the chapters on detectability methods associated in particular with ecological sampling, sampling design is emphasized. In the chapter on the prediction or kriging methods associated with geostatistics, the connection to regression estimation results is noted. In the chapter on network sampling, originally associated with epidemiological surveys, the notation has been simplified and connections to basic unequal probability sampling estimators are observed.

Although the range of topics in this book is for the above-noted reasons considerably wider than has been traditional for sampling texts, it has been necessary, in order to keep the book of the desired size, to be selective in what to include. To the reader for whom an additional topic would have been particularly helpful, I can only offer the recompense of the references cited throughout the text to give access to the wider literature in sampling.

My immediate purposes in writing this book were to provide a text for graduate and upper-level undergraduate courses in sampling at the University of Alaska Fairbanks and at the University of Auckland and to provide a manual of useful sampling and estimation methods for researchers with whom I had worked on various projects in a variety of scientific fields. No available manual or text covered the range of topics of interest to these people.

In my experience the backgrounds of the researchers and students interested in sampling topics have been extremely diverse: While some are in statistics or mathematics, many others are in the natural and social sciences and other fields. In writing this book I have assumed the same diversity of backgrounds; the only common factor I feel I can take for granted is some previous course in statistics. The chapters are for the most part organized so that the basic methods and worked examples come first, with generalizations and key derivations following for those interested.

A basic one-semester course in sampling can consist of Chapters 1 through 8 and 11 through 13 or 14, with one or more topics from the remainder of the book added, depending on time and interest. For a graduate class in which many of the students are interested in the special topics of the last three parts of the book, the instructor may wish to cover the basic ideas and methods of the first three parts quite quickly, drawing on them for background later, and spend most of the time on the second half of the book.

I would like to give my thanks to the many people who have influenced and enriched the contents of this book through conversations, joint work, and other interactions on sampling and statistics. In particular, I would like to express appreciation to Fred Ramsey, P. X. Quang, Dana Thomas, and Lyle Calvin. Also, I am grateful to Lyman McDonald, David Siegmund, Richard Cormack, Stephen Buckland, Bryan Manly, Scott Overton, and Tore Schweder for enlightening conversations on statistical sampling methods. I would like to thank my colleagues at Auckland—George Seber, Alastair Scott, Chris Wild, Chris Triggs, Alan Lee, Peter

Danaher, and Ross Ihaka—for the benefits of our collaborations, discussions, and daily interactions through which my awareness of relevant and interesting issues in sampling has been increased. I thank my sabbatical hosts at the Institute of Mathematical Statistics at the University of Copenhagen, where some of the sampling designs of this book were first seen as sketches on napkins in the lunch room: Søren Johansen, Tue Tjur, Hans Brøns, Martin Jacobsen, Inge Henningsen, Søren Tolver Jensen, and Steen Andersson. Among the many friends and associates around Alaska who have shared their experiences and ideas on sampling to the benefit of this book are Pat Holmes, Peter Jackson, Jerry McCrary, Jack Hodges, Hal Geiger, Dan Reed, Earl Becker, Dave Bernard, Sam Harbo, Linda Brannian, Allen Bingham, Alan Johnson, Terry Quinn, Bob Fagen, Don Marx, and Daniel Hawkins. Questions and comments leading to rethinking and rewriting of sampling topics have been contributed by many students, to each of whom I offer my thanks and among whom I would particularly like to mention Cheang Wai Kwong, Steve Fleischman, Ed Berg, and Heather McIntyre.

I would like to give a special thanks to my editor, Kate Roach, at John Wiley & Sons for her encouragement and enthusiasm. Research support provided by two grants from the National Science Foundation (DMS-8705812, supported by the Probability and Statistics Program and DMS-9016708, jointly supported by the Probability and Statistics Program and the Environmental Biology Division) resulted in a better book than would have otherwise been possible. I wish to thank Mary for, among many other things, her supportive sense of humor; when on a trip through Norway I could not find a certain guide book after ransacking the luggage jumble from one end of our vehicle to the other, she reminded me to "use adaptive sampling" and, starting with the location of another book randomly discovered amidst the chaos, soon produced the wanted volume. Finally, I thank Jonathan, Lynn, Daniel, and Christopher for an environment of enthusiasm and innovativeness providing inspiration all along the way.

*Auckland, New Zealand*                                              STEVEN K. THOMPSON

Sampling

CHAPTER 1

# Introduction

*Sampling* consists of selecting some part of a population to observe so that one may estimate something about the whole population. Thus, to estimate the amount of lichen available as food for caribou in Alaska, a biologist collects lichen from selected small plots within the study area. Based on the dry weight of these specimens, the available biomass for the whole region is estimated. Similarly, to estimate the amount of recoverable oil in a region, a few (highly expensive) sample holes are drilled. The situation is similar in a national opinion survey, in which only a sample of the people in the population is contacted, and the opinions in the sample are used to estimate the proportions with the various opinions in the whole population. To estimate the prevalence of a rare disease, the sample might consist of a number of medical institutions, each of which has records of patients treated. To estimate the abundance of a rare and endangered bird species, the abundance of birds in the population is estimated based on the pattern of detections from a sample of sites in the study region. In a study of risk behaviors associated with the transmission of the human immunodeficiency virus (HIV), a sample of injecting drug users is obtained by following social links from one member of the population to another.

Some obvious questions for such studies are how best to obtain the sample and make the observations and, once the sample data are in hand, how best to use them to estimate the characteristic of the whole population. Obtaining the observations involves questions of sample size, how to select the sample, what observational methods to use, and what measurements to record. Getting good estimates with observations means picking out the relevant aspects of the data, deciding whether to use auxiliary information in estimation, and choosing the form of the estimator.

Sampling is usually distinguished from the closely related field of *experimental design*, in that in experiments one deliberately perturbs some part of a population in order to see what the effect of that action is. In sampling, more often one likes to find out what the population is like without perturbing or disturbing it. Thus, one hopes that the wording of a questionnaire will not influence the respondents' opinions or that observing animals in a population will not significantly affect the distribution or behavior of the population.

1

Sampling is also usually distinguished from *observational studies*, in which one has little or no control over how the observations on the population were obtained. In sampling one has the opportunity to deliberately select the sample, thus avoiding many of the factors that make data observed by happenstance, convenience, or other uncontrolled means "unrepresentative."

More broadly, the field of sampling concerns every aspect of how data are selected, out of all the possibilities that might have been observed, whether the selection process has been under the control of investigators or has been determined by nature or happenstance, and how to use such data to make inferences about the larger population of interest. Surveys in which there is some control over the procedure by which the sample is selected turn out to have considerable advantages for purposes of inference about the population from which the sample comes.

## 1.1 BASIC IDEAS OF SAMPLING AND ESTIMATION

In the basic sampling setup, the population consists of a known, finite number $N$ of units—such as people or plots of ground. With each unit is associated a value of a variable of interest, sometimes referred to as the *y-value* of that unit. The $y$-value of each unit in the population is viewed as a fixed, if unknown quantity—not a random variable. The units in the population are identifiable and may be labeled with numbers $1, 2, \ldots, N$.

Only a sample of the units in the population are selected and observed. The data collected consist of the $y$-value for each unit in the sample, together with the unit's label. Thus, for each hole drilled in the oil reserve, the data not only record how much oil was found but also identify, through the label, the location of the hole. In addition to the variable of interest, any number of auxiliary variables, such as depth and substrate types, may be recorded. In a lichen survey, auxiliary variables recorded could include elevation, presence of other vegetation, or even "eyeball" estimates of the lichen biomass. In an opinion poll, auxiliary variables such as gender, age, or income class may be recorded along with the opinions.

The procedure by which the sample of units is selected from the population is called the *sampling design*. With most well-known sampling designs, the design is determined by assigning to each possible sample $s$ the probability $P(s)$ of selecting that sample. For example, in a simple random sampling design with sample size $n$, a possible sample $s$ consists of a set of $n$ distinct units from the population, and the probability $P(s)$ is the same for every possible sample $s$. In actual practice, the design may equivalently be described as a step-by-step procedure for selecting units rather than the resulting probabilities for selecting whole samples. In the case of simple random sampling, a step-by-step procedure consists of selecting a unit label at random from $\{1, 2, \ldots, N\}$, selecting the next unit label at random from the remaining numbers between 1 and $N$, and so on until $n$ distinct sample units are selected.

The entire sequence $y_1, y_2, \ldots, y_N$ of $y$-values in the population is considered a fixed characteristic or parameter of the population in the basic sampling view. The usual inference problem in sampling is to estimate some summary characteristic of the population, such as the mean or the total of the $y$-values, after observing only the sample. Additionally, in most sampling and estimation situations, one would like to be able to assess the accuracy or confidence associated with estimates; this assessment is most often expressed with a confidence interval.

In the basic sampling view, if the sample size were expanded until all $N$ units of the population were included in the sample, the population characteristic of interest would be known exactly. The uncertainty in estimates obtained by sampling thus stems from the fact that only part of the population is observed. While the population characteristic remains fixed, the estimate of it depends on which sample is selected. If for every possible sample the estimate is quite close to the true value of the population characteristic, there is little uncertainty associated with the sampling strategy; such a strategy is considered desirable. If, on the other hand, the value of the estimate varies greatly from one possible sample to another, uncertainty is associated with the method. A trick performed with many of the most useful sampling designs—more clever than it may appear at first glance—is that this variability from sample to sample is estimated using only the single sample selected.

With careful attention to the sampling design and using a suitable estimation method, one can obtain estimates that are unbiased for population quantities, such as the population mean or total, without relying on any assumptions about the population itself. The estimate is unbiased in that its expected value over all possible samples that might be selected with the design equals the actual population value. Thus, through the design and estimation procedure, an unbiased estimate of lichen biomass is obtained whether lichens are evenly distributed throughout the study area or are clumped into a few patches. Additionally, the random or probability selection of samples removes recognized and unrecognized human sources of bias, such as conscious or unconscious tendencies to select units with larger (or smaller) than average values of the variable of interest. Such a procedure is especially desirable when survey results are relied on by persons with conflicting sets of interests—a fish population survey that will be used by fishery managers, commercial fishermen, and environmentalists, for instance. In such cases, it is unlikely that all parties concerned could agree on the purposive selection of a "representative" sample.

A probability design such as simple random sampling thus can provide unbiased estimates of the population mean or total and also an unbiased estimate of variability, which is used to assess the reliability of the survey result. Unbiased estimates and estimates of variance can also be obtained from unequal probability designs, provided that the probability of inclusion in the sample is known for each unit and for pairs of units.

Along with the goal of unbiased or nearly unbiased estimates from the survey come goals of precise or low-variance estimates and procedures that are convenient or cost-effective to carry out. The desire to satisfy as many of these goals as possible under a variety of circumstances has led to the development of widely used sampling designs and estimation methods, including simple random and unequal probability sampling; the use of auxiliary information; stratified, systematic, cluster, multistage, and double sampling; and other techniques.


## 1.2 SAMPLING UNITS

With many populations of people and institutions, it is straightforward to identify the type of units to be sampled and to conceive of a list or frame of the units in the population, whatever the practical problems of obtaining the frame or observing the selected sample. The units may be people, households, hospitals, or businesses. A complete list of the people, households, medical institutions, or firms in the target population would provide an ideal frame from which the sample units could be selected. In practice, it is often difficult to obtain a list that corresponds exactly to the population of interest. A telephone directory does not list people without telephones or with unlisted numbers. The set of all possible telephone numbers, which may be sampled by random dialing, still does not include households without telephones. A list of public or private institutions may not be up-to-date.

With many other populations, it is not so clear what the units should be. In a survey of a natural resource or agricultural crop in a region, the region may be divided into a set of geographic units (*plots* or *segments*) and a sample of units may be selected using a map. However, one is free to choose alternative sizes and shapes of units, and such choices may affect the cost of the survey and the precision of estimators. Further, with a sampling procedure in which a point location is chosen at random in a study region and sample units are then centered around the selected points, the sample units can potentially overlap, and hence the number of units in the population from which the sample is selected is not finite.

For an elusive population with detectability problems, the role of units or plots may be superseded by that of detectability functions, which are associated with the methods by which the population is observed and the locations are selected for making the observations. For example, in selecting the locations of line transects in a bird survey and choosing the speed at which they are traversed, one determines the "effective areas" observed within the study area in place of traditional sampling units or plots.

In some sampling situations the variable of interest may vary continuously over a region. For example, in a survey to assess the oil reserves in a region, the variable measured may be the depth or core volume of oil at a location. The value of such a variable is not necessarily associated with any of a finite set of units in the region, but rather, may be measured or estimated either at a point or as a total over a subregion of any size or shape.

Although the foregoing sampling situations go beyond the framework of a population divided uniquely into a finite collection of units from which the sample is selected, basic sampling design considerations regarding random sampling, stratified sampling, and other designs, and estimation results on design-unbiased estimation, ratio estimation, and other methods still apply.

## 1.3 SAMPLING AND NONSAMPLING ERRORS

The basic sampling view assumes that the variable of interest is measured on every unit in the sample without error, so that errors in the estimates occur only because just part of the population is included in the sample. Such errors are referred to as *sampling errors*. But in real survey situations, nonsampling errors may arise also. Some people in a sample may be away from home when phoned or may refuse to answer a question on a questionnaire, and such nonrespondents may not be typical of the population as a whole, so that the sample tends to be unrepresentative of the population and the estimates are biased. In a fish survey, some selected sites may not be observed due to rough weather conditions; sites farthest from shore, which may not be typical of the study region as a whole, are the most likely to have such weather problems.

The problem of nonresponse is particularly pronounced in a survey with a very low response rate, in which the probability of responding is related to the characteristic to be measured—magazine readership surveys of sexual practices exemplify the problem. The effect of the nonresponse problem may be reduced through additional sampling effort to estimate the characteristics of the nonresponse stratum of the population, by judicious use of auxiliary information available on both responding and nonresponding units, or by modeling of the nonresponse situation. But perhaps the best advice is to strive to keep nonresponse rates as low as possible.

Errors in measuring or recording the variable of interest may also occur. Quality-control effort throughout every stage of a survey is needed to keep errors to a minimum. In some situations, it may be possible to model measurement errors separately from sampling issues in order to relate the observations to population characteristics.

Detectability problems are a type of nonsampling error that occurs with a wide range of elusive populations. On a bird survey, the observer is typically unable to detect every individual of the species in the vicinity of a sampling site. In a trawl survey of fish, not every fish in the path of the net is caught. Nor is every homeless person in a society counted in a census. A number of special techniques, including line transect, capture–recapture, and related methods, have been developed for estimating population quantities when detectability problems are a central issue.

## 1.4 MODELS IN SAMPLING

In the basic sampling view the population is a finite set of units, each with a fixed value of the variable of interest, and probability enters only through the design, that is, the procedure by which the sample of units is selected. But for some populations it may be realistic and of practical advantage to consider a probability model for the population itself. The model might be based on knowledge of the natural phenomena influencing the distribution of the type of population or on a pragmatic statistical model summarizing some basic characteristics of such populations.

For example, a regression model may empirically describe a relationship between a variable of interest, the yield of a horticultural crop, say, with an auxiliary variable, such as the median level of an air pollutant. The model relating the variable of interest with the auxiliary variable has implications both for how to design the survey and how to make estimates.

In spatial sampling situations, the existence of correlations between values of the variable of interest at different sites, depending on the distance between the sites, has implications for choices regarding sampling design, estimation or prediction, and observational method. A model-based approach utilizing such correlation patterns has been particularly influential in geological surveys of mineral and fossil-fuel resources. In ecological surveys, such correlation patterns have implications not only for the spatial selection of observational sites, but for the observational methods (including plot shapes) used.

Ideally, one would like to be able to use a model of the population without having all conclusions of the survey depend on the model's being exactly true. A "robust" approach to sampling uses models to suggest efficient procedures while using the design to protect against departures from the model.

## 1.5 ADAPTIVE AND NONADAPTIVE DESIGNS

Surveys of rare, clustered populations motivate a further advance beyond the basic view of a sampling design. In adaptive sampling designs, the procedure for selecting sites or units on which to make observations may depend on observed values of the variable of interest. For example, in a survey for estimating the abundance of a natural resource, additional sites may be added to the sample during the survey in the vicinity of high observed abundance. Such designs have important applications to surveys of animal, plant, mineral, and fossil-fuel resources and may also have applications to other fields such as epidemiology and quality control.

The main purpose of adaptive procedures is to achieve gains in precision or efficiency, compared to conventional designs of equivalent sample size, by taking advantage of observed characteristics of the population. Adaptive procedures include such procedures as sequential stopping rules and sequential allocation among strata—procedures that have been rather heavily studied outside the finite-population context in the field of sequential analysis. With the population units identifiable as in the sampling situation, the possibilities for adaptive procedures are even greater, since it is possible to decide during a survey not just how many units to sample next but exactly which units or group of units to sample next.

In adaptive cluster sampling, whenever an observed value of the variable of interest satisfies a given criterion—for example, high abundance of animals observed at a site—units in the neighborhood of that unit (site) are added to the sample. A number of variations on this type of design are described in the final chapters of this book. For some populations, the designs produce remarkable increases in efficiency and appear to be particularly effective for sampling rare, clustered populations.

The sampling design is given for a conventional or nonadaptive design by a probability $P(s)$ of selecting any particular sample $s$. For an adaptive design, the probability of selecting a given sample of units is $P(s \mid \mathbf{y})$, that is, the probability of selecting sample $s$ is conditional on the set $\mathbf{y}$ of values of the variable of interest in the population. Of course, in practice, the selection procedure can depend only on those values already observed.

Many natural populations tend to aggregate into fairly small portions of the study region, but the locations of these concentrations cannot be predicted prior to the survey. An effective adaptive design for such a population can result in higher selection probabilities assigned to samples that have a preponderance of units in those concentration areas. While the primary purpose of such a design may be to obtain a more precise estimate of the population total, a secondary benefit can be a dramatic increase in the yield of interesting observations—for example, more animals seen or more of a mineral obtained. Once adaptive designs are considered, the scope and potential of sampling methodology widens considerably.

## 1.6   SOME SAMPLING HISTORY

In the earliest known European nonfiction book, *The Histories* (ca. 440 B.C.), the author Herodotus describes a sampling method used by a Persian king to estimate the number of his troops during an invasion of Greece. A sample group of a fixed number of soldiers was instructed to stand as close together as possible and the area in which they had stood was enclosed by a short fence. Then the entire army was marched through, filling the enclosure group by group, and the number of groups required was tabulated. Multiplying the number of groups by the number in the sample group gave the estimated size of the whole force. No attempt was made to assess the accuracy of the estimate, and no description is given of how the initial sample group was selected. In fact, historians believe that the estimate reported, 1,700,000, was a gross overestimate based on present knowledge regarding feasible sizes of populations and armies at that time. Even so, the sampling strategy appears to be a fairly sensible use of an expansion estimator, and the recorded overestimate may have more to do with military propagandizing or to Herodotus's enthusiasm for large numbers than to sampling variability or bias.

> This place seemed to Xerxes a convenient spot for reviewing and numbering his soldiers; which things accordingly he proceeded to do....What the exact number of the troops of each nation was I cannot say with certainty—for it is not mentioned by any one—but the whole land army together was found to amount to one million seven hundred thousand men. The manner in which the numbering took place was the following. A body of ten thousand men was brought to a certain place, and the men were made to stand as close together as possible; after which a circle was drawn around them, and the men were let go: then where the circle had been, a fence was built about the height of a man's middle; and the enclosure was filled continually with fresh troops, till the whole army had in this way been numbered. When the numbering was over, the troops were drawn up according to their several nations. (*The History of Herodotus, Book VII*, translated by George

Rawlingson, The Internet Classics Archive by Daniel C. Stevenson, Web Atomics, 1994–2000, http:classics.mit.edu/Herodotus/history.html)

Many of the specific sampling designs and estimation methods in wide use today were developed in the twentieth century. Early in the twentieth century there was considerable debate among survey practitioners on the merits of random sampling versus purposively trying to select the most "representative" sample possible. The basic methods and formulas of simple random sampling were worked out in the first two decades of the century. An article by Neyman (1934) compared the two methods and laid out the conceptual basis for probability sampling, in which the sample is selected at random from a known distribution. Most standard sampling designs— stratified sampling, systematic sampling, cluster sampling, multistage sampling, and double or multiphase sampling—had been introduced by the end of the 1930s. The U.S. Census introduced probability sampling methods when it took over the sample survey of unemployment in the early 1940s. Unequal probability designs were introduced in the 1940s and 1950s.

The theory and methods of sampling have continued to develop and expand throughout the second half of the twentieth and the early twenty-first centuries. Studies in the theory of sampling by Godambe and others from the early 1950s forward have helped clarify the inference issues in sampling and have opened the way for subsequent development of new methods. A number of new designs and inference methods have been introduced in response to difficult problems in studies of natural and human populations, with contributing developments coming from many fields. Differences of opinion over design-based versus model-based approaches in sampling have lead to the development of methods that combine both approaches. Recent developments in the field of missing data analysis have opened up new analysis methods and underscored the importance of how observed data are selected from the potential observations.

More detailed notes on the history of sampling are found in Bellhouse (1988b), Hansen et al. (1985), and Kruskal and Mosteller (1980). Some general references to sampling or specific aspects of sampling include Barnett (1991), Bart et al. (1990), Bolfarine and Zacks (1992), Chaudhuri and Stenger (1992), Cochran (1977), Foreman (1991), Ghosh and Meeden (1997), Govindaragulu (1999), Hansen et al. (1953), Hedayat and Sinha (1991), Kish (1965), Lohr (1999), Orton (2000), Raj (1968), Rubin (1987), Sampath (2001), Särndal et al. (1992), Schreuder et al. (1993), Sukhatme and Sukhatme (1970), M. E. Thompson (1997), Thompson and Seber (1996), Tryfos (1996), and Yates (1981).

PART I

# Basic Sampling

CHAPTER 2

# Simple Random Sampling

*Simple random sampling*, or *random sampling without replacement*, is a sampling design in which $n$ distinct units are selected from the $N$ units in the population in such a way that every possible combination of $n$ units is equally likely to be the sample selected. The sample may be obtained through $n$ selections in which at each step every unit of the population not already selected has equal chance of selection. Equivalently, one may make a sequence of independent selections from the whole population, each unit having equal probability of selection at each step, discarding repeat selections and continuing until $n$ distinct units are obtained.

A simple random sample of $n = 40$ units from a population of $N = 400$ units is depicted in Figure 2.1. Another simple random sample, just as likely as the first to be selected, is shown in Figure 2.2. Each such combination of 40 units has equal probability of being the sample selected. With simple random sampling, the probability that the $i$th unit of the population is included in the sample is $\pi_i = n/N$, so that the inclusion probability is the same for each unit. Designs other than simple random sampling may give each unit equal probability of being included in the sample, but only with simple random sampling does each possible *sample* of $n$ units have the same probability.

## 2.1 SELECTING A SIMPLE RANDOM SAMPLE

A simple random sample may be selected by writing the numbers 1 through $N$ on $N$ pieces of paper, putting the pieces of paper in a hat, stirring them thoroughly, and, without looking, selecting $n$ of the pieces of paper without replacing any. The sample consists of the set of population units whose labels correspond to the numbers selected. To reduce the labor of the selection process and to avoid such problems as pieces of paper sticking together, the selection is more commonly made using a random number table or a computer "random number" generator.

To select a simple random sample of $n$ units from the $N$ in the population using a random number table, one may read down columns of digits in the table starting from a haphazard location. As many columns of the table as $N - 1$ has digits are

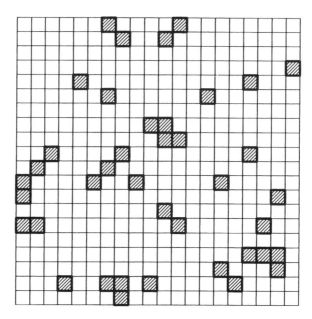

**Figure 2.1.** Simple random sample of 40 units from a population of 400 units.

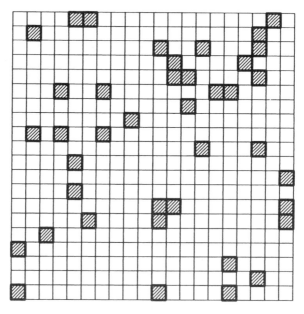

**Figure 2.2.** Another simple random sample of 40 units.

used. When using three columns, the digits "000" would be interpreted as unit 1000. When using the table, repeat selections and numbers greater than $N$ are ignored, and selection continues until $n$ distinct units are obtained.

The basic random number generator on most computers produces decimal fractions uniformly distributed between zero and 1. The first few digits after the decimal place in such numbers can be used to represent unit label numbers.

Table 2.1 lists 285 uniform random numbers, each shown to 10 digits after the decimal point, produced by a computer random number generator. Suppose that we wish to select a simple random sample of $n = 10$ units from a population of $N = 67$ units. Starting from the first entry in Table 2.1 and reading pairs of digits downward (it would also be valid to read across rows of the table, to use pairs of digits other than the first after the decimal point, or to start at a point other than the beginning of the list), the first pair is 99. Since there is no unit 99 in the population, this entry is passed over, and the first unit selected to be in the sample is unit 21. Continuing down the column, the sample selected consists of the units 21, 12, 1, 15, 29, 43, 30, 63, 2, and 8. In making the selection, note that entries 68, 76, 86, 97, and 100 (represented by the pair 00) were passed over since each is larger than $N$. Entry 43 was passed over the second time it appeared, so the sample contains 10 distinct units. Many computer systems include facilities for direct random selection without replacement of $n$ integers between 1 and $N$, eliminating the tedium of passing over repeat values or values larger than $N$. For example, in either of the statistical systems S-PLUS or R, the command "s_sample(1:27502, 12500, replace$=$F)" selects $n = 12,500$ integers at random without replacement from the set of integers from 1 to 27,502 and stores the selected numbers in "s."

## 2.2   ESTIMATING THE POPULATION MEAN

With simple random sampling, the sample mean $\bar{y}$ is an unbiased estimator of the population mean $\mu$. The population mean $\mu$ is the average of the $y$-values in the whole population:

$$\mu = \frac{1}{N}(y_1 + y_2 + \cdots + y_N) = \frac{1}{N}\sum_{i=1}^{N} y_i \tag{1}$$

The sample mean $\bar{y}$ is the average of the $y$-values in the sample:

$$\bar{y} = \frac{1}{n}(y_1 + y_2 + \cdots + y_n) = \frac{1}{n}\sum_{i=1}^{n} y_i \tag{2}$$

Also with simple random sampling, the sample variance $s^2$ is an unbiased estimator of the *finite population variance* $\sigma^2$. The finite population variance is defined as

$$\sigma^2 = \frac{1}{N-1}\sum_{i=1}^{N}(y_i - \mu)^2 \tag{3}$$

**14**

**Table 2.1. Uniform Random Numbers**

| | | | | |
|---|---|---|---|---|
| .9915338159 | .3376058340 | .1529208720 | .0008221702 | .3645994067 |
| .2110764831 | .4482254982 | .0259101614 | .1159885451 | .5011445284 |
| .1215928346 | .4434396327 | .1677099317 | .5284986496 | .9135305882 |
| .0125039294 | .2536827028 | .1724499613 | .5171836615 | .5422329903 |
| .1583184451 | .4694896638 | .9516881704 | .3874872923 | .0451180041 |
| .2974444926 | .9606751800 | .2988916636 | .7681296468 | .3288438320 |
| .4321415126 | .9025109410 | .6112304330 | .4916386008 | .8434410095 |
| .3065150678 | .5485164523 | .6078377366 | .1443793625 | .7657701969 |
| .6806892753 | .0791656822 | .7079550028 | .7407252192 | .7297828197 |
| .7614942193 | .4598654807 | .8545978069 | .4847860932 | .7846541405 |
| .8696339726 | .2160511613 | .5071278811 | .0302107912 | .3910638690 |
| .4398060441 | .0101473443 | .0496022329 | .2955447733 | .6359770298 |
| .9754472375 | .0900140777 | .9543433189 | .7030580044 | .6982350349 |
| .6345051527 | .9645981193 | .4215144813 | .8500274420 | .4303097129 |
| .0047403700 | .9751796722 | .6224800944 | .4581535459 | .3851253986 |
| .0205896683 | .2392801940 | .0118337637 | .6197799444 | .9798330665 |
| .0894387960 | .1349214613 | .0790547207 | .1108237952 | .1181035042 |
| .6207187772 | .4988264143 | .9772401452 | .2934628427 | .7792176604 |
| .8887537122 | .3153925836 | .4549961388 | .3680104315 | .8818087578 |
| .3764214814 | .6713073850 | .9082747102 | .3485270441 | .7828890681 |
| .9147837162 | .4565998316 | .2507463396 | .8603917360 | .3503700197 |
| .7551217675 | .6151723266 | .6706758142 | .9292267561 | .7541347742 |
| .4477638602 | .4369836152 | .4551322758 | .8340566158 | .6796288490 |
| .8799548149 | .5218108892 | .2309677154 | .6433401108 | .0874217674 |
| .6529608965 | .9821792245 | .8369561434 | .8693770766 | .3227941990 |
| .9485814571 | .7658874393 | .5788805485 | .8377626538 | .1910941452 |
| .9316777587 | .5495033860 | .7132855058 | .9236876369 | .1685705334 |
| .6445560455 | .1993282586 | .1627506465 | .0411975421 | .0192697253 |
| .0773160681 | .6400896907 | .4214436412 | .5431558490 | .5692960024 |
| .8540129066 | .5267632008 | .6384039521 | .4066059291 | .0482674502 |
| .6418970227 | .2250400186 | .6437576413 | .2099322975 | .3629093170 |
| .1715016663 | .1052204221 | .6630748510 | .1328498721 | .1639286429 |
| .1240955144 | .0937742889 | .4384917915 | .4143532813 | .8565336466 |
| .9962730408 | .1046832651 | .1845341027 | .7540032864 | .8298202157 |
| .1585547477 | .7293077707 | .7993465066 | .7446641326 | .5463740826 |
| .7089923620 | .1290157437 | .8575667739 | .0251938123 | .7664318085 |
| .3898053765 | .9139558077 | .3378374279 | .2337769121 | .4814206958 |
| .7222445011 | .6537817717 | .1274980158 | .0039445930 | .3522033393 |
| .1698853821 | .5726385117 | .7305127382 | .2965210974 | .2888952196 |
| .5344746709 | .2255166918 | .0169686452 | .5906063914 | .9546776414 |
| .7548384070 | .0843338221 | .8771440983 | .7653347254 | .5916480422 |
| .1039589792 | .6858401299 | .6389055848 | .9076186419 | .8857548237 |
| .5081589222 | .2550631166 | .1969931573 | .0558514856 | .6456795335 |
| .3169104457 | .5660375357 | .6318614483 | .1304887086 | .4802035689 |
| .6693667173 | .9299270511 | .8694118261 | .2035958767 | .9613003135 |
| .3214286268 | .8198484778 | .8971202970 | .0275031179 | .1577183455 |
| .1545569003 | .2482915521 | .7872648835 | .4376204610 | .2435218245 |

**Table 2.1** (*Continued*)

| | | | | |
|---|---|---|---|---|
| .5372928381 | .5366832614 | .4940558970 | .5881735682 | .5513799191 |
| .0131097753 | .9373838305 | .9739696383 | .5421801805 | .3240519464 |
| .3482980430 | .7070090175 | .6941514015 | .1654081792 | .3356401920 |
| .4537515640 | .8378376961 | .3140848875 | .5731232762 | .2575304508 |
| .3538932502 | .5364976525 | .0633419156 | .2484393269 | .7877063751 |
| .6873268485 | .3285647929 | .7112956643 | .5748419762 | .8346126676 |
| .1625820547 | .6026779413 | .9953029752 | .7957111597 | .2106933594 |
| .9141720533 | .6276242733 | .7062586546 | .0587451383 | .3998769820 |
| .4099894762 | .7787652612 | .3133662939 | .8499189615 | .0682335123 |
| .4036674798 | .4339759648 | .7664646506 | .0310811996 | .7275006175 |

The sample variance is defined as

$$s^2 = \frac{1}{n-1}\sum_{i=1}^{n}(y_i - \bar{y})^2 \tag{4}$$

The variance of the estimator $\bar{y}$ with simple random sampling is

$$\mathrm{var}(\bar{y}) = \left(\frac{N-n}{N}\right)\frac{\sigma^2}{n} \tag{5}$$

An unbiased estimator of this variance is

$$\widehat{\mathrm{var}}(\bar{y}) = \left(\frac{N-n}{N}\right)\frac{s^2}{n} \tag{6}$$

The square root of the variance of the estimator is its standard error; the estimated standard error is in general not an unbiased estimator of the actual standard error.

The quantity $(N-n)/N$, which may alternatively be written $1-(n/N)$, is termed the *finite population correction factor*. If the population is large relative to the sample size, so that the sampling fraction $n/N$ is small, the finite population correction factor will be close to 1, and the variance of the sample mean $\bar{y}$ will be approximately equal to $\sigma^2/n$. Omitting the finite population correction factor in estimating the variance of $\bar{y}$ in such a situation will tend to give a slight overestimate of the true variance. In sampling small populations, however, the finite population correction factor may have an appreciable effect in reducing the variance of the estimator, and it is important to include it in the estimate of that variance. Note that as sample size $n$ approaches the population size $N$ in simple random sampling, the finite population correction factor approaches zero, so that the variance of the estimator $\bar{y}$ approaches zero.

## 2.3   ESTIMATING THE POPULATION TOTAL

To estimate the population total $\tau$, where

$$\tau = \sum_{i=1}^{N} y_i = N\mu \tag{7}$$

the sample mean is multiplied by $N$. An unbiased estimator of the population total is

$$\hat{\tau} = N\bar{y} = \frac{N}{n}\sum_{i=1}^{n} y_i \tag{8}$$

Since the estimator $\hat{\tau}$ is $N$ times the estimator $\bar{y}$, the variance of $\hat{\tau}$ is $N^2$ times the variance of $\bar{y}$. Thus,

$$\text{var}(\hat{\tau}) = N^2 \ \text{var}(\bar{y}) = N(N-n)\frac{\sigma^2}{n} \tag{9}$$

An unbiased estimator of this variance is

$$\widehat{\text{var}}(\hat{\tau}) = N^2\widehat{\text{var}}(\bar{y}) = N(N-n)\frac{s^2}{n} \tag{10}$$

***Example 1: Estimates from Survey Data.***   In an experimental survey of caribou on the Arctic Coastal Plain of Alaska, caribou were counted from an aircraft flying over selected lines across the study region (Davis et al. 1979; Valkenburg 1990). All caribou within $1/2$ mile to either side of each line flown are recorded, so that each unit is a 1-mile-wide strip. A simple random sample of 15 north–south strips were selected from the 286-mile-wide study region, so that $n = 15$ and $N = 286$. The numbers of caribou in the 15 sample units were 1, 50, 21, 98, 2, 36, 4, 29, 7, 15, 86, 10, 21, 5, and 4.

The sample mean [using Equation (2)] is

$$\bar{y} = \frac{1 + 50 + \cdots + 4}{15} = 25.9333$$

The sample variance [using Equation (4)] is

$$s^2 = \frac{(1 - 25.93)^2 + (50 - 25.93)^2 + \cdots + (4 - 25.93)^2}{15 - 1} = 919.0667$$

The estimated variance of the sample mean [using Equation (6)] is

$$\widehat{\text{var}}\,(\bar{y}) = \left(\frac{286 - 15}{286}\right)\frac{919.07}{15} = 58.0576$$

so that the estimated standard error is $\sqrt{58.06} = 7.62$.

An estimate of the total number of caribou in the study region [using Equation (8)] is

$$\hat{\tau} = 286(25.9333) = 7417$$

The estimated variance associated with the estimate of the total [using Equation (10)] is

$$\widehat{\mathrm{var}}(\hat{\tau}) = 286^2(58.0576) = 4{,}748{,}879$$

giving an estimated standard error of $\sqrt{4{,}748{,}879} = 2179$. $\qquad\qquad\square$

## 2.4  SOME UNDERLYING IDEAS

The estimator $\bar{y}$ is a random variable, the outcome of which depends on which sample is selected. With any given sample, the value of $\bar{y}$ may be either higher or lower than the population mean $\mu$. But the expected value of $\bar{y}$, taken over all possible samples, equals $\mu$. Thus, the estimator $\bar{y}$ is said to be *design-unbiased* for the population quantity $\mu$, since the probability with respect to which the expectation is evaluated arises from the probabilities, due to the design, of selecting different samples.

Therefore, the unbiasedness of the sample mean of the population mean with simple random sampling does not depend on any assumptions about the population itself.

The variance estimates are similarly design-unbiased for their population counterparts. The actual variance of the estimator $\bar{y}$ depends on the population through the population variance $\sigma^2$. For a given population, however, a larger sample size $n$ will always produce a lower variance for the estimators $\bar{y}$ and $\hat{\tau}$.

***Example 2: All Possible Samples.***   The ideas underlying simple random sampling can be illustrated with the sampling of a very small population. The object of the sampling is to estimate the number of persons attending a lecture. To make a very quick estimate, a random sample of $n = 2$ of the $N = 4$ seating sections in the lecture theater were selected, and the number of persons in each section selected were counted. The units (seating sections) were labeled 1, 2, 3, and 4, starting from the entrance.

Using random digits generated on a computer (four numbered pieces of paper in a hat would have done as well), the sample $\{1, 3\}$ was selected. There were 10 people in unit 1 and 13 people in unit 3. The data, which include the unit labels as well as the $y$-values in the sample, are $\{(i, y_i), i \in s\} = \{(1, 10), (3, 13)\}$.

The sample mean is $\bar{y} = (10 + 13)/2 = 11.5$. The estimate of the population total $\tau$, the number of people attending the lecture, is $\hat{\tau} = N\bar{y} = 4(11.5) = 46$. The sample variance [using Equation (4)] is $s^2 = [(10 - 11.5)^2 + (13 - 11.5)^2]/(2 - 1) = 4.5$. The estimated variance of $\hat{\tau}$ [using Equation (10)] is $\widehat{\mathrm{var}}(\hat{\tau}) = [(4)(4 - 2)(4.5)]/2 = 18$.

Had another sample of two units been selected, different values would have been obtained for each of these statistics. Since the population is so small, it is possible to look at every possible sample and the estimates obtained with each. Counting the number of people in the remaining seating sections, the population $y$-values were determined as summarized in the following table:

| Unit, $i$ | 1 | 2 | 3 | 4 |
|---|---|---|---|---|
| People, $y_i$ | 10 | 17 | 13 | 20 |

The population parameters are $\tau = 60$ people attending the lecture and $\mu = 15$ people per section on average; the finite population variance is $\sigma^2 = 19.33$. With $N = 4$ and $n = 2$, there are $\binom{4}{2} = 6$ possible samples. Table 2.2 lists each of the possible samples $s$ along the $y$-values $\mathbf{y}_s$, and the estimates and the confidence interval (c.i.) obtained with each sample.

Because of the simple random sampling used, each possible sample has probability $P(s) = 1/6$ of being the one selected. An estimator such as $\hat{\tau}$ is a random variable whose value depends on the sample selected. The expected value of $\hat{\tau}$ with respect to the design is the sum, over all possible samples, of the value of the estimator for that sample times the probability of selecting that sample. Thus, the expected value of $\hat{\tau}$ is

$$E(\hat{\tau}) = 54\left(\frac{1}{6}\right) + 46\left(\frac{1}{6}\right) + 60\left(\frac{1}{6}\right) + 60\left(\frac{1}{6}\right) + 74\left(\frac{1}{6}\right) + 66\left(\frac{1}{6}\right) = 60$$

demonstrating for this population that the estimator $\hat{\tau}$ is indeed unbiased for the parameter $\hat{\tau}$ under simple random sampling. Similarly, one can show directly for the other estimators that $E(\bar{y}) = \mu$, $E(s^2) = \sigma^2$, and $E[\widehat{\text{var}}(\hat{\tau})] = \text{var}(\hat{\tau})$. On the other hand, direct computation of the expected value, over all possible samples, of the sample standard deviation $s = \sqrt{s^2}$ gives $E(s) = 4.01$, while the population standard deviation is $\sigma = \sqrt{19.33} = 4.40$, so the sample standard deviation is not unbiased for the population standard deviation under simple random sampling.

**Table 2.2. Data for Example 2**

| Sample | $\mathbf{y}_s$ | $\bar{y}$ | $\hat{\tau}$ | $s^2$ | $\widehat{\text{var}}(\hat{\tau})$ |
|---|---|---|---|---|---|
| $(1,2)$ | $(10,17)$ | 13.5 | 54 | 24.5 | 98 |
| $(1,3)$ | $(10,13)$ | 11.5 | 46 | 4.5 | 18 |
| $(1,4)$ | $(10,20)$ | 15.0 | 60 | 50.0 | 200 |
| $(2,3)$ | $(17,13)$ | 15.0 | 60 | 8.0 | 32 |
| $(2,4)$ | $(17,20)$ | 18.5 | 74 | 4.5 | 18 |
| $(3,4)$ | $(13,20)$ | 16.5 | 66 | 24.5 | 98 |

The variance of $\hat{\tau}$ is the sum, over all possible samples, of the value of $(\hat{\tau} - \tau)^2$ times the probability of that sample. Thus, direct computation of the variance of $\hat{\tau}$ (using the data in Table 2.2) gives

$$\text{var}(\hat{\tau}) = (54 - 60)^2 \left(\frac{1}{6}\right) + (46 - 60)^2 \left(\frac{1}{6}\right) + (60 - 60)^2 \left(\frac{1}{6}\right)$$

$$+ (60 - 60)^2 \left(\frac{1}{6}\right) + (74 - 60)^2 \left(\frac{1}{6}\right) + (66 - 60)^2 \left(\frac{1}{6}\right)$$

$$= 77.333 \qquad \qquad \square$$

## 2.5   RANDOM SAMPLING WITH REPLACEMENT

Imagine drawing $n$ poker chips from a bowl of $N$ numbered chips one at a time, returning each chip to the bowl before selecting the next. With such a procedure, any of the chips may be selected more than once. A sample of $n$ units selected by such a procedure from a population of $N$ units is called a *random sample with replacement*. The $n$ selections are independent, and each unit in the population has the same probability of inclusion in the sample. Simple random sampling with replacement is characterized by the property that each possible *sequence* of $n$ units—distinguishing order of selection and possibly including repeat selections—has equal probability under the design.

One practical advantage of sampling with replacement is that in some situations, it is an important convenience not to have to determine whether any unit in the data is included more than once. However, for a given sample size $n$, simple random sampling with replacement is inherently less efficient than simple random sampling without replacement.

Let $\bar{y}_n$ denote the sample mean of the $n$ observations; that is,

$$\bar{y}_n = \frac{1}{n} \sum_{i=1}^{n} y_i \qquad (11)$$

Note that if a unit is selected more than once, its $y$-value is utilized more than once in the estimator.

The variance of $\bar{y}_n$ is

$$\text{var}(\bar{y}_n) = \frac{1}{nN} \sum_{i=1}^{N} (y_i - \mu)^2 = \frac{N-1}{nN} \sigma^2 \qquad (12)$$

Thus, the variance of the sample mean with simple random sampling without replacement is lower, since it is $(N - n)/(N - 1)$ times that of the sample mean of all the observations when the sampling is with replacement.

An unbiased estimate of the variance of $\bar{y}_n$ is

$$\widehat{\mathrm{var}}(\bar{y}_n) = \frac{s^2}{n} \tag{13}$$

The estimator $\bar{y}_n$ depends on the number of times each unit was selected, so that two surveys observing exactly the same set of distinct units, but with different repeat selections, would in general yield different estimates. This situation can be avoided by using the sample mean of the distinct observations.

The number of distinct units contained in the sample, termed the *effective sample size*, is denoted $\nu$. Let $\bar{y}_\nu$ be the sample mean of the distinct observations:

$$\bar{y}_\nu = \frac{1}{\nu}\sum_{i=1}^{\nu} y_i \tag{14}$$

The estimator $\bar{y}_\nu$ is an unbiased estimator of the population mean. The variance of $\bar{y}_\nu$ can be shown to be less than that of $\bar{y}_n$. However, it is still not as small as the variance of the sample mean under simple random sampling without replacement (see Cassel et al. 1977, p. 41). Even so, in some survey situations the practical convenience of sampling with replacement could allow a larger sample size to be used, resulting in improved precision for a given amount of time or expense.

***Example 3: Random Sampling with Replacement.*** In a simple random sample with replacement with nominal sample size $n = 5$, the following $y$-values are obtained: 2, 4, 0, 4, 5. However, examination of the labels of the units in the sample reveals that one unit, the one with $y_i = 4$, was selected twice. The estimate of the population mean based on the sample mean of the five observations, not all of which are distinct [using Equation (11)], is

$$\bar{y}_n = \frac{2+4+0+4+5}{5} = 3.0$$

The estimate based only on the four distinct units in the sample [using Equation (14)] is

$$\bar{y}_\nu = \frac{2+4+0+5}{4} = 2.75 \qquad\qquad \square$$

## 2.6 DERIVATIONS FOR RANDOM SAMPLING

Since the number of combinations of $n$ distinct units from a population of size $N$ is

$$\binom{N}{n} = \frac{N!}{n!\,(N-n)!} \tag{15}$$

the design simple random sampling assigns probability $1 \Big/ \binom{N}{n}$ to each possible sample $s$ of $n$ distinct units. The probability $\pi_i$ that a given unit $i$ is included in the sample is the same for every unit in the population and is given by $\pi_i = n/N$.

It is customary in sampling to write the $y$-values in the population as $y_1, y_2, \ldots, y_N$ and the $y$-values in the sample as $y_1, y_2, \ldots, y_n$, and for most purposes no confusion results from this simple notation. A more precise notation lists the $y$-values in sample $s$ as $y_{s1}, y_{s2}, \ldots, y_{sn}$, distinguishing, for example, that the first unit in the sample is not necessarily the same unit as the first unit in the population. With the more careful notation, the sample mean for sample $s$ is written $\bar{y}_s = (1/n) \sum_{i=1}^{n} y_{si}$.

The expected value of the sample mean $\bar{y}$ in simple random sampling is defined as $E(\bar{y}) = \sum \bar{y}_s P(s)$, where the summation is over all possible samples $s$ of size $n$, and $\bar{y}_s$ denotes the value of the sample mean for the sample $s$. This expectation may be computed directly, since $P(s) = 1 \Big/ \binom{N}{n}$ for every sample. The number of samples that include a given unit $i$ is $\binom{N-1}{n-1}$. Thus,

$$E(\bar{y}) = \sum \bar{y}_s P(s) = \frac{1}{n} \sum_{i=1}^{N} y_i \binom{N-1}{n-1} \Big/ \binom{N}{n} = \frac{1}{N} \sum_{i=1}^{N} y_i \qquad (16)$$

so the sample mean is an unbiased estimator of the population mean under simple random sampling.

Alternatively, the expectation of the sample mean under simple random sampling can be derived using a device that proves useful in many more complicated designs as well. For each unit $i$ in the population, define an indicator variable $z_i$ such that $z_i = 1$ if unit $i$ is included in the sample and $z_i = 0$ otherwise. Then the sample mean can be written in the alternative form

$$\bar{y} = \frac{1}{n} \sum_{i=1}^{N} y_i z_i \qquad (17)$$

Each of the $z_i$ is a (Bernoulli) random variable, with expected value $E(z_i) = P(z_i = 1) = n/N$. Hence the expected value of the sample mean is

$$E(\bar{y}) = \frac{1}{n} \sum_{i=1}^{N} y_i E(z_i) = \frac{1}{n} \sum_{i=1}^{N} y_i \frac{n}{N} = \frac{1}{N} \sum_{i=1}^{N} y_i = \mu \qquad (18)$$

The variance of the sample mean under simple random sampling can be derived similarly by either method. Using the indicator-variable method, the variance is

$$\text{var}(\bar{y}) = \text{var}\left( \frac{1}{n} \sum_{i=1}^{N} y_i z_i \right) = \frac{1}{n^2} \left[ \sum_{i=1}^{N} y_i^2 \text{var}(z_i) + \sum_{i=1}^{N} \sum_{j \neq i}^{N} y_i y_j \, \text{cov}(z_i, z_j) \right]$$

Since $z_i$ is a Bernoulli random variable, $\text{var}(z_i) = (n/N)(1 - n/N)$.

The number of samples containing both units $i$ and $j$, when $i \neq j$, is $\binom{N-2}{n-2}$, so that the probability that both units are included is $\binom{N-2}{n-2} / \binom{N}{n} = n(n-1)/[N(N-1)]$. The product $z_i z_j$ is zero except when both $i$ and $j$ are included in the sample, so

$$E(z_i z_j) = P(z_i = 1, z_j = 1) = \frac{n(n-1)}{N(N-1)}$$

The covariance is

$$\text{cov}(z_i, z_j) = E(z_i z_j) - E(z_i)E(z_j) = \frac{n(n-1)}{N(N-1)} - \left(\frac{n}{N}\right)^2 = \frac{-n(1 - n/N)}{N(N-1)}$$

Thus, the variance of the sample mean is

$$\text{var}(\bar{y}) = \frac{1}{n^2} \left(\frac{n}{N}\right) \left(1 - \frac{n}{N}\right) \left[\sum_{i=1}^{N} y_i^2 - \frac{1}{N-1} \sum_{i=1}^{N} \sum_{i \neq j} y_i y_j \right]$$

Using the identity

$$\sum_{i=1}^{N} (y_i - \mu)^2 = \sum_{i=1}^{N} y_i^2 - \frac{\left(\sum y_i\right)^2}{N} = \frac{1}{N} \left[(N-1) \sum_{i=1}^{N} y_i^2 - \sum_{i=1}^{N} \sum_{j \neq i} y_i y_j \right]$$

the variance expression simplifies to

$$\text{var}(\bar{y}) = \frac{1}{n} \left(1 - \frac{n}{N}\right) \frac{\sum (y_i - \mu)^2}{N-1} = \left(1 - \frac{n}{N}\right) \frac{\sigma^2}{n}$$

For simple random sampling with replacement, the expected value and variance of the sample mean and the expected value of the sample variance are obtained from the usual statistical properties of the sample mean of independent and identically distributed random variables. On any draw, unit $i$ has probability $p_i = 1/N$ of being selected. The probability that unit $i$ is included (one or more times) in the sample is $\pi_i = 1 - (1 - N^{-1})^n$. The expected number of times unit $i$ is included in the sample is $n/N$.

## 2.7   MODEL-BASED APPROACH TO SAMPLING

In the fixed-population or design-based approach to sampling, the values $y_1, y_2, \ldots, y_N$ of the variable of interest in the population are considered as fixed but unknown constants. Randomness or probability enters the problem only through the deliberately imposed design by which the sample of units to observe is selected. In the design-based approach, with a design such as simple random sampling the sample mean is a random variable only because it varies from sample to sample. One sample gives a value of the sample mean that is greater than the population mean, another sample gives a value of the sample mean that is lower than the population mean.

In the stochastic-population or model-based approach to sampling, the values of the variable of interest, denoted $Y_1, Y_2, \ldots, Y_N$, are considered to be random variables. The population model is given by the joint probability distribution or density function $f(y_1, y_2, \ldots, y_N; \theta)$, which may depend on one or more unknown parameters $\theta$. The population values $y_1, y_2, \ldots, y_N$ realized represent just one outcome of many possible outcomes under the model for the population.

Suppose that the object is to estimate the population mean: for example, mean household expenditure for a given month in a geographical region. Economic theory may suggest a statistical model, such as a normal or lognormal distribution, for the amount a household might spend. The amount that the household spends that month is then one realization among the many possible under the assumed distribution.

As a very simple population model, assume that the population variables $Y_1, Y_2, \ldots, Y_N$ are independent, identically distributed (i.i.d.) random variables from a distribution having a mean $\theta$ and a variance $\gamma^2$. That is, for any unit $i$, the variable of interest $Y_i$ is a random variable with expected value $E(Y_i) = \theta$ and variance $\mathrm{var}(Y_i) = \gamma^2$, and for any two units $i$ and $j$, the variables $Y_i$ and $Y_j$ are independent.

Suppose that we have a sample $s$ of $n$ distinct units from the population and the object is to estimate the parameter $\theta$ of the distribution from which the population comes. For the given sample $s$, the sample mean

$$\bar{Y} = \frac{1}{n} \sum_{i \in s} Y_i$$

is a random variable, whether or not the sample is selected at random, because for each unit $i$ in the sample $Y_i$ is a random variable that can take on different outcomes. With the assumed model, the expected value of the sample mean is $E(\bar{Y}) = \theta$ and its variance is $\mathrm{var}(\bar{Y}) = \gamma^2/n$. Thus, $\bar{Y}$ is a model-unbiased estimator of the parameter $\theta$. An approximate $1 - \alpha$ confidence interval for the parameter $\theta$, based on the central limit theorem for the sample mean of independent, identically distributed random variables, is given by

$$\bar{Y} \pm tS/\sqrt{n}$$

where $S$ is the sample standard deviation and $t$ is the upper $\alpha/2$ point of the $t$ distribution with $n - 1$ degrees of freedom. If additionally the $Y_i$ are assumed to have a normal distribution, then the confidence level is exact, even with a small sample size.

In the study of household expenditure the focus of interest may not be on the parameter $\theta$ of the model, however, but on the actual average amount spent by households in the community that month. That is, the object is to estimate (or predict) the value of the random quantity

$$Z = \frac{1}{N} \sum_{i=1}^{N} Y_i$$

The difference between inference about the random variable $Z$ and the model parameter $\theta$ can be appreciated by considering a survey in which every household in a community is included in the sample, so that $n = N$. Then, with the expenditure $Y_i$ measured for every household, there is no uncertainty about the value of the population mean $Z = (1/N) \sum_{i=1}^{N} Y_i$. However, even with the whole population observed, there is still uncertainty about the parameter $\theta$ of the model that produced the population values, since we have observed only one realization of the $N$ values from that distribution. That is, with the entire population of households observed, there is no uncertainty about the household expenditure realized in that population (assuming no measurement error), but there is uncertainty about the exact distribution or process that produced the expenditure pattern realized. In reality, the more common situation is that the sample size is much smaller than the population size, so that there is uncertainty both about the population values realized and the parameters of their distribution.

To estimate or predict the value of the random variable $Z = (1/N) \sum_{i=1}^{N} Y_i$ from the sample observations, an intuitively reasonable choice is again the sample mean $\hat{Z} = \bar{Y} = \sum_{i \in s} Y_i / n$. Both $Z$ and $\hat{Z}$ have expected value $\theta$, since the expected value of each of the $Y_i$ is $\theta$. Because $E(\hat{Z}) = E(Z)$, with the expectations evaluated under the assumed model distribution, the predictor $\hat{Z}$ is said to be "model unbiased" for the population quantity $Z$. More precisely, a predictor $\hat{Z}$ is said to be *model unbiased* for $Z$ if for any given sample $s$, the conditional expectations are equal, that is,

$$E(\hat{Z} \mid s) = E(Z \mid s)$$

Additionally, for the type of designs we are considering, the expectation of the population quantity $Z$ does not depend on the sample $s$ selected, so that $E(Z \mid s) = E(Z)$.

Note that the design unbiasedness of the sample mean for the population mean under our assumed model does not depend on how the sample was selected, that is, does not depend on the design. Under the assumed model, the predictor is unbiased with the specific sample selected.

In estimating or predicting the value of a random variable $Z$ with a predictor $\hat{Z}$, one measure of the uncertainty is the mean square prediction error

$$E(\hat{Z} - Z)^2$$

If the predictor $\hat{Z}$ is model unbiased for Z, then $E(\hat{Z} - Z) = 0$ and the mean square prediction error is the variance of the difference,

$$E(\hat{Z} - Z)^2 = \text{var}(\hat{Z} - Z)$$

In the case of the sample mean $\hat{Z} = \bar{Y}$ as a predictor of the population mean $Z = \sum_{i=1}^{N} Y_i/N$, with the model in which $Y_1, \ldots, Y_N$ are i.i.d. from a distribution with mean $\theta$ and variance $\gamma^2$, the mean square prediction error is

$$E(\bar{Y} - Z)^2 = \left(\frac{N-n}{N}\right)\frac{\gamma^2}{n}$$

*Proof:*   Because $E(\bar{Y}) = E(Z)$,

$$E(\bar{Y} - Z)^2 = \text{var}\left(\frac{1}{n}\sum_{i \in s} Y_i - \frac{1}{N}\sum_{i=1}^{N} Y_i\right)$$

Separating the terms for units in the sample $s$ from the units in $\bar{s}$ outside the sample yields

$$\text{var}\left(\bar{Y} - \frac{1}{N}\sum_{i=1}^{N} Y_i\right) = \text{var}\left[\left(\frac{1}{n} - \frac{1}{N}\right)\sum_{i \in s} Y_i - \frac{1}{N}\sum_{i \in \bar{s}} Y_i\right]$$

Since the values in the sample are independent of the values outside the sample, the variance of the difference between the two independent terms is

$$\text{var}\left[\left(\frac{1}{n} - \frac{1}{N}\right)\sum_{i \in s} Y_i - \frac{1}{N}\sum_{i \in \bar{s}} Y_i\right] = \left(\frac{1}{n} - \frac{1}{N}\right)^2 n\gamma^2 + \frac{1}{N^2}(N-n)\gamma^2$$

$$= \left[\frac{(N-n)^2}{nN^2} + \frac{n(N-n)}{nN^2}\right]\gamma^2$$

$$= \frac{N-n}{nN}\gamma^2$$

Notice that in the model framework, the finite population variance

$$V = \frac{\sum_{i=1}^{N}(Y_i - Z)^2}{N-1}$$

is itself a random variable. The notation Z in place of $\mu$ and V in place of $\sigma^2$ is used only to emphasize the model-based viewpoint in which these population quantities are themselves random variables. With the i.i.d. model,

$$E(V) = \gamma^2$$

since by standard results in statistics, the expectation of a sample variance of i.i.d. random variables equals the variance of the distribution from which the variables come.

An unbiased estimator or predictor of the mean square prediction error is

$$\hat{E}(\hat{Z} - Z)^2 = \frac{N - n}{N} \frac{S^2}{n}$$

since $E(S^2) = \gamma^2$ since the $Y_i$ are i.i.d. with variance $\gamma^2$.

Further, an approximate $1 - \alpha$ prediction interval for $Z$ is given by

$$\bar{Y} \pm t \sqrt{\hat{E}(\hat{Z} - Z)^2}$$

where $t$ is the upper $\alpha/2$ point of the $t$ distribution with $n - 1$ degrees of freedom. If, additionally, the distribution of the $Y_i$ is assumed to be normal, the confidence level is exact.

Thus, with the assumed i.i.d. model, the estimation and assessment of uncertainty are carried out using exactly the same calculations from the sample data as those used with simple random sampling in the design-based approach. The validity of the inference in the model-based approach does not require that the sample be selected by random sampling, but does depend on the realism of the assumed model.

## EXERCISES

1. In Figure 2.3, the locations of objects (e.g., trees, mines, dwellings) in a study region are given by the centers of "+" symbols. The goal is to estimate the number of the objects in the study region.

   (a) A random sample without replacement of $n = 10$ units has been selected from the $N = 100$ units in the population. Units selected are indicated by shading in Figure 2.3. List the sample data. Use the sample to estimate the number of objects in the figure. Estimate the variance of your estimator.

   (b) Repeat part (a), selecting another sample of size 10 by simple random sampling (without replacement) and making new estimates. Indicate the positions of the units of the samples on the sketch.

   (c) Give the inclusion probability for the unit in the upper left-hand corner. How many possible samples are there? What is the probability of selecting the sample you obtained in part (a)?

2. A simple random sample of 10 households is selected from a population of 100 households. The numbers of people in the sample households are 2, 5, 1, 4, 4, 3, 2, 5, 2, 3.

   (a) Estimate the total number of people in the population. Estimate the variance of your estimator.

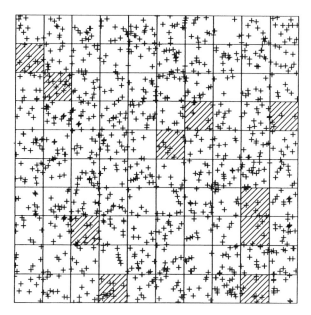

**Figure 2.3.** Simple random sample of 10 units from a population of 100 units. The variable of interest is the number of point objects within each unit. (See Exercise 1.)

   **(b)** Estimate the mean number of people per household and estimate the variance of that estimator.

**3.** Consider a small population of $N = 5$ units, labeled 1, 2, 3, 4, 5, with respective $y$-values 3, 1, 0, 1, 5. Consider a simple random sampling design with a sample size $n = 3$. For your convenience, several parts of the following may be combined into a single table.
   **(a)** Give the values of the population parameters $\mu$, $\tau$, and $\sigma^2$. List every possible sample of size $n = 3$. For each sample, what is the probability that it is the one selected?
   **(b)** For each sample, compute the sample mean $\bar{y}$ and the sample median $m$. Demonstrate that the sample mean is unbiased for the population mean and determine whether the sample median is unbiased for the population median.

**4.** Show that $E(s^2) = \sigma^2$ in simple random sampling, where the sample variance $s^2$ is defined with $n - 1$ in the denominator and the population variance $\sigma^2$ is defined with $N - 1$ in the denominator. [*Hint*: Write $y_i - \bar{y}$ as $y_i - \mu - (\bar{y} - \mu)$, verify that

$$\sum_{i=1}^{n}(y_i - \bar{y})^2 = \sum_{i=1}^{n}(y_i - \mu)^2 - n(\bar{y} - \mu)^2$$

and either take expectation over all possible samples or define an indicator variable for each unit, indicating whether it is included in the sample.]

5. The best way to gain understanding of a sampling and estimation method is to carry it out on some real population of interest to you. If you are not already involved in such a project professionally, choose a population and set out to estimate the mean or total by taking a simple random sample. Examples include estimating the number of trees on a university campus by conceptually dividing the campus into plots, estimating the number of houses in a geographic area by selecting a simple random sample of blocks, or estimating the mean number of people per vehicle during rush hour. In the process of carrying out the survey and making the estimates, think about or discuss with others the following:

   (a) What practical problems arise in establishing a frame, such as a map or list of units, from which to select the sample?

   (b) How is the sample selection actually carried out?

   (c) What special problems arise in observing the units selected?

   (d) Estimate the population mean and total.

   (e) Estimate the variance of the estimators used in part (d).

   (f) How would you improve the survey procedure if you were to do it again?

# CHAPTER 3

# Confidence Intervals

Having selected a sample and used the sample data to make an estimate of the population mean or total, it is desirable in addition to make an assessment regarding the accuracy of the estimate. This is most often done by constructing a confidence interval within which one is sufficiently sure that the true population value lies or, equivalently, placing a bound on the probable error of the estimate. A confidence interval procedure uses the data to determine an interval with the property that—viewed before the sample is selected—the interval has a given high probability of containing the true population value.

## 3.1 CONFIDENCE INTERVAL FOR THE POPULATION MEAN OR TOTAL

Let $I$ represent a confidence interval for the population mean $\mu$. Choosing some small number $\alpha$ as the allowable probability of error, the procedure should have the property that $P(\mu \in I) = 1 - \alpha$. The random quantity in this expression is the interval $I$. The endpoints of the interval $I$ vary from sample to sample, whereas the parameter $\mu$, although unknown, is fixed.

The quantity $1 - \alpha$ is termed the *confidence coefficient*, and the interval is called a $100(1 - \alpha)\%$ confidence interval. Typical (arbitrary but conventional) choices for the value of $\alpha$ are 0.01, 0.05, and 0.1. With $\alpha = 0.05$, for instance, the confidence coefficient is 0.95. Under simple random sampling, a 95% confidence interval procedure has the interpretation that for 95% of the possible samples of size $n$, the interval covers the true value of the population mean $\mu$.

Approximate $100(1 - \alpha)\%$ confidence intervals for the population mean and total can be constructed based on a normal approximation for the distribution of the sample mean under simple random sampling. An approximate $100(1 - \alpha)\%$ confidence interval for the population mean $\mu$ is

$$\bar{y} \pm t\sqrt{\left(\frac{(N-n)}{N}\right)\left(\frac{s^2}{n}\right)}$$

where $t$ is the upper $\alpha/2$ point of Student's $t$ distribution with $n-1$ degrees of freedom.

An approximate $100(1-\alpha)\%$ confidence interval for the population total $\tau$ is

$$\hat{\tau} \pm t\sqrt{N(N-n)\frac{s^2}{n}}$$

For sample sizes larger than 50, the upper $\alpha/2$ point of the standard normal distribution may be used for the value of $t$ in the confidence intervals above.

More generally, if $\hat{\theta}$ is a normally distributed, unbiased estimator for a population parameter $\theta$, then a $1-\alpha$ confidence interval for $\theta$ is given by

$$\hat{\theta} \pm z\sqrt{\text{var}(\hat{\theta})}$$

where $z$ is the upper $\alpha/2$ point of the normal distribution. In practice, the estimator may have a distribution which is approximately normal (based on the central limit theorem) even if the original $y$-values are not. Also, the variance of the estimator is typically estimated from the sample data, and the confidence interval is

$$\hat{\theta} \pm z\sqrt{\widehat{\text{var}}(\hat{\theta})}$$

In such cases, the coverage probability of the confidence interval is only approximately $1-\alpha$. For sample sizes less than 50, it is advisable to use the upper $\alpha/2$ point $t$ from the $t$-distribution with $n-1$ degrees of freedom, giving a somewhat wider interval than that obtained with the normal value $z$.

***Example 1.*** An approximate 90% confidence interval for the total number of caribou in the study area of Example 1 of Chapter 2 with sample size $n=15$, estimator $\hat{\tau} = 7417$, and variance estimate $\widehat{\text{var}}(\hat{\tau}) = 4,748,879$, is

$$7417 \pm 1.761\sqrt{4,748,879} = 7417 \pm 3838 = (3578, 11,255)$$

where the value 1.761 is the upper 0.05 point from the $t$-table with 14 degrees of freedom.

Since the number of units in the population was $N=286$, a confidence interval for the mean could be obtained by dividing each endpoint of the interval for the total by 286, giving a 90% interval of $(12.5, 39.4)$.                                    □

## 3.2   FINITE-POPULATION CENTRAL LIMIT THEOREM

When the individual observations $y_1, y_2, \ldots, y_n$ are not normally distributed, the approximate confidence levels of the usual confidence intervals depend on the approximate normal distribution of the sample mean $\bar{y}$. If $y_1, y_2, \ldots, y_n$ are a sequence of *independent* and *identically distributed* random variables with finite mean and variance, the distribution of

$$\frac{\bar{y} - \mu}{\sqrt{\operatorname{var}(\bar{y})}}$$

approaches a standard normal distribution as $n$ gets large, by the central limit theorem. The result also holds if the variance is replaced by a reasonable estimator of variance.

When a finite population is sampled using random sampling with replacement, the $n$ observations are indeed independent and identically distributed, so that the usual central limit theorem applies. With random sampling without replacement, however, the sample observations are not independent—selecting a unit with a large $y$-value on the first draw, for instance, removes that unit from the selection pool and hence reduces the probability of obtaining large $y$-value subsequent draws. A special version of the central limit theorem applies to random sampling without replacement from a finite population (Erdös and Rényi 1959; Hájek 1960, 1961; Lehmann 1975, p. 353; Madow 1948; Scott and Wu 1981; M. E. Thompson 1997).

For sampling without replacement from a finite population, it is necessary to think of a sequence of populations, with population size $N$ becoming large along with sample size $n$. For the population with a given size $N$ in the sequence, let $\mu_N$ be the population mean and $\bar{y}_N$ be the sample mean of a simple random sample selected from that population. According to the finite-population central limit theorem, the distribution of

$$\frac{\bar{y}_N - \mu_N}{\sqrt{\operatorname{var}(\bar{y}_N)}}$$

approaches the standard normal distribution as both $n$ and $N - n$ become large.

The result also holds with the estimated variance $\widehat{\operatorname{var}}(\bar{y}_N)$ of the sample mean of a simple random sample of size $n$ from a population of size $N$ substituted for $\operatorname{var}(\bar{y}_N)$. A technical condition in the theorem requires that in the progression of hypothetical populations of increasing size, the proportion of the population variance contributed by any one unit is not too large.

For more complicated survey design and estimation methods, it is often still possible to use confidence intervals of the basic form described in this chapter, with approximate coverage probability based on some form of the central limit theorem. Estimation of the variance of an estimator may be more involved in the more complex survey situations, however. A number of approaches to variance estimation are introduced as needed in this book. Further discussion of variance estimation in survey sampling can be found in J. N. K. Rao (1988), Skinner et al. (1989), and Wolter (1985).

**Table 3.1. Data for Example 2**

| Sample | $y_s$ | $\bar{y}$ | $\hat{\tau}$ | $s^2$ | $\widehat{\mathrm{var}}(\hat{\tau})$ | c.i. |
|--------|-------|-----------|--------------|-------|--------------------------------------|------|
| (1, 2) | (10, 17) | 13.5 | 54 | 24.5 | 98 | $54 \pm 126$ |
| (1, 3) | (10, 13) | 11.5 | 46 | 4.5 | 18 | $46 \pm 54$ |
| (1, 4) | (10, 20) | 15.0 | 60 | 50.0 | 200 | $60 \pm 180$ |
| (2, 3) | (17, 13) | 15.0 | 60 | 8.0 | 32 | $60 \pm 72$ |
| (2, 4) | (17, 20) | 18.5 | 74 | 4.5 | 18 | $74 \pm 54$ |
| (3, 4) | (13, 20) | 16.5 | 66 | 24.5 | 98 | $66 \pm 126$ |

Confidence intervals that avoid the normal approximation are available for some sampling situations. When the variable of interest indicates the presence or absence of some attribute (see Chapter 5), confidence intervals based on the exact distribution are possible with simple random sampling. Nonparametric confidence intervals for the population median and other quantiles may also be constructed [see Sedransk and Smith (1988) for a review].

***Example 2: All Possible Samples.*** For the sample $\{1, 3\}$ of Example 2 of Chapter 2, having $y$-values 10 and 13, the estimate of the population total was $\hat{\tau} = 46$ and the estimated variance of $\hat{\tau}$ was $\widehat{\mathrm{var}}(\hat{\tau}) = 18$. A nominal 95% confidence interval, constructed mechanically but not justified by the sample size, is $46 \pm 12.706 \sqrt{18} = 46 \pm 54$, where 12.706 is the upper 0.025 point of the $t$-distribution with 1 degree of freedom.

The exact coverage probability for the confidence interval procedure can also be determined from Table 3.1, as the proportion of samples for which the Interval includes the true value 60. While the nominal confidence level was 95%, the actual coverage is 100% for this population.        □

### Some Details

The finite population central limit theorem requires the concept of a sequence of populations $U_1, U_2, \ldots$, the $N$th population in the sequence has $N$ units and $y$-values $y_{1N}, y_{1N}, \ldots, y_{NN}$. The sample size $n_N$ of the simple random sample selected from the $N$th population also depends on $N$, and the sample mean of this sample is $\bar{y}_N = \sum_{i \in s} y_{iN}$. For any positive constant $\epsilon$, let the set of units with $y$-values farthest from the mean in the $N$th population be denoted

$$A_N = \left\{ i : |y_{iN} - \mu_N| > \epsilon \sqrt{\left(\frac{n}{N}\right)\left(1 - \frac{n}{N}\right) \sum_{j=1}^{N} (y_{jN} - \mu_N)^2} \right\}$$

A neccessary and sufficient condition for the distribution of

$$\frac{\bar{y}_N - \mu_N}{\sqrt{\mathrm{var}(\bar{y}_N)}}$$

to approach the normal (0,1) distribution as $N \to \infty$ is that

$$\lim_{N \to \infty} \frac{\sum_{A_N} (y_{iN} - \mu_N)^2}{\sum_{i=1}^{N} (y_{iN} - \mu_N)^2} = 0$$

for any $\epsilon > 0$. An additional description of these results, dating back to Erdös and Renyi (1959) and Hájek (1960), is given in the book by M. E. Thompson (1997, pp. 56–61).

## EXERCISES

1. From the data of parts (a) and (b) of Exercise 1 of Chapter 2, construct approximate 95% confidence intervals for the population total. Also construct approximate 95% confidence intervals for the population mean per unit. On what assumptions or results is the confidence interval procedure based, and how well does the method apply here?

2. For Exercise 2 of Chapter 2, give an approximate 90% confidence interval (a) for the population total and (b) for the mean.

3. For the population of $N = 5$ units of Exercise 3 of Chapter 2:
   (a) Compute directly the variance $\mathrm{var}(\bar{y})$ of the sample mean and the variance $\mathrm{var}(m)$ of the sample median.
   (b) From each sample, compute the sample variance $s^2$ and the estimate $\widehat{\mathrm{var}}(\bar{y})$ of the variance of the sample mean. Show that the sample variance $s^2$ is unbiased for the finite population variance $\sigma^2$ but that the sample standard deviation $s = \sqrt{s^2}$ is not unbiased for the population standard deviation $\sigma = \sqrt{\sigma^2}$.
   (c) For each sample, construct a standard 95% confidence interval for the population mean. What is the actual coverage probability for the method with this population and design?

4. Using the data from your own survey (see Exercise 5 of Chapter 2) give 95% and 99% confidence intervals for the population mean and total.

# CHAPTER 4

# Sample Size

The first question asked when a survey is being planned is, more often than not: What sample size should be used? Obtaining the answer is not always as simple as desired.

Suppose that one wishes to estimate a population parameter $\theta$—for example, the population mean or total—with an estimator $\hat{\theta}$. One would wish the estimate to be close to the true value with high probability. Specifying a maximum allowable difference $d$ between the estimate and the true value and allowing for a small probability $\alpha$ that the error may exceed that difference, the object is to choose a sample size $n$ such that

$$P(|\hat{\theta} - \theta| > d) < \alpha \tag{1}$$

If the estimator $\hat{\theta}$ is an unbiased, normally distributed estimator of $\theta$, then

$$\frac{\hat{\theta} - \theta}{\sqrt{\operatorname{var}(\hat{\theta})}}$$

has a standard normal distribution. Letting $z$ denote the upper $\alpha/2$ point of the standard normal distribution yields

$$P\left(\frac{|\hat{\theta} - \theta|}{\sqrt{\operatorname{var}(\hat{\theta})}} > z\right) = P\left(|\hat{\theta} - \theta| > z\sqrt{\operatorname{var}(\hat{\theta})}\right) = \alpha \tag{2}$$

The variance of the estimator $\hat{\theta}$ decreases with increasing sample size $n$, so that the inequality above will be satisfied if we can choose $n$ large enough to make $z\sqrt{\operatorname{var}(\hat{\theta})} \leq d$.

**35**

## 4.1   SAMPLE SIZE FOR ESTIMATING A POPULATION MEAN

With simple random sampling, the sample mean $\bar{y}$ is an unbiased estimator of the population mean $\mu$ with variance $\text{var}(\bar{y}) = (N - n)\sigma^2/Nn$. Setting

$$z\sqrt{\left(\frac{N - n}{N}\right)\frac{\sigma^2}{n}} = d$$

and solving for $n$ gives the necessary sample size:

$$n = \frac{1}{d^2/z^2\sigma^2 + 1/N} = \frac{1}{1/n_0 + 1/N} \tag{3}$$

where

$$n_0 = \frac{z^2\sigma^2}{d^2} \tag{4}$$

If the population size $N$ is large relative to the sample size $n$, so that the finite population correction factor can be ignored, the formula for sample size simplifies to $n_0$.

For sampling designs more elaborate than simple random sampling, sample size can generally be chosen in much the same way, with sample size (or sizes) determined so that the half-width of the confidence interval equals the specified distance. The weak point in the system is usually the estimate of population variance used.

## 4.2   SAMPLE SIZE FOR ESTIMATING A POPULATION TOTAL

For estimating the population total $\tau$, the equation to be solved for $n$ is

$$z\sqrt{N(N - n)\frac{\sigma^2}{n}} = d \tag{5}$$

which gives the necessary sample size as

$$n = \frac{1}{d^2/N^2z^2\sigma^2 + 1/N} = \frac{1}{1/n_0 + 1/N} \tag{6}$$

where

$$n_0 = \frac{N^2z^2\sigma^2}{d^2} \tag{7}$$

Ignoring the finite population correction factor, the formula for sample size reduces to $n_0$.

A bothersome aspect of sample size formulas such as these is that they depend on the population variance, which generally is unknown. In practice, one may be able to estimate the population variance using a sample variance from past data from the same or similar population.

**Example 1.**    What sample size would be necessary to estimate the caribou population total to within $d = 2000$ animals of the true total with 90% confidence $(\alpha = 0.10)$?

Ignoring the finite size of the population and using the sample variance $s^2 = 919$ from the preliminary survey (Example 1 of Chapter 2) as an estimate of the population variance $\sigma^2$, the simpler sample size formula [Equation (7)] would give

$$n_0 = \frac{286^2(1.645)^2(919)}{2000^2} = 50.9 \approx 51$$

The constant 1.645 is the upper $\alpha = 0.05$ point of the standard normal distribution. Although simpler to compute, the formula ignoring the finite population correction always gives a larger sample size to meet given criteria. In fact, for small populations, the simple formula can give a sample size larger than the number of units in the population.

Taking the finite population size into account produces the following choice of sample size [Equation (3)] for any future survey of this population:

$$n = \frac{1}{1/50.9 + 1/286} = 43.2 \approx 44$$

with the choice to round up considered conservative.                          □

## 4.3   SAMPLE SIZE FOR RELATIVE PRECISION

If instead of controlling the absolute error $d$, one is concerned with the relative error $r$—that is, the difference between the estimate and the true value, divided by the true value—the criterion to be met is

$$p\left(\frac{|\hat{\theta} - \theta|}{\theta} > r\right) < \alpha$$

or, equivalently,

$$p(|\hat{\theta} - \theta| > r\theta) < \alpha$$

where $\theta$ represents either the population mean or the population total.

To estimate the population mean $\mu$ to within $r\mu$ of the true value or to estimate the population total $\tau$ to within $r\tau$ of the true value, with probability $1 - \alpha$, the sample size formula is

$$n = \frac{1}{r^2\mu^2/z^2\sigma^2 + 1/N}$$

This result may be obtained either by substituting $r\mu$ for $d$ in the sample size formula for estimating the mean or by substituting $r\tau = rN\mu$ for $d$ in the sample size formula for estimating the population total.

Letting $\gamma$ denote the coefficient of variation for the population (i.e., $\gamma = \sigma/\mu$), the sample size formula may be written

$$n = \frac{1}{r^2/z^2\gamma^2 + 1/N}$$

Thus, the coefficient of variation is the population quantity on which sample size depends when the desire is to control relative precision.


**EXERCISES**

1. A botanical researcher wishes to design a survey to estimate the number of birch trees in a study area. The study area has been divided into 1000 units or plots. From previous experience, the variance in the number of stems per plot is known to be approximately $\sigma^2 \approx 45$. Using simple random sampling, what sample size should be used to estimate the total number of trees in the study area to within 500 trees of the true value with 95% confidence? To within 1000 trees? To within 2000 trees?

2. Compute the sample sizes for Exercise 1 when the finite population correction factor is ignored. What do you conclude about the importance of the finite population correction factor for this population?

3. Using the sample variance from the data of your own survey to estimate the population variance, specify a desired precision and calculate the sample size necessary to achieve it with 95% confidence in a future survey of the same population.

CHAPTER 5

# Estimating Proportions, Ratios, and Subpopulation Means

In some sampling situations, the object is to estimate the proportion of units in the population having some attribute. For example, one may wish to estimate the proportion of voters favoring a proposition, the proportion of females in an animal population, the proportion of plots in a study area in which a certain plant species is present, or the proportion of a specific mineral in the composition of a rock. In such a situation, the variable of interest is an indicator variable: $y_i = 1$ if unit $i$ has the attribute, and $y_i = 0$ if it does not.

With the $y$-variables taking on only the values zero and 1, the population total $\tau$ is the number of units in the population with the attribute, and the population mean $\mu$ is the proportion of units in the population with the attribute. Thus, to estimate a population proportion using simple random sampling, the usual methods associated with estimating a population mean, forming confidence intervals based on the normal approximation and determining sample size could be used. However, several special features are worth noting: (1) the formulas simplify considerably with attribute data; (2) exact confidence intervals are possible; and (3) a sample size sufficient for a desired absolute precision may be chosen without any information on population parameters.

A section on estimating a ratio is included in this chapter mainly to distinguish that situation from estimating a proportion. The statistical properties of estimates of ratios are covered in later chapters. Estimating proportions, means, and totals for subpopulations is treated as a special topic because when the simple random sample is selected from the whole population, the sample size for the subpopulation is a random variable.

## 5.1 ESTIMATING A POPULATION PROPORTION

Writing $p$ for the proportion in the population with the attribute

$$p = \frac{1}{N} \sum_{i=1}^{N} y_i = \mu$$

**39**

the finite population variance is

$$\sigma^2 = \frac{\sum_{i=1}^{N}(y_i - p)^2}{N-1} = \frac{\sum y_i^2 - Np^2}{N-1} = \frac{Np - Np^2}{N-1} = \frac{N}{N-1}p(1-p)$$

Letting $\hat{p}$ denote the proportion in the sample with the attribute

$$\hat{p} = \frac{1}{n}\sum_{i=1}^{n} y_i = \bar{y}$$

the sample variance is

$$s^2 = \frac{\sum_{i=1}^{n}(y_i - \bar{y})^2}{n-1} = \frac{\sum y_i^2 - n\hat{p}^2}{n-1} = \frac{n}{n-1}\hat{p}(1-\hat{p})$$

Thus, the relevant statistics can be computed from the sample proportion alone.

Since the sample proportion is the sample mean of a simple random sample, it is unbiased for the population proportion and has variance

$$\text{var}(\hat{p}) = \left(\frac{N-n}{N-1}\right)\frac{p(1-p)}{n}$$

An unbiased estimator of this variance is

$$\widehat{\text{var}}(\hat{p}) = \left(\frac{N-n}{N}\right)\frac{\hat{p}(1-\hat{p})}{n-1}$$

## 5.2 CONFIDENCE INTERVAL FOR A PROPORTION

An approximate confidence interval for $p$ based on a normal distribution is given by

$$\hat{p} \pm t\sqrt{\widehat{\text{var}}(\hat{p})}$$

where $t$ is the upper $\alpha/2$ point of the $t$-distribution with $n-1$ degrees of freedom. The normal approximation on which this interval is based improves the larger the sample size and the closer $p$ is to 0.5.

Confidence limits may also be obtained based on the exact hypergeometric distribution of the number of units in the sample with the attribute. The exact method is conceptually simple but computationally complex. Let $a = \sum_{i=1}^{n} y_i$ be the number of units with the attribute in the sample obtained in the survey. For exact limits, one equates the situation to an urn model. The urn contains $\tau$ red balls (units with the attribute in the population) and $N - \tau$ white balls. A random sample of $n$ balls is selected without replacement from the urn. Let the random variable $X$ denote the number of red balls in the sample from the urn. Given $\tau$ red balls in the urn, the probability that the number of red balls in the sample from the urn is $j$ is

$$P(X = j \mid \tau) = \binom{\tau}{j}\binom{N-\tau}{n-j} \bigg/ \binom{N}{n}$$

For a desired $100(1 - \alpha)\%$ confidence limit for the number $\tau$ of units in the population with the attribute, an upper limit $\tau_U$ is determined as the number of red balls in the urn giving probability $\alpha_1$ of obtaining $a$ or fewer red balls in the sample, where $\alpha_1$ is approximately equal to half the desired $\alpha$. That is, $\tau_U$ satisfies

$$P(X \leq a \mid \tau_U) = \sum_{i=0}^{a} \binom{\tau_U}{i}\binom{N-\tau_U}{n-i} \bigg/ \binom{N}{n} = \alpha_1$$

The lower limit $\tau_L$ is the number of red balls in the urn giving probability $\alpha_2$ of obtaining $a$ or more red balls in the sample, where $\alpha_2$ is approximately equal to half the desired $\alpha$. That is, $\tau_L$ satisfies

$$P(X \geq a \mid \tau_L) = \sum_{i=a}^{n} \binom{\tau_L}{i}\binom{N-\tau_L}{n-i} \bigg/ \binom{N}{n} = \alpha_2$$

Confidence limits for the population proportion $p$ are $p_L = \tau_L/N$ and $p_U = \tau_U/N$.

If $\alpha_1$ and $\alpha_2$ are chosen in advance, one should choose $\tau_U$ as the largest whole number such that $P(X \leq a \mid \tau_U) > \alpha_1$ and choose $\tau_L$ as the smallest whole number such that $P(X \geq a \mid \tau_L) > \alpha_2$. This procedure ensures a coverage probability of at least $1 - \alpha_1 - \alpha_2$ and gives a slightly narrower interval than that given by the related method prescribed in Cochran (1977). The requirement that $\tau$ be a whole number may preclude the possibility of attaining precisely a preselected confidence coefficient $1 - \alpha$. Approximate confidence intervals can also be constructed based on the binomial distribution.

## 5.3 SAMPLE SIZE FOR ESTIMATING A PROPORTION

To obtain an estimator $\hat{p}$ having probability at least $1 - \alpha$ of being no farther than $d$ from the population proportion, the sample size formula based on the normal approximation gives

$$n = \frac{Np(1-p)}{(N-1)(d^2/z^2) + p(1-p)} \tag{1}$$

where $z$ is the upper $\alpha/2$ point of the normal distribution. When the finite population correction can be ignored, the formula reduces to

$$n_0 = \frac{z^2 p(1-p)}{d^2} \tag{2}$$

For computational purposes, the exact sample size formula (1) may be written

$$n = \frac{1}{(N-1)/Nn_0 + 1/N} \approx \frac{1}{1/n_0 + 1/N} \tag{1a}$$

Note that the formulas depend on the unknown population proportion $p$. If no estimate of $p$ is available prior to the survey, a worst-case value of $p = 0.5$ can be used in determining sample size. The quantity $p(1-p)$, and hence the value of $n$ required by the formula, assumes its maximum value when $p$ is 1/2.

***Example 1.*** A geologist wishes to estimate the proportion of gold in a thin section of rock by taking a simple random sample of $n$ points and noting the presence or absence of the mineral. How large a sample is needed to obtain an estimate within $d = 0.05$ of the true proportion with probability 0.95 ($\alpha = 0.05$)?

The finite population correction factor can be ignored, since the sample "points" (units) are small in area so that the sample size $n$ is small relative to $N$. Using the simple formula [Equation (2)]

$$n = \frac{(1.96^2)(0.5)(0.5)}{0.05^2} = 384.16$$

where 1.96 is the upper 0.025 point of the normal distribution. Thus, a sample size of 384 or 385 would be sufficient to meet the criteria no matter what the actual population proportion $p$ is.  □

## 5.4  SAMPLE SIZE FOR ESTIMATING SEVERAL PROPORTIONS SIMULTANEOUSLY

Suppose that a biologist needs to collect a sample of fish to estimate the proportion of the population in each age class and that she would like the probability to be at least 0.95 that all of the estimates are simultaneously within 0.05 of the population proportions. She does not have prior knowledge of the population proportions and is not even sure how many age classes there are in the population. How large a sample size should she take?

The worst-case population would be one with the combination of $k$ proportions that give the maximum probability of a sample for which at least one of the sample proportions was unacceptably far from the corresponding population proportion. Since the proportions must sum to 1, the situation can never be quite so bad as every population proportion equal to $\frac{1}{2}$. The worst case to be considered in such a situation is given in a theorem in Fitzpatrick and Scott (1987) and Thompson (1987). Typically, the worst case is a situation in which virtually the whole population is equally distributed between two or three categories (age classes). The exact case that is worst depends on the $\alpha$-level chosen, but does not depend on the number $k$ of categories in the population.

The sample size needed may be obtained from Table 5.1. For Example 1, the biologist has chosen $d = 0.05$ and $\alpha = 0.05$. In the row of the table for $\alpha = 0.05$ the entry 1.27359 for $d^2 n_0$ is divided by $d^2 = 0.05^2$ to determine the sufficient sample size of $n_0 = 510$ fish. This sample size ensures that all the estimates will *simultaneously* be as close as desired to their respective population proportions. This sample size will be adequate whether the population of fish is divided into three age classes or 15.

The sample size $n_0$ from Table 5.1 for estimating multinomial proportions is appropriate when the population size $N$ is large relative to sample size $n$, so that finite population correction factors in the variances can be ignored. When the population size $N$ is small, a smaller sample size $n$ will be adequate to meet the same precision criteria, $d$ and $\alpha$.

Obtaining $n_0$ from Table 5.1, the smaller sample size $n$ may be obtained from the relationship

$$\frac{(N - 1)n}{N - n} = n_0$$

**Table 5.1. Sample Size $n_0$ for Simultaneously Estimating Several Proportions within Distance $d$ of the True Values at Confidence Level $1 - \alpha$**

| $\alpha$ | $d^2 n_0$ | $n_0$ with $d = 0.05$ | $m$ |
|---|---|---|---|
| 0.50 | 0.44129 | 177 | 4 |
| 0.40 | 0.50729 | 203 | 4 |
| 0.30 | 0.60123 | 241 | 3 |
| 0.20 | 0.74739 | 299 | 3 |
| 0.10 | 1.00635 | 403 | 3 |
| 0.05 | 1.27359 | 510 | 3 |
| 0.025 | 1.55963 | 624 | 2 |
| 0.02 | 1.65872 | 664 | 2 |
| 0.01 | 1.96986 | 788 | 2 |
| 0.005 | 2.28514 | 915 | 2 |
| 0.001 | 3.02892 | 1212 | 2 |
| 0.0005 | 3.33530 | 1342 | 2 |
| 0.0001 | 4.11209 | 1645 | 2 |

*Source:* Thompson (1987). With permission from the American Statistical Association.

Solving for $n$ gives

$$n = \frac{1}{(N-1)/Nn_0 + 1/N} \approx \frac{1}{1/n_0 + 1/N}$$

For example, if the population size is $N = 1000$ and one wishes each estimated proportion to be simultaneously within $d = 0.05$ of the true value with probability $1 - \alpha = 0.95$, an adequate worst-case sample size is

$$n = \frac{1}{1/510 + 1/1000} = 338$$

*Comments*
The sample sizes for simultaneously estimating several proportions—that is, $k$ proportions which add up to 1—are based on the normal approximation together with Bonferroni's inequality, which states that the probability that any one or more of the estimates misses the corresponding true value (by more than $d$) is less than or equal to the sum of the $k$ individual probabilities of missing them. The theorem in Fitzpatrick and Scott (1987) and Thompson (1987) showed that the worst case—that is, the population values demanding the largest sample size—occurs when some $m$ of the proportions in the population are equal and the rest are zero. The value of $m$ depends on $\alpha$. For example, when $\alpha = 0.01$, $m = 2$, and when $\alpha = 0.05$, $m = 3$. Sample size does not depend on the number $k$ of categories in the population as long as $k$ is greater than or equal to $m$. The sample sizes are "conservative" in that they may actually be somewhat larger than necessary to attain the precision desired.

## 5.5   ESTIMATING A RATIO

Suppose that a biologist studying an animal population selects a simple random sample of plots in the study region and in each selected plot counts the number $y_i$ of young animals and the number $x_i$ of adult females, with the object of estimating the ratio of young to adult females in the population. In a household survey to estimate the number of television sets per person in the region, a random sample of households is selected and for each household selected the number $y_i$ of television sets and the number $x_i$ of people are recorded.

In such cases the population ratio is commonly estimated by dividing the total of the $y$-values by the total of the $x$-values in the sample. The estimator may be written $r = \sum_{i=1}^{n} y_i / \sum_{i=1}^{n} x_i = \bar{y}/\bar{x}$ and is called a *ratio estimator*. Because the denominator $\bar{x}$ as well as the numerator $\bar{y}$ is a random variable, ratio estimators are not unbiased with the design simple random sampling, although they may be nearly so. Properties of ratio estimators under simple random sampling are described in Chapters 7 and 12.

## 5.6   ESTIMATING A MEAN, TOTAL, OR PROPORTION OF A SUBPOPULATION

In a sample survey to estimate the proportion of registered voters who intend to vote for a particular candidate, a simple random sample of $n$ people in the voting district are selected. Of the people in the sample, $n_1$ are found to be registered voters, and of these, $a_1$ intend to vote for the candidate. The proportion of registered voters favoring the candidate is estimated with the sample proportion $\hat{p}_1 = a_1/n_1$. The proportion to be estimated is not a proportion of the whole population but of a subpopulation (or domain of study), the subpopulation of registered voters. The estimator $\hat{p}_1$ differs from the sample proportions considered in the first part of this chapter in that the denominator $n_1$, the number of registered voters in the sample, is a random variable. A similar situation arises when a random sample of households is selected for estimating mean or total expenditures, but estimates are wanted not just for the whole population but for subpopulations based on family size, geographic area, or other factors. Subpopulation estimation is of interest in an ecological survey of the abundance of an animal species when it is desired, using a simple random sample of plots in a study region, to estimate not only the abundance or density of the species in the region, but also the abundance in the subregion of the study area that has been most strongly affected by pollution.

### Estimating a Subpopulation Mean

Of the $N$ units in the population, let $N_k$ be the number that belong to the $k$th subpopulation. The variable of interest for the $i$th unit in the $k$th subpopulation is $y_{ki}$. The subpopulation total is $\tau_k = \sum_{i=1}^{N_k} y_{ki}$ and the subpopulation mean is $\mu_k = \tau_k/N_k$. A simple random sample of $n$ units is selected from the $N$ units in the population, and $n_k$ of the sample units are found to be from subpopulation $k$. The sample mean of those $n_k$ units is $\bar{y}_k = (1/n_k)\sum_{i=1}^{n_k} y_{ki}$.

The domain sample mean $\bar{y}_k$ is an unbiased estimator of the subpopulation mean $\mu_k$. This can be shown by conditioning on the domain sample size $n_k$. Given $n_k$, every possible combination of $n_k$ of the $N_k$ subpopulation units has equal probability of being included in the sample. Thus, conditional on $n_k$, $\bar{y}_k$ behaves as the sample mean of a simple random sample of $n_k$ units from $N_k$, so that the conditional expectation is $\mathrm{E}(\bar{y}_k \mid n_k) = \mu_k$. The unconditional expectation is $\mathrm{E}(\bar{y}_k) = \mathrm{E}[\mathrm{E}(\bar{y}_k \mid n_k)] = \mu_k$, so $\bar{y}_k$ is unbiased. The variance of $\bar{y}_k$ is obtained using the well-known decomposition $\mathrm{var}(\bar{y}_k) = \mathrm{E}[\mathrm{var}(\bar{y}_k \mid n_k)] + \mathrm{var}[\mathrm{E}(\bar{y}_k \mid n_k)]$. The second term is zero since the conditional expectation is constant. The conditional variance of $\bar{y}_k$, from simple random sampling, is

$$\mathrm{var}(\bar{y}_k \mid n_k) = \frac{N_k - n_k}{N_k n_k}\sigma_k^2 = \sigma_k^2\left(\frac{1}{n_k} - \frac{1}{N_k}\right)$$

where $\sigma_k^2$ is the population variance for the units in the $k$th subpopulation,

$$\sigma_k^2 = \frac{1}{N_k - 1} \sum_{i=1}^{N_k} (y_{ki} - \mu_k)^2$$

The unconditional variance is thus

$$\operatorname{var}(\bar{y}_k) = \sigma_k^2 \left[ \operatorname{E}\left(\frac{1}{n_k}\right) - \frac{1}{N_k} \right]$$

The sample size $n_k$ is a random variable with expected value $nN_k/N$. A design with a fixed sample size $nN_k/N$ would give lower variance, since $\operatorname{E}(1/n_k)$ is greater than $N/(nN_k)$ (by Jensen's inequality from probability theory).

For an estimated variance with which to construct a confidence interval, one can use

$$\widehat{\operatorname{var}}(\bar{y}_k) = \frac{N_k - n_k}{N_k n_k} s_k^2$$

where $s_k^2$ is the sample variance of the sample units in the $k$th subpopulation,

$$s_k^2 = \frac{1}{n_k - 1} \sum_{i=1}^{n_k} (y_{ki} - \bar{y}_k)^2$$

This is an unbiased estimate of the conditional variance of $\bar{y}_k$ given $n_k$.

An approximate $100(1 - \alpha)\%$ confidence interval for $\mu_k$ is provided by

$$\bar{y}_k \pm t \sqrt{\widehat{\operatorname{var}}(\bar{y}_k)}$$

where $t$ is the upper $\alpha/2$ point of the $t$-distribution with $n_k - 1$ degrees of freedom. Since the confidence interval procedure has the desired approximate coverage probability conditional on domain sample size $n_k$, it has the desired approximate coverage probability unconditionally as well.

If the subpopulation size $N_k$ is unknown, as is often the case, the finite population correction factor $(N_k - n_k)/N_k$ can be replaced with its expected value $(N - n)/N$.

### Estimating a Proportion for a Subpopulation

A subpopulation proportion is a special case of a subpopulation mean, with the variable of interest $y_{ki} = 1$ if the $i$th unit of the $k$th subpopulation has the attribute and $y_{ki} = 0$ otherwise. The number of units in the $k$th subpopulation having the attribute is denoted $A_k$ and the number of those in the sample is denoted $a_k$. The subpopulation sample proportion $\hat{p}_k = a_k/n_k$, as a subpopulation sample mean, is unbiased for the subpopulation proportion $p = A_k/N_k$. The estimate of variance with attribute data reduces to

$$\widehat{\text{var}}(\hat{p}_k) = \left(\frac{N_k - n_k}{N_k}\right) \frac{\hat{p}_k(1 - \hat{p}_k)}{n_k - 1}$$

with $(N_k - n_k)/N_K$ replaced by $(N - n)/N$ when $N_k$ is unknown.

## Estimating a Subpopulation Total

If the subpopulation size $N_K$ is known, the subpopulation total $\tau_k$ can be unbiasedly estimated with $N_k \bar{y}_k$. The variance of the estimator is $\text{var}(N_k \bar{y}_k) = N_k^2 \text{var}(\bar{y}_k)$ and the estimate of (conditional) variance $\widehat{\text{var}}(N_k \bar{y}_k) = N_k^2 \widehat{\text{var}}(\bar{y}_k)$ may be used.

If $N_k$ is not known, the unbiased estimator

$$\hat{\tau}_k = \frac{N}{n} \sum_{i=1}^{n_k} y_{ki}$$

may be used.

To show that $\hat{\tau}_k$ is unbiased for $\tau_k$, define a new variable of interest $y'$ that is identical to $y$ for every unit in the subpopulation and is zero for all other units. The population total for the new variable is $\sum_{i=1}^{N} y'_i = \tau_k$. The population variance for the new variable is $\sigma'^2 = \sum_{i=1}^{N}(y'_i - \tau_k/N)^2/(N - 1)$.

With a simple random sample of $n$ units from the population, the sample mean of the new variables is $\bar{y}' = \sum_{i=1}^{n} y'_i/n$. The estimator $\hat{\tau}_k$ can be written $\hat{\tau}_k = N\bar{y}'$. By the usual results of simple random sampling, $\hat{\tau}_k$ is an unbiased estimator of $\tau_k$ with variance

$$\text{var}(\hat{\tau}_k) = N^2 \left(\frac{N - n}{Nn}\right) \sigma'^2$$

An unbiased estimate of this variance is

$$\widehat{\text{var}}(\hat{\tau}_k) = N^2 \left(\frac{N - n}{Nn}\right) s'^2$$

where $s'^2$ is the sample variance computed with the new variables,

$$s'^2 = \frac{1}{n - 1} \sum_{i=1}^{n} (y'_i - \bar{y}')^2$$

When the subpopulation size $N_k$ is known, the first estimator $N_k \bar{y}_k$ is preferable to $\hat{\tau}_k$ as the variance of $N_k \bar{y}_k$ is smaller.

In many surveys it is important to make estimates not just for the overall population or large subgroups, but also for small geographic areas and small subpopulations. For example, in a national health survey it may be important to have estimates for small geographic subdivisions and for specific ethnic groups or for recent immigrants from a given region. Because of practical and budgetary limitations, the number of sample units in a given domain of interest may be very small, so that direct estimates based just on those units will have high variances. To obtain better estimates for small areas, a number of techniques have been developed that "borrow strength" from surrounding areas. One approach, referred to as *synthetic estimation*, makes use of estimates and auxiliary information from the larger area to adjust estimates for the small area of interest. Composite estimators use a weighted average of the synthetic and direct estimators. A variety of model-based methods using auxiliary information from the small area and larger population have also been introduced. Ghosh and Rao (1994) provide a review and appraisal of small-area estimation methods.

## EXERCISES

1. To estimate the proportion of voters in favor of a controversial proposition, a simple random sample of 1200 eligible voters was contacted and questioned. Of these, 552 reported that they favored the proposition. Estimate the population proportion in favor and give a 95% confidence interval for population proportion. The number of eligible voters in the population is approximately 1,800,000.

2. A market researcher plans to select a simple random sample of households from a population of 30,000 households to estimate the proportion of households with one or more videocassette recorders. The researcher asks a statistician what sample size to use and is told to use the formula

$$n = \frac{Np(1-p)}{(N-1)(d^2/z^2) + p(1-p)}$$

   Describe exactly what the researcher needs to know or decide to fill in values for $d$, $z$, and $p$ in the equation, and what will be accomplished by that choice of sample size.

3. What sample size is required to estimate the proportion of people with blood type O in a population of 1500 people to be within 0.02 of the true proportion with 95% confidence? Assume no prior knowledge about the proportion.

4. For the population of Exercise 3, what sample size is required to simultaneously estimate the proportion with each blood type to within 0.02 of the true proportion with 95% confidence?

5. Carry out a survey to estimate a proportion in a population of your choice. Examples include the proportion of vehicles with only one occupant and the proportion of a university campus covered by tree canopy. For the latter, the campus can be partitioned into extremely small plots, so that each plot location may be determined to be either covered or not covered and the finite population correction factor will be negligible. In the process of carrying out the survey and making the estimate, think about or discuss with others the following:

   (a) What practical problems arise in establishing a frame, such as a map or list of units, from which to select the sample?

   (b) How is the sample selection actually carried out?

   (c) What special problems arise in observing the units selected?

   (d) Estimate the population proportion.

   (e) Estimate the variance of the estimated proportion.

   (f) Give a 95% confidence interval for the population proportion.

   (g) How large a sample size would be needed to estimate the proportion with 0.02 of the true value with 95% confidence? (Answer this first using your estimate proportion and second for the worst case.)

   (h) How would you improve the survey procedure if you were to do it again?

CHAPTER 6

# Unequal Probability Sampling

With some sampling procedures, different units in the population have different probabilities of being included in a sample. The differing inclusion probabilities may result from some inherent feature of the sampling procedure, or they may be imposed deliberately to obtain better estimates by including "more important" units with higher probability. In either case, the unequal inclusion probabilities must be taken into account in order to come up with reasonable estimates of population quantities.

Examples of unequal probability designs include line-intercept sampling of vegetation cover, in which the size of a patch of vegetation is measured whenever a randomly selected line intersects it. The larger the patch, the higher the probability of inclusion in the sample. Forest surveys to assess the board feet in a stand of trees are often done in such a way as to give higher inclusion probabilities to larger trees.

If a study area is divided into plots of unequal sizes, it may be desired to assign larger inclusion probabilities to larger plots; this may be done by selecting points in the study area with equal probability and including a plot whenever a point selected is within it. More generally, unequal probability selections may be carried out by assigning to each unit an interval whose length is equal to the desired probability and selecting random numbers from the uniform distribution: A unit is included if the random number is in its interval.

## 6.1 SAMPLING WITH REPLACEMENT: THE HANSEN–HURWITZ ESTIMATOR

Suppose that sampling is with replacement and that on each draw the probability of selecting the $i$th unit of the population is $p_i$, for $i = 1, 2, \ldots, N$. Then an unbiased estimator of the population total $\tau$ is

$$\hat{\tau}_p = \frac{1}{n} \sum_{i=1}^{n} \frac{y_i}{p_i} \tag{1}$$

The variance of this estimator is

$$\text{var}(\hat{\tau}_p) = \frac{1}{n} \sum_{i=1}^{N} p_i \left( \frac{y_i}{p_i} - \tau \right)^2 \tag{2}$$

An unbiased estimator of this variance is

$$\widehat{\text{var}}(\hat{\tau}_p) = \frac{1}{n(n-1)} \sum_{i=1}^{n} \left( \frac{y_i}{p_i} - \hat{\tau}_p \right)^2 \tag{3}$$

An unbiased estimator of the population mean $\mu$ is $\hat{\mu}_p = (1/N)\hat{\tau}_p$, having variance $\text{var}(\hat{\mu}_p) = (1/N^2)\text{var}(\hat{\tau}_p)$ and estimated variance $\widehat{\text{var}}(\hat{\mu}_p) = (1/N^2)\widehat{\text{var}}(\hat{\tau}_p)$.

An approximate $(1 - \alpha)$ 100% confidence interval for the population total, based on the large-sample normal approximation for the estimator $\hat{\tau}_p$, is

$$\hat{\tau}_p \pm z \sqrt{\widehat{\text{var}}(\hat{\tau}_p)}$$

where $z$ is the upper $\alpha/2$ point of the standard normal distribution. For sample sizes less than about 50, the $t$-distribution with $n - 1$ degrees of freedom may be used in place of the normal distribution.

This estimator was introduced by Hansen and Hurwitz (1943). Note that because of the with-replacement sampling, a unit may be selected on more than one draw. The estimator utilizes that unit's value as many times as the unit is selected.

If the selection probabilities $p_i$ were proportional to the variables $y_i$, the ratio $y_i/p_i$ would be constant and the Hansen–Hurwitz estimator would have zero variance. The variance would be low if the selection probabilities could be set approximately proportional to the $y$-values. Of course, the population $y$-values are unknown prior to sampling. If it is believed that the $y$-values are approximately proportional to some known variable such as the sizes of the units, the selection probabilities can be chosen proportional to the value of that known variable.

***Example 1.*** Large mammals in open habitat are often surveyed from aircraft. As the aircraft flies over a selected strip, all animals of the species within a prescribed distance of the aircraft path are counted, the distance sometimes being determined by markers on the wing struts of the aircraft. Because of the irregularities in the shape of the study area, the strips to be flown may be of varying lengths. One may select units (strips) with probability proportional to their lengths by randomly selecting $n$ points on a map of the study region and including in the sample any strip that contains a selected point. The draw-by-draw selection probability for any strip equals its length times its width divided by the area of the study region. A strip is selected more than once if it contains more than one of the points selected. It is immaterial whether the unit is physically observed more than once. The design

**Table 6.1. Sample Obervations for Example 1**

| $y_i$ | Length | $p_i$ |
|---|---|---|
| 60 | 5 | 0.05 |
| 60 | 5 | 0.05 |
| 14 | 2 | 0.02 |
| 1 | 1 | 0.01 |

unbiasedness of the estimator depends on the observation receiving a weight in the estimator equal to the number of times the unit is selected.

For simplicity, consider a study area of 100 square kilometers partitioned into strips 1 kilometer wide but varying in length. A sample of $n = 4$ strips is selected by this method. One strip, with $y_i = 60$ animals, selected twice; its length was 5 kilometers, and hence its selection probability was $p_i = 0.05$. The set of sample observations is listed in Table 6.1, with the repeat selection listed twice.

The *Hansen–Hurwitz estimator*, also known as the *probability-proportional-to-size* (PPS) *estimator* in this type of design situation [using Equation (1)], is

$$\hat{\tau}_p = \frac{1}{4}\left(\frac{60}{0.05} + \frac{60}{0.05} + \frac{14}{0.02} + \frac{1}{0.01}\right)$$

$$= \frac{1}{4}(1200 + 1200 + 700 + 100) = 800 \text{ animals}$$

To compute the estimated variance, note that an ordinary sample variance of the numbers $1200, \ldots, 100$, the mean of which is 800, is involved [Equation (3)]:

$$\widehat{\text{var}}(\hat{\tau}_p) = \frac{1}{4(3)}[(1200 - 800)^2 + \cdots + (100 - 800)^2] = \frac{273{,}333}{4} = 68{,}333$$

The standard error is $\sqrt{\widehat{\text{var}}(\hat{\tau}_p)} = \sqrt{68{,}333} = 261$.     □

## 6.2  ANY DESIGN: THE HORVITZ–THOMPSON ESTIMATOR

With any design, with or without replacement, giving probability $\pi_i$ that unit $i$ is included in the sample, for $i = 1, 2, \ldots, N$, an unbiased estimator of the population total $\tau$, introduced by Horvitz and Thompson (1952), is

$$\hat{\tau}_\pi = \sum_{i=1}^{\nu} \frac{y_i}{\pi_i} \tag{4}$$

where $\nu$ is the effective sample size—the number of distinct units in the sample—and the summation is over the distinct units in the sample.

This estimator does not depend on the number of times a unit may be selected. Each distinct unit of the sample is utilized only once.

Let the probability that *both* unit $i$ and unit $j$ are included in the sample be denoted by $\pi_{ij}$. The variance of the estimator is

$$\text{var}(\hat{\tau}_\pi) = \sum_{i=1}^{N}\left(\frac{1-\pi_i}{\pi_i}\right)y_i^2 + \sum_{i=1}^{N}\sum_{j\neq i}\left(\frac{\pi_{ij}-\pi_i\pi_j}{\pi_i\pi_j}\right)y_iy_j \tag{5}$$

An unbiased estimator of this variance is

$$\widehat{\text{var}}(\hat{\tau}_\pi) = \sum_{i=1}^{\nu}\left(\frac{1-\pi_i}{\pi_i^2}\right)y_i^2 + \sum_{i=1}^{\nu}\sum_{j\neq i}\left(\frac{\pi_{ij}-\pi_i\pi_j}{\pi_i\pi_j}\right)\frac{y_iy_j}{\pi_{ij}}$$

$$= \sum_{i=1}^{\nu}\left(\frac{1}{\pi_i^2}-\frac{1}{\pi_i}\right)y_i^2 + 2\sum_{i=1}^{\nu}\sum_{j>i}\left(\frac{1}{\pi_i\pi_j}-\frac{1}{\pi_{ij}}\right)y_iy_j \tag{6}$$

if all of the joint inclusion probabilities $\pi_{ij}$ are greater than zero. An unbiased estimator of the population mean $\mu$ is $\hat{\mu}_\pi = (1/N)\hat{\tau}_\pi$, having variance $\text{var}(\hat{\mu}_\pi) = (1/N^2)\text{var}(\hat{\tau}_\pi)$ and estimated variance $\widehat{\text{var}}(\hat{\mu}_\pi) = (1/N^2)\widehat{\text{var}}(\hat{\tau}_\pi)$.

An approximate $(1-\alpha)100\%$ confidence interval for the population total, based on the large-sample normal approximation for the estimator $\hat{\tau}_\pi$, is

$$\hat{\tau}_\pi \pm z\sqrt{\widehat{\text{var}}(\hat{\tau}_\pi)} \tag{7}$$

where $z$ is the upper $\alpha/2$ point of the standard normal distribution. As a pragmatic rule, the $t$-distribution with $\nu-1$ degrees of freedom could be substituted for the normal distribution with small sample sizes (less than about 50).

Although unbiased, the variance estimator $\widehat{\text{var}}(\hat{\tau}_\pi)$ is somewhat tedious to compute and with some designs can give negative estimates, whereas the true variance must be nonnegative. A very simple approximate variance estimation formula, which though biased is considered "conservative" (tending to be larger than the actual variance) and is invariably nonnegative, has been suggested as an alternative (see Brewer and Hanif 1983, p. 68). For the $i$th of the $\nu$ distinct units in the sample, a variable $t_i = \nu y_i/\pi_i$ is computed, for $i = 1,\ldots,\nu$. Each of the $t_i$ is an estimate of the population total, and their average is the Horvitz–Thompson estimate. The sample variance of the $t_i$ is defined by $s_t^2 = [1/(\nu-1)]\sum_{i=1}^{\nu}(t_i - \hat{\tau}_\pi)^2$. The alternative variance estimator is

$$\widetilde{\text{var}}(\hat{\tau}_\pi) = \left(\frac{N-\nu}{N}\right)\frac{s_t^2}{\nu} \tag{8}$$

If the inclusion probabilities $\pi_i$ could be chosen approximately proportional to the values $y_i$ the variance of the Horvitz–Thompson estimator would be low. It is not easy, however, to devise without-replacement sampling schemes that obtain a desired set of unequal inclusion probabilities, nor is it easy to compute the inclusion probabilities for given without-replacement sampling schemes. Reviews of methods developed for these purposes are found in Brewer and Hanif (1983) and Chaudhuri and Vos (1988).

For a design in which the effective sample size $\nu$ is fixed rather than random, the variance can be written in the form

$$\mathrm{var}(\hat{\tau}_\pi) = \sum_{i=1}^{N} \sum_{j<i} (\pi_i \pi_j - \pi_{ij}) \left( \frac{y_i}{\pi_i} - \frac{y_j}{\pi_j} \right)^2$$

and an unbiased variance estimator is given by

$$\widehat{\mathrm{var}}(\hat{\tau}_\pi) = \sum_{i=1}^{\nu} \sum_{j<i} \left( \frac{\pi_i \pi_j - \pi_{ij}}{\pi_{ij}} \right) \left( \frac{y_i}{\pi_i} - \frac{y_j}{\pi_j} \right)^2$$

provided that all of the joint inclusion probabilities $\pi_{ij}$ are greater than zero (Sen 1953; Yates and Grundy 1953).

***Example 2.*** The Horvitz–Thompson estimator can be used with the PPS design of Example 1, so that the estimate will depend only on the distinct units in the sample, not on numbers of repeat selections. On each of the $n$ draws, the probability of selecting the $i$th unit is $p_i$, and, because of the with-replacement selection, the $n$ draws are independent. Thus, the probability of unit $i$ not being included in the sample is $(1 - p_i)^n$, so that the inclusion probability is $\pi_i = 1 - (1 - p_i)^n$.

For the first unit in the sample, with $p_1 = 0.05$, the inclusion probability is $\pi_1 = 1 - (1 - 0.05)^4 = 1 - 0.8145 = 0.1855$. (See Table 6.1 for data.) Similarly, for the second distinct unit in the sample, with $p_2 = 0.02$, we get $\pi_2 = 1 - (1 - 0.02)^4 = 1 - 0.9224 = 0.0776$. For the third unit, $\pi_3 = 1 - (1 - 0.01)^4 = 1 - 0.9606 = 0.0394$.

The Horvitz–Thompson estimator [using Equation (4)] is

$$\hat{\tau}_\pi = \frac{60}{0.1855} + \frac{14}{0.0776} + \frac{1}{0.0394} \approx 529 \text{ animals}$$

Using the relation between probabilities of intersections and unions of events, the probability of including both unit $i$ and unit $j$ in the sample is the probability of including unit $i$ plus the probability of including unit $j$ minus the probability of including either $i$ or $j$. Thus, the joint inclusion probability is

$$\pi_{ij} = \pi_i + \pi_j - [1 - (1 - p_i - p_j)^n]$$

For the three units in the sample, this gives $\pi_{12} = 0.0112$, $\pi_{13} = 0.0056$, and $\pi_{23} = 0.0023$. The estimated variance [using Equation (6)] is

$$
\begin{aligned}
\widehat{\mathrm{var}}(\hat{\tau}_\pi) = {}& \left(\frac{1}{0.1855^2} - \frac{1}{0.1855}\right)60^2 + \left(\frac{1}{0.0776^2} - \frac{1}{0.0776}\right)14^2 \\
& + \left(\frac{1}{0.0394^2} - \frac{1}{0.0394}\right)1^2 \\
& + 2\left[\frac{1}{0.1855(0.0776)} - \frac{1}{0.0112}\right](60)(14) \\
& + 2\left[\frac{1}{0.1855(0.0394)} - \frac{1}{0.0056}\right](60)(1) \\
& + 2\left[\frac{1}{0.0776(0.0394)} - \frac{1}{0.0023}\right](14)(1) \\
= {}& 74{,}538
\end{aligned}
$$

so that the estimated standard error is $\sqrt{74{,}538} = 273$ animals.

For the alternative variance estimate, an estimate of the total based on the first of the $\nu = 3$ distinct units in the sample is $t_1 = 3(60)/0.1855 = 970.35$. For the other two units, the values are $t_2 = 3(14)/0.0776 = 541.24$ and $t_3 = 3(1)/0.0394 = 76.14$. The sample variance of these three numbers is $s_t^2 = 200{,}011$. The estimated variance [using Equation (8)] is

$$
\widetilde{\mathrm{var}}(\hat{\tau}_\pi) = \left(\frac{100 - 3}{100}\right)\left(\frac{200{,}011}{3}\right) = 64{,}670
$$

giving an estimated standard error of 254. Note that because the effective sample size is not fixed for this design, the Sen–Yates–Grundy variance estimator does not apply. ☐

## 6.3 GENERALIZED UNEQUAL-PROBABILITY ESTIMATOR

The Horvitz–Thompson estimator is unbiased but can have a large variance for populations in which the variables of interest and inclusion probabilities are not well related. A generalized unequal-probability estimator of the population mean is

$$
\hat{\mu}_g = \frac{\sum_{i \in s} y_i/\pi_i}{\sum_{i \in s} 1/\pi_i}
$$

In the numerator of this estimator is the ordinary Horvitz–Thompson estimator, which gives an unbiased estimate of the population total $\tau$. The denominator can be viewed as another Horvitz–Thompson estimator, giving an unbiased estimate of the population size $N$. Thus, the estimator $\hat{\mu}_g$ estimates $\tau/N = \overline{\tau}$. As the ratio of two unbiased estimators the estimator $\hat{\mu}_g$ is not precisely unbiased, but the bias tends to be small and to decrease with increasing sample size.

The variance or mean square error of $\hat{\mu}_g$ is approximately

$$
\text{var}(\hat{\mu}_g) = \frac{1}{N^2}\left[\sum_{i=1}^{N}\left(\frac{1-\pi_i}{\pi_i}\right)(y_i-\mu)^2 + \sum_{i=1}^{N}\sum_{j\neq i}\left(\frac{\pi_{ij}-\pi_i\pi_j}{\pi_i\pi_j}\right)(y_i-\mu)(y_j-\mu)\right]
$$

$$(9)$$

An estimator of this variance is given by

$$
\widehat{\text{var}}(\hat{\mu}_g) = \frac{1}{N^2}\left[\sum_{i=1}^{\nu}\left(\frac{1-\pi_i}{\pi_i^2}\right)(y_i-\hat{\mu}_g)^2 + \sum_{i=1}^{\nu}\sum_{j\neq i}\left(\frac{\pi_{ij}-\pi_i\pi_j}{\pi_i\pi_j}\right)\frac{(y_i-\hat{\mu}_g)(y_j-\hat{\mu}_g)}{\pi_{ij}}\right]
$$

$$(10)$$

provided that all of the joint inclusion probabilities $\pi_{ij}$ are greater than zero. The value of $N$ in this estimator may be replaced by its estimator, $\hat{N} = \sum_{i\in s}(1/\pi_i)$.

For the population total, the generalized estimator is $\hat{\tau}_g = N\hat{\mu}_g$ and its variance is $\text{var}(\hat{\tau}_g) = N^2\text{var}(\hat{\mu}_g)$ with variance estimator $\widehat{\text{var}}(\hat{\tau}_g) = N^2\widehat{\text{var}}(\hat{\mu}_g)$.

**Example 3.** With the data from Example 1, the generalized estimator of the population total is

$$
\hat{\tau}_g = (50)\frac{60/0.1855 + 14/0.0776 + 1/0.0394}{1/0.1855 + 1/0.0776 + 1/0.0394} \approx 606 \text{ animals} \qquad \square
$$

## 6.4  SMALL POPULATION EXAMPLE

To illustrate the design-based properties of the two unbiased estimators with an unequal probability design, consider the following artificially small example. The population consists of $N = 3$ farms, of which $n = 2$ are selected with probabilities proportional to size with replacement. The variable of interest is production of wheat in metric tons. The population values are given in the following table.

| Unit (Farm), $i$ | 1 | 2 | 3 |
|---|---|---|---|
| Selection Probability, $p_i$ | 0.3 | 0.2 | 0.5 |
| Wheat Produced, $y_i$ | 11 | 6 | 25 |

Every possible *ordered sample*, with units listed in the order selected, is listed in Table 6.2, together with the probability $p(s)$ that the ordered sample is the one selected. The sample values $\mathbf{y}_s$ of the variable of interest and the values of the two estimators $\hat{\tau}_p$ and $\hat{\tau}_\pi$ are also listed for each of the samples. Because sampling

**Table 6.2. Samples for Small Population Example**

| $s$ | $p(s)$ | $y_s$ | $\hat{\tau}_p$ | $\hat{\tau}_\pi$ | $\hat{\tau}_g$ |
|------|------------------------|----------|-------|--------|--------|
| 1, 1 | 0.3(0.3) = 0.09 | (11, 11) | 36.67 | 21.57 | 33.00 |
| 2, 2 | 0.2(0.2) = 0.04 | (6, 6) | 30.00 | 16.67 | 18.00 |
| 3, 3 | 0.5(0.5) = 0.25 | (25, 25) | 50.00 | 33.33 | 75.00 |
| 1, 2 | 0.3(0.2) = 0.06 | (11, 6) | 33.33 | 38.24 | 24.21 |
| 2, 1 | 0.2(0.3) = 0.06 | (6, 11) | 33.33 | 38.24 | 24.21 |
| 1, 3 | 0.3(0.5) = 0.15 | (11, 25) | 43.33 | 54.90 | 50.00 |
| 3, 1 | 0.5(0.3) = 0.15 | (25, 11) | 43.33 | 54.90 | 50.00 |
| 2, 3 | 0.2(0.5) = 0.10 | (6, 25) | 40.00 | 50.00 | 36.49 |
| 3, 2 | 0.5(0.2) = 0.10 | (25, 6) | 40.00 | 50.00 | 36.49 |
| Mean: | | | 42 | 42 | 47.64 |
| Bias: | | | 0 | 0 | 5.64 |
| Variance: | | | 34.67 | 146.46 | 334.02 |
| Std. Dev.: | | | 5.89 | 12.10 | 18.28 |

is with replacement, the draws are independent, so the probability $p(s)$ of selecting a given sample of units is the product of the draw-by-draw selection probabilities for the units in the ordered sample.

The inclusion probabilities $\pi_i$ can be obtained directly from the table using

$$\pi_i = P(i \in s) = \sum_{\{s:i\in s\}} P(s)$$

Thus,

$$\pi_1 = 0.09 + 0.06 + 0.06 + 0.15 + 0.15 = 0.51$$
$$\pi_2 = 0.04 + 0.06 + 0.06 + 0.10 + 0.10 = 0.36$$
$$\pi_3 = 0.25 + 0.15 + 0.15 + 0.10 + 0.10 = 0.75$$

Alternatively, the inclusion probabilities can be computed using the analytical formula for *this particular design* as $\pi_i = 1 - (1 - p_i)^n$. For example, $p_1 = 1 - (1 - 0.3)^2 = 1 - 0.49 = 0.51$.

For computing the sample values of the Hanson–Hurwitz estimator $\hat{\tau}_p = (1/n)\sum_{i=1}^n y_i/p_i$, it is first useful to compute

$$\frac{y_1}{p_1} = \frac{11}{0.3} = 36.37$$

$$\frac{y_2}{p_2} = \frac{6}{0.2} = 30.00$$

$$\frac{y_3}{p_3} = \frac{25}{0.5} = 50.00$$

For each sample, $\hat{\tau}_p$ is the average of two of these values.

For computing the sample values of the Horvitz–Thompson estimator $\hat{\tau}_\pi = \sum_{i=1}^{\nu} y_i/\pi_i$, it is first useful to compute

$$\frac{y_1}{\pi_1} = \frac{11}{0.51} = 21.57$$

$$\frac{y_2}{\pi_2} = \frac{6}{0.36} = 16.67$$

$$\frac{y_3}{\pi_3} = \frac{25}{0.75} = 33.33$$

For each sample, $\hat{\tau}_\pi$ is the sum of these values for the $\nu$ distinct units in the sample. The expectation of an estimator $\hat{\tau}$ under the design is

$$E(\hat{\tau}) = \sum_s \hat{\tau}_s P(s)$$

where $\hat{\tau}_s$ is the value of the estimator for sample $s$ and the summation is over every possible sample.

The expected value of the estimator $\hat{\tau}_p$ for this population with the design used is thus

$$E(\hat{\tau}_p) = 36.67(0.09) + 30.00(0.04) + 50.00(0.25) + 2(33.33)(0.06)$$
$$+ 2(43.33)(0.15) + 2(40.00)(0.10) = 42$$

The expected value of the estimator $\hat{\tau}_\pi$ for this population with the design used is thus

$$E(\hat{\tau}_\pi) = 21.57(0.09) + 16.67(0.04) + 33.33(0.25) + 2(38.24)(0.06)$$
$$+ 2(54.90)(0.15) + 2(50.00)(0.10) = 42$$

The expected value of the generalized estimator $\hat{\tau}_g$, computed in similar fashion, is $E(\hat{\tau}_g) = 47.64$, so the bias of this estimator with this population total is $E(\hat{\tau}_g) - \tau = 47.64 - 42 = 5.64$. The variance is $\text{var}(\hat{\tau}_g) = E(\hat{\tau}_g - 47.64)^2 = 334.02$. The mean square error is $E(\hat{\tau}_g - \tau)^2 = E(\hat{\tau}_g - 42)^2 = 365.85$.

A small population example with small sample size such as this illustrates conceptually how each of the estimators works with the given design but does not provide guidance on which estimator to use in a given real situation. In this particular population, there is a roughly proportional relationship between the $y_i$ and the $p_i$ values, and the estimator $\hat{\tau}_p$ performs well, whereas for many real populations with roughly proportional relationships, the estimator $\hat{\tau}_\pi$ has been found to perform better. The estimator $\hat{\tau}_g$, which is recommended for cases in which the $y_i$ and $\pi_i$ values are not well related, performs least well of the three estimators here. None of the three estimators is uniformly better than the others for every type of population or design.

## 6.5  DERIVATIONS AND COMMENTS

When sampling is with replacement, the selections are independent. Consider a single draw. The $j$th unit of the population is selected with probability $p_j$, for $j = 1, \ldots, N$. Let $y_s$ denote the sample $y$-value and $p_s$ the selection probability for the unit in the sample. The Hansen–Hurwitz estimator for the sample of one size may be written $t_s = y_s/p_s$. Its expected value is

$$E(t_s) = \sum_s t_s p_s = \sum_{j=1}^N y_j = \tau$$

so that $t_s$ is an unbiased estimator of the population total.

The variance of $t_s$ is

$$\text{var}(t_s) = \sum_s (t_s - \tau)^2 p_s = \sum_{j=1}^N \left( \frac{y_j}{p_j} - \tau \right)^2 p_j$$

With $n$ independent draws, in which unit $j$ has selection probability $p_j$ on each draw, the Hansen–Hurwitz estimator is the sample mean of $n$ independent and identically distributed (i.i.d.) random variables $t_{s1}, t_{s2}, \ldots, t_{sn}$, each with the mean and variance above, so that one may write $\hat{\tau}_p = (1/n) \sum_{i=1}^n t_{si} = \bar{t}_s$. Thus, using the properties of a sample mean of i.i.d. random variables, $E(\hat{\tau}_p) = \tau$ and $\text{var}(\hat{\tau}_p) = (1/n) \sum_{j=1}^N p_j [(y_j/p_j) - \tau]^2$.

Further, since the $t_{si}$ are i.i.d., a sample variance based on them is an unbiased estimator of their variance, so that

$$E \left[ \frac{1}{n(n-1)} \sum_{i=1}^n (t_{si} - \hat{\tau}_p)^2 \right] = E[\widehat{\text{var}}(\hat{\tau}_p)] = \frac{1}{n} \text{var}(t_s) = \text{var}(\hat{\tau}_p)$$

The above derivation method is useful for any with-replacement sampling strategy, in which the estimator may be written as sample mean of independent, unbiased estimators.

For the Horvitz–Thompson estimator, define the indicator variable $z_i$ to be 1 if the $i$th unit of the population is included in the sample and zero otherwise, for $i = 1, 2, \ldots, N$. For any unit $i$, $z_i$ is a Bernoulli random variable with $E(z_i) = P(z_i = 1) = P(i \in s) = \pi_i$ and variance $\text{var}(z_i) = \pi_i(1 - \pi_i)$. For two distinct units $i$ and $j$, with $i \neq j$, the covariance of $z_i$ and $z_j$ is

$$\text{cov}(z_i, z_j) = E(z_i z_j) - E(z_i)E(z_j) = P(z_i = 1, z_j = 1) - \pi_i \pi_j = \pi_{ij} - \pi_i \pi_j$$

The Horvitz–Thompson estimator may be written in terms of the random variables $z_i$ as

$$\hat{\tau}_\pi = \sum_{i=1}^N \frac{y_i z_i}{\pi_i}$$

in which the only random quantities are the $z_i$. The expected value is thus

$$E(\hat{\tau}_\pi) = \sum_{i=1}^{N} \frac{y_i E(z_i)}{\pi_i} = \sum_{i=1}^{N} \frac{y_i \pi_i}{\pi_i} = \sum_{i=1}^{N} y_i = \tau$$

so that the Horvitz–Thompson estimator is unbiased.

Similarly, the variance of the estimator is

$$var(\hat{\tau}_\pi) = \sum_{i=1}^{N} \left(\frac{y_i}{\pi_i}\right)^2 var(z_i) + \sum_{i=1}^{N} \sum_{j \neq i} \frac{y_i y_j}{\pi_i \pi_j} cov(z_i, z_j)$$

$$= \sum_{i=1}^{N} \left(\frac{1 - \pi_i}{\pi_i}\right) y_i^2 + \sum_{i=1}^{N} \sum_{j \neq i} \left(\frac{\pi_{ij} - \pi_i \pi_j}{\pi_i \pi_j}\right) y_i y_j$$

To see that $\widehat{var}(\hat{\tau}_\pi)$ is unbiased for $var(\hat{\tau}_\pi)$, define $z_{ij}$ to be 1 if both units $i$ and $j$ are included in the sample and zero otherwise. The estimator of variance may be written

$$\widehat{var}(\hat{\tau}_\pi) = \sum_{i=1}^{N} \left(\frac{1 - \pi_i}{\pi_i^2}\right) y_i^2 z_i + \sum_{i=1}^{N} \sum_{j \neq i} \left(\frac{\pi_{ij} - \pi_i \pi_j}{\pi_i \pi_j}\right) \frac{y_i y_j z_{ij}}{\pi_{ij}}$$

Since $E(z_{ij}) = P(z_i = 1, z_j = 1) = \pi_{ij}$, unbiasedness follows immediately.

Alternative estimators of the variance of the Horvitz–Thompson estimator, based on a simple sum of squared deviations, have been investigated by a number of authors. Such estimators are to be desired for the computational simplicity and for favorable properties under specific population models. More general forms for such variance estimators are described in Brewer and Hanif (1983), Hájek (1981), and Kott (1988).

Asymptotic normality of the Hansen–Hurwitz estimator under sampling with replacement follows from the usual central limit theorem for the sample mean of independent, identically distributed random variables. For the Horvitz–Thompson estimator the situation is more complicated, since the terms $y_i/\pi_i$ of the distinct units in the sample are not in general independent. Results on the asymptotic normality of the Horvitz–Thompson estimator under certain designs are given in Rosen (1972a,b) and Sen (1988).

The generalized unequal probability estimator $\hat{\mu}_g$ can be obtained using an "estimating equation" approach. The population mean is defined implicitly by the population estimation equation

$$\sum_{i=1}^{N} (y_i - \mu) = 0$$

Notice that solving this equation for $\mu$ gives $\mu = \sum_{i=1}^{N} y_i/N$, the usual definition of the population mean. The left-hand side of the estimating equation can itself be viewed as a kind of population total, although it depends on an unknown parameter. Denote this total by $\phi(\mu) = \sum_{i=1}^{N}(y_i - \mu)$. From a sample $s$ an unbiased estimator of this total, of Horvitz–Thompson form, is

$$\hat{\phi}(\mu) = \sum_{i \in s} \frac{y_i - \mu}{\pi_i}$$

Setting this sample quantity to zero, since the total it estimates equals zero, gives the sample estimating equation

$$\sum_{i \in s} \frac{y_i - \mu}{\pi_i} = 0$$

Now solving this equation for $\mu$ gives the generalized estimator, $\hat{\mu}_g$.

Since the generalized estimator $\hat{\mu}_g$ is the ratio of two random variables, Taylor series approximation can be used to give

$$\hat{\mu}_g - \mu \approx \frac{1}{N} \sum_{i \in s} \frac{y_i - \mu}{\pi_i}$$

which has the form of a Horvitz–Thompson estimator. The approximate variance and variance estimator formulas are then obtained from the usual formulas for a Horvitz–Thompson estimator. Taylor series approximations of ratios of random variables are described in more detail in Chapter 7. For designs to which a finite population central limit theorem applies for the Horvitz–Thompson estimator, approximate $(1 - \alpha)$ confidence intervals may be based directly on the unbiased variance estimator for the sample estimation function. The interval consists of those values of $\mu$ for which $-t\sqrt{\widehat{\text{var}}(\hat{\phi}(\mu))} \le \hat{\phi}(\mu) \le t\sqrt{\widehat{\text{var}}(\hat{\phi}(\mu))}$. The estimated variance $\widehat{\text{var}}(\hat{\phi}(\mu))$ is obtained from the Horvitz–Thompson variance estimator with $\hat{\mu}_g$ substituted for $\mu$. The estimating equation approach to inference in survey sampling is discussed in Binder and Patak (1994), Godambe (1995), Godambe and Thompson (1986), and M. E. Thompson (1997). Further discussion on this type of estimator can be found in Särndal et al. (1992, pp. 176–184) and M. E. Thompson (1997, pp. 93–111). A more general form of this estimator is described in Chapter 7.

As mentioned earlier, it is difficult to devise without-replacement designs with inclusion probabilities exactly proportional to unit size or another specified variable. One simple design consists of selecting the first unit with probability proportional to size and, at each step thereafter, selecting the next unit with probability proportional to size among those units not already selected, until $n$ distinct units are included in the sample. The inclusion probabilities for this design are not easy to compute, but the conditional probability of selection, given the units selected previously, is easy. Das (1951) and Raj (1956) introduced unbiased estimators based on these conditional selection probabilities. The value of the estimator depends on the order in which the units in the sample were selected. Murthy (1957) improved the estimator by removing the dependence on order—but the improved estimator is not easy to compute. Comparisons of these estimators with others are summarized in Brewer and Hanif (1983, pp. 90–91).

## EXERCISES

1. An unequal probability sample of size 3 is selected from a population of size 10 with replacement. The $y$-values of the selected units are listed along with their draw-by-draw selection probabilities: $y_1 = 3$, $p_1 = 0.06$; $y_2 = 10$, $p_2 = 0.20$; $y_3 = 7$, $p_3 = 0.10$.
   (a) Estimate the population total using the Hansen–Hurwitz estimator.
   (b) Estimate the variance of the estimator.

2. For the design and data of Exercise 1:
   (a) Estimate the population total using the Horvitz–Thompson estimator.
   (b) Give an unbiased estimate of the variance of the estimator. (This may be compared to the value of the simpler alternative estimator.)

3. In a water pollution study, a sample of lakes is selected from the 320 lakes in a study region by the following procedure. A rectangle of length $l$ and width $w$ was drawn around the study region on a map. Pairs of random numbers between 0 and 1 were generated from the uniform distribution using a random number generator on a computer. The first random number of a pair was multiplied by $l$ and the second by $w$ to give location coordinates within the study region. If the location was in a lake, that lake was selected. This process was continued until four of the points had fallen on lakes. The first lake in the sample was selected twice by this process, while each of the other two were selected just once. The pollutant concentrations (in parts per million) for the three lakes in the sample were 2, 5, and 10. The respective sizes of the three lakes (in km²) were 1.2, 0.2, and 0.5. In all, 80 km² of the study region was covered by lakes.
   (a) Describe concisely the type of design used.
   (b) Give an unbiased estimate of the mean pollution concentration per lake in the population.
   (c) Estimate the variance of the estimator above.

4. Use sampling with probability proportional to size, with replacement, to estimate the mean or total of a population of your choice. One example would be the average elevation of lakes on a topographic map. In the process of carrying out the survey and making the estimate, think about or discuss with others the following:

(a) What practical problems arise in establishing a frame, such as a map or list of units, from which to select the sample?

(b) How is the sample selection actually carried out?

(c) What special problems arise in observing the units selected?

(d) Estimate the population mean or total.

(e) Estimate the variance of the estimator used above.

(f) Give a 95% confidence interval for the population proportion.

(g) How would you improve the survey procedure if you were to do it again?

# PART II

# Making the Best Use of Survey Data

# CHAPTER 7

# Auxiliary Data and Ratio Estimation

In addition to the variable of interest $y_i$, one or more auxiliary variables $x_i$ may be associated with the $i$th unit of the population. For example, if the variable of interest is the volume of a tree, the breast-height diameter or an "eyeball" estimate of volume may serve as an auxiliary variable. If the variable of interest is the number of animals in a plot, auxiliary variables could include the area of the plot, the vegetation type, or the average elevation of the plot. In many surveys of human populations, the value of the variable of interest from a previous census may serve as an auxiliary variable.

Auxiliary information may be used either in the sampling design or in estimation. Stratification based on vegetation type or elevation represents a use of auxiliary information in the design. Sampling with replacement with selection probabilities proportional to size—for instance, size of the plot or size of the tree—is another use of auxiliary information in the design.

At the estimation stage, the relationship between $y_i$ and $x_i$ can sometimes be exploited to produce more precise estimates than can be obtained from the $y$-data alone. Ratio and regression estimators are examples of the use of auxiliary information in estimation. In some situations, the $x$-values may be known for the entire population, while in other situations, the $x$'s are known only for units included in the sample.

In this chapter the ratio estimator is examined first in the traditional, design-based context, in which the $y$- and $x$-values for each unit in the population are viewed as fixed and probability is involved only through the sample selection procedure. The ratio estimator is considered first when the design is simple random sampling and then more generally under unequal probability designs. In many sampling situations, it may be reasonable to assume a statistical relationship between the $x$- and $y$-values that goes beyond the sampling design, that is, to assume a regression-through-the-origin model. With such a model, one thinks of the $y$-values as random variables that can take on different values even for a given sample, depending on the whims of nature. The implications of the model-based approach to both estimation and design are discussed in the latter part of this chapter.

## 7.1 RATIO ESTIMATOR

Suppose that the $x$-values are known for the whole population and that the relationship between the $x$'s and the $y$'s is linear and it is reasonable to assume that when $x_i$ is zero, $y_i$ will be zero. For example, as plot size goes to zero, the number of animals on the plot will almost certainly go to zero; as tree diameter approaches zero, so will tree volume.

Let $\tau_x$ denote the population total of the $x$'s (i.e., $\tau_x = \sum_{i=1}^{N} x_i$). Let $\mu_x$ denote the population mean of the $x$'s (i.e., $\mu_x = \tau_x/N$). These population quantities for the $x$-variable are assumed known; the object of inference is to estimate the population mean $\mu$ or total $\tau$ of the $y$-values.

For a simple random sample of $n$ of the units, the sample $y$-values are recorded along with the associated $x$-values. The population ratio $R$ is defined to be

$$R = \frac{\sum_{i=1}^{N} y_i}{\sum_{i=1}^{N} x_i} = \frac{\tau_y}{\tau_x} \tag{1}$$

and the sample ratio $r$ is

$$r = \frac{\sum_{i=1}^{n} y_i}{\sum_{i=1}^{n} x_i} = \frac{\bar{y}}{\bar{x}} \tag{2}$$

The *ratio estimate* of the population mean $\mu$ is

$$\hat{\mu}_r = r\mu_x \tag{3}$$

Since the ratio estimator is not unbiased, its mean square error will be of interest for comparing its efficiency relative to other estimators. The mean square error of the ratio estimator is by definition $\mathrm{mse}(\hat{\mu}_r) = \mathrm{E}(\hat{\mu}_r - \mu_r)^2$. For an unbiased estimator the mean square error equals the variance, but for a biased estimator the mean square error equals the variance plus the bias squared, that is, $\mathrm{mse}(\hat{\mu}_r) = \mathrm{var}(\hat{\mu}_r) + [\mathrm{E}(\hat{\mu}) - \mu]^2$. With the ratio estimator, the squared bias is small relative to the variance, so the first-order approximation to the mean square error is the same as for the variance.

A formula for the approximate mean square error or variance of the ratio estimator is

$$\mathrm{var}(\hat{\mu}_r) \approx \left( \frac{N-n}{N} \right) \frac{\sigma_r^2}{n} \tag{4}$$

where

$$\sigma_r^2 = \frac{1}{N-1} \sum_{i=1}^{N} (y_i - Rx_i)^2$$

The ratio estimator thus tends to be more precise than the sample mean of the $y$-values for populations for which $\sigma_r^2$ is less than $\sigma^2$. This is the case for populations for which the $y$'s and $x$'s are highly correlated, with roughly a linear relationship through the origin.

The traditional estimator of the mean square error or variance of the ratio estimator is

$$\widehat{\mathrm{var}}(\hat{\mu}_r) = \left(\frac{N-n}{N}\right)\frac{s_r^2}{n} \tag{5}$$

where

$$s_r^2 = \frac{1}{n-1}\sum_{i=1}^{n}(y_i - rx_i)^2 \tag{6}$$

The estimator $\widehat{\mathrm{var}}(\hat{\mu}_r)$ tends to give high values with samples having high values of $\bar{x}$ and low values with samples having low values of $\bar{x}$. The adjusted estimator

$$\widetilde{\mathrm{var}}(\hat{\mu}_r) = \left(\frac{\mu_x}{\bar{x}}\right)^2 \widehat{\mathrm{var}}(\hat{\mu}_r) \tag{7}$$

has therefore been suggested (see Cochran 1977, p. 155; P. S. R. S. Rao 1988; Robinson 1987).

An approximate $100(1-\alpha)\%$ confidence interval for $\mu$, based on the normal approximation, is given by

$$\hat{\mu}_r \pm t_{n-1}(\alpha/2)\sqrt{\widehat{\mathrm{var}}(\hat{\mu}_r)} \tag{8}$$

where $t_{n-1}(\alpha/2)$ denotes the upper $\alpha/2$ point of the Student $t$-distribution with $n-1$ degrees of freedom. The alternative variance estimate $\widetilde{\mathrm{var}}(\hat{\mu}_r)$ may be substituted in the confidence interval expression.

The ratio estimate of the population total $\tau$ is

$$\hat{\tau}_r = N\hat{\mu}_r = r\tau_x \tag{9}$$

for which the variance expressions above are multiplied by $N^2$ and the confidence interval endpoints by $N$.

For estimating the population ratio $R$, the sample ratio $r$ may be used. Although not unbiased, it is approximately so with large sample sizes. The approximate variance is

$$\mathrm{var}(r) \approx \left(\frac{N-n}{N\mu_x^2}\right)\frac{\sigma_r^2}{n}$$

An estimate of this variance is

$$\widehat{\text{var}}(r) = \left(\frac{N-n}{N\mu_x^2}\right)\frac{s_r^2}{n}$$

or the adjusted estimator

$$\widetilde{\text{var}}(r) = \left(\frac{N-n}{N\bar{x}^2}\right)\frac{s_r^2}{n}$$

***Example 1: Survey Data with Auxiliary Information.*** In surveys of financial variables, exact or audited figures may be difficult to obtain, whereas less accurate reported figures may be readily available. In such surveys, the actual figures are the variables of interest and the reported figures provide auxiliary information. A small-scale version of such a survey was carried out to estimate the average amount of money—in coins only—carried by $N = 53$ persons in a lecture theater. First, each person was asked to write down a guess $x_i$ of the amount of money that she or he was carrying in change. These amounts were available for all 53 people in the population. Then a simple random sample of $n = 10$ people was selected, and each person in the sample counted the actual amount $y_i$ of money that he or she had in coins.

The average of the values guessed for the whole population was $\mu_x = \$3.33$. (The experiment was carried out in New Zealand, where one- and two-dollar coins had recently been added to the currency; all figures are in New Zealand dollars.) The $(x_i, y_i)$ values for the 10 people in the sample were (\$8.35, \$8.75), (\$1.50, \$2.55), (\$10.00, \$9.00), (\$0.60, \$1.10), (\$7.50, \$7.50), (\$7.95, \$5.00), (\$0.95, \$1.15), (\$4.40, \$3.40), (\$1.00, \$2.00), and (\$0.50, \$1.25).

The sample mean of the amounts guessed is $\bar{x} = \$4.275$, and the sample mean of the actual amounts is $\bar{y} = \$4.17$. The sample ratio is $r = \$4.17/\$4.275 = 0.975$. The ratio estimate of the population mean $\mu$ [using Equation (3)] is

$$\hat{\mu}_r = 0.975(\$3.33) = \$3.25$$

The estimated variance of the ratio estimator [using Equation (5)] is

$$\begin{aligned}
\widehat{\text{var}}(\hat{\mu}_r) &= \left(\frac{53-10}{53}\right)\frac{1}{(10)9}\{[\$8.75 - 0.975(\$8.35)]^2 \\
&\quad + [\$2.55 - 0.975(\$1.50)]^2 + \cdots + [\$1.25 - 0.975(\$0.50)]^2\} \\
&= \frac{43}{53}\left(\frac{1.387}{10}\right) = \$0.1125
\end{aligned}$$

An approximate 95% confidence interval [using Equation (8)] is

$$\$3.25 \pm 2.262\sqrt{0.1125} = \$3.25 \pm \$0.76 = (\$2.49, \$4.01)$$

in which 2.262 is the upper 0.025 point of the $t$-distribution with 9 degrees of freedom.

The adjusted variance estimator for this example [from Equation (7)] is

$$\widetilde{\text{var}}(\hat{\mu}_r) = \left(\frac{\$3.33}{\$4.275}\right)^2 (0.1125) = 0.0683$$

and the 95% confidence interval based on this variance estimator is $3.25 ± $0.59 = ($2.66, $3.84)$.

The estimate of the population mean based on the sample $y$-values only, ignoring the auxiliary data, is $\bar{y} = \$4.17$. The sample variance of the $y$-values is $s^2 = \$10.12$, and the estimated variance of $\bar{y}$ is $\widehat{\text{var}}(\bar{y}) = (53 - 10)\ (\$10.12)/[53(10)] = \$0.8211$. The 95% confidence interval is $4.17 ± $2.05 = ($2.12, $6.22)$.

In essence, the ratio estimator adjusts for our sample with higher-than-average $x$-values, since $\bar{x} = 4.275$ is greater than $\mu_x = 3.33$. At the end of the survey, the remaining members of the class were asked to count their money, and the true population mean was determined to be $\mu = \$3.52$. The $x$- and $y$-values for the entire population is plotted in Figure 7.1. The slope of the line in the figure is the population ratio $R$.                                                              □

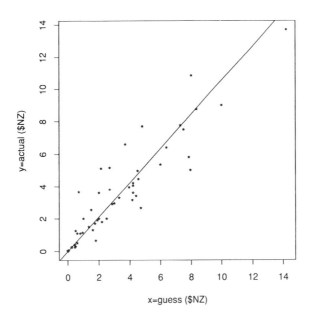

**Figure 7.1.** Plot of the amounts guessed $(x)$ and the actual amounts $(y)$ of money ($NZ) in coins on 53 people. The fitted ratio line is shown.

## 7.2   SMALL POPULATION ILLUSTRATING BIAS

The bias and mean square error under simple random sampling of the ratio estimator can be illustrated by imagining the sampling of a very small population and
looking at the sample space, that is, the set of all possible samples. Imagine that
it is desired to estimate the total number of fish caught, in a given day, along a river
on which fish are caught in nets fixed adjacent to established fishing sites. Suppose
that there are $N = 4$ sites along the river and suppose that the number of nets $x_i$ at
each site in the population is readily observed, say from an aircraft flying the length
of the river. The number $y_i$ of fish caught can be supposed to be obtained only with
more difficulty, by visiting a given site at the end of the fishing day. A simple random sample of $n = 2$ sites will be selected and ratio estimation used to estimate the
total number of fish caught.

Values for the entire population are listed in the following table:

| Site, $i$ | 1 | 2 | 3 | 4 |
|---|---|---|---|---|
| Nets, $x_i$ | 4 | 5 | 8 | 5 |
| Fish, $y_i$ | 200 | 300 | 500 | 400 |

The actual population total $\tau$—the number to be estimated by the samplers—is
1400 fish caught. The population total for the auxiliary variable—which is known
to the samplers—is 22 nets. If, for example, the samplers selected $s = \{1, 2\}$, consisting of the first and second sites, the ratio estimator [Equation (9)] would be
$\hat{\tau}_r = (22)(200 + 300)/(4 + 5) = 1222$.

The number of possible samples [from Equation (15) of Chapter 2] is

$$\binom{N}{n} = \binom{4}{2} = \frac{4!}{2!\,(4-2)!} = 6$$

Table 7.1 lists every possible sample, along with the value of the ratio estimator for
that sample of sites.

Since each possible sample has the same probability $P(s) = 1/6$ of being the one
selected with simple random sampling, the expected value of the ratio estimator

**Table 7.1.  Samples for Small Population with Bias**

| Sample | $\hat{\tau}_r$ |
|---|---|
| $(1, 2)$ | 1222 |
| $(1, 3)$ | 1283 |
| $(1, 4)$ | 1467 |
| $(2, 3)$ | 1354 |
| $(2, 4)$ | 1540 |
| $(3, 4)$ | 1523 |

$E(\hat{\tau}_r) = \sum_s \hat{\tau}_{rs}P(s)$ is the arithmetic average of the six possible values of the estimator. Thus, $E(\hat{\tau}_r) = 1398.17$, while $\tau = 1400$, so the ratio estimator is slightly biased under simple random sampling with this population.

The mean square error for the ratio estimator with this population is

$$\text{mse}(\hat{\tau}_r) = \sum_s (\hat{\tau}_{rs} - \tau)^2 p(s)$$

$$= (1222 - 1400)^2 \left(\frac{1}{6}\right) + \cdots + (1523 - 1400)^2 \left(\frac{1}{6}\right)$$

$$= 14{,}451.2$$

and the root mean square error is $\sqrt{14{,}451} = 120$. The variance would be obtained by subtracting 1398.17, rather than 1400, from each term before squaring, giving $\text{var}(\hat{\tau}_r) = 14{,}447.8$. The bias is $1398.17 - 1400 = -1.83$, and the squared bias is 3.4.

With the same sampling design, the samplers could use the unbiased estimator $\hat{\tau} = N\bar{y}$ based on the sample $y$-values only. The variance of this estimator, by the standard formula, is 66,667, with square root (standard error) 258. Thus, although the ratio estimator is slightly biased with this population, it is considerably more precise than the unbiased estimator based on the $y$-values alone.

## 7.3 DERIVATIONS AND APPROXIMATIONS FOR THE RATIO ESTIMATOR

Because the sample mean $\bar{x}$ of the $x$'s, as well as the sample mean $\bar{y}$ of the $y$'s, varies from sample to sample, the sample ratio $r = \bar{y}/\bar{x}$ is not unbiased, over all possible samples under simple random sampling, for the population ratio $R = \mu/\mu_x$, even though $E(\bar{y}) = \mu$ and $E(\bar{x}) = \mu_x$. In fact, an exact expression for the bias of $r$ can be obtained (Hartley and Ross 1954) as follows:

$$\text{cov}(r, \bar{x}) = E(r\bar{x}) - E(r)E(\bar{x})$$
$$= \mu - \mu_x E(r)$$

so that the expectation of $r$ is

$$E(r) = \frac{\mu - \text{cov}(r, \bar{x})}{\mu_x} = R - \text{cov}(r, \bar{x})$$

Thus the bias is $E(r) - R = -\text{cov}(r, \bar{x})/\mu_x$.

Since the covariance of two random variables cannot exceed in absolute value the product of their standard deviations,

$$|E(r) - R| \leq \sqrt{\frac{\text{var}(r)\text{var}(\bar{x})}{\mu_x}}$$

so that

$$\frac{|E(r) - R|}{\sqrt{\text{var}(r)}} \leq \frac{\sqrt{\text{var}(\bar{x})}}{\mu_x}$$

That is, the magnitude of bias relative to the standard deviation of the estimator is no greater than the coefficient of variation of $\bar{x}$.

The approximate mean square error of the ratio estimator $\hat{\mu}_r$ is obtained by using a linear approximation for the (nonlinear) function $f(\bar{x}, \bar{y}) = \bar{y}/\bar{x}$. The linear approximating function is $(y_i - Rx_i)/\mu_x$, obtained as the first term in a Taylor series of $f$ about the point $(\mu_x, \mu)$. The expected value of $y_i - Rx_i$ under simple random sampling is zero, since $E(y_i) = \mu$ and $E(x_i) = \mu_x$ and $R\mu_x = \mu$. Thus, the mean square error of the ratio estimator $\hat{\mu}_r$ is approximated by the variance of the variables $y_i - Rx_i$, which under simple random sampling is given by Equation (4).

Proceeding in more detail, the Taylor series expansion of a function $g(x, y)$ about a point $(a, b)$ is

$$g(x, y) = g(a, b) + g_x(a, b)(x - a) + g_y(a, b)(y - b)$$
$$+ \frac{1}{2}g_{xx}(a, b)(x - a)^2 + g_{xy}(a, b)(x - a)(y - b) + \frac{1}{2}g_{yy}(a, b)(y - b)^2$$
$$+ \cdots + \frac{1}{p! \, q!}g_{x^p y^q}(a, b)(x - a)^p(y - b)^q + \cdots$$

where $g_x(a, b)$ is the partial derivative of $g$ with respect to $x$ evaluated at $(a, b)$ and, more generally,

$$g_{x^p y^q}(a, b) = \frac{\partial^{p+q}}{\partial x^p \, \partial y^q} g(x, y)$$

evaluated at $x = a$, $y = b$.

For the sample ratio $r = \bar{y}/\bar{x}$, let

$$g(x, y) = \frac{y}{x}$$

and approximate $r = g(\bar{x}, \bar{y})$ by expanding about the point $(\mu_x, \mu_y)$. The first term of the expansion is $g(\mu_x, \mu_y) = \mu_y/\mu_x = R$. The partial derivative of $g$ with respect to $x$ is $(\partial/\partial x)g(x, y) = -y/x^2$, which evaluated at $(\mu_x, \mu_y)$ gives $g_x(\mu_x, \mu_y) = -\mu_y/\mu_x^2$. Similarly, $g_y(\mu_x, \mu_y) = 1/\mu_x$. The first-order approximation for the sample ratio $r$ is thus

$$r = g(\bar{x}, \bar{y})$$
$$\approx \frac{\mu_y}{\mu_x} - \frac{\mu_y}{\mu_x^2}(\bar{x} - \mu_x) + \frac{1}{\mu_x}(\bar{y} - \mu_y)$$
$$= R - \frac{\mu_y}{\mu_x^2}(\bar{x} - \mu_x) + \frac{1}{\mu_x}(\bar{y} - \mu_y)$$

Under simple random sampling, the expected value of the sample ratio is, to a first-order approximation,

$$E(r) \approx R$$

since $E(\bar{x} - \mu_x) = 0$ and $E(\bar{y} - \mu_y) = 0$. The approximate variance from the first-order approximation is

$$\text{var}(r) \approx \left(\frac{\mu_y}{\mu_x^2}\right)^2 \text{var}(\bar{x}) + \frac{1}{\mu_x^2}\text{var}(\bar{y}) + 2\frac{\mu_y}{\mu_x^3}\text{cov}(\bar{x}, \bar{y})$$

$$= \frac{1}{\mu_x^2}\text{var}(\bar{y} - R\bar{x})$$

With simple random sampling, $\text{var}(\bar{x}) = [(N - n)/Nn]\sigma_x^2$ and $\text{var}(\bar{y}) = [(N - n)/Nn]\sigma_y^2$. The covariance can be shown to be

$$\text{cov}(\bar{x}, \bar{y}) = \frac{N - n}{Nn}\sum_{i=1}^{N}\frac{(x_i - \mu_x)(y_i - \mu_y)}{N - 1}$$

This is shown either by using the inclusion indicator variable $z_i$ as in the derivation for $\text{var}(\bar{y})$ under simple random sampling, or by defining for each unit a variable $u_i = x_i + y_i$ and noting that under simple random sampling,

$$\text{var}(\bar{u}) = \frac{N - n}{Nn(N - 1)}\sum_{i=1}^{N}(x_i - \mu_x + y_i - \mu_y)^2$$

$$= \frac{N - n}{Nn(N - 1)}\sum_{i=1}^{N}[(x_i - \mu_x)^2 + (y_i - \mu_y)^2 + 2(x_i - \mu_x)(y_i - \mu_y)]$$

$$= \text{var}(\bar{y}) + \text{var}(\bar{x}) + 2\frac{N - n}{Nn(N - 1)}\sum_{i=1}^{N}(x_i - \mu_x)(y_i - \mu_y)$$

Since also

$$\text{var}(\bar{u}) = \text{var}(\bar{x} + \bar{y}) = \text{var}(\bar{x}) + \text{var}(\bar{y}) + 2\,\text{cov}(\bar{x}, \bar{y})$$

the expression for $\text{cov}(\bar{x}, \bar{y})$ follows by subtraction.

Approximations using more terms than one of Taylor's formula are used in examining the bias, mean square error, and higher moments of the ratio estimator. It is worth noting in passing that for most populations, the Taylor series for $\bar{y}/\bar{x}$ does not converge for some samples, specifically those for which $|\bar{x} - \mu_x| > \mu_x$. For those samples, the approximation becomes worse, not better, as more terms are added to the series. Even so, however, expectations under simple random sampling of the approximations based on the initial one to several terms of the series do provide consistent and useful approximations, which get closer to the true values the larger the sample size, as shown in David and Sukhatme (1974), who in addition gave bounds for the absolute difference between the actual and the approximating bias and mean square error.

## 7.4  FINITE-POPULATION CENTRAL LIMIT THEOREM FOR THE RATIO ESTIMATOR

The finite-population central limit theorem for the ratio estimator is proved in Scott and Wu (1981). As with the finite-population central limit theorem for the sample mean, one conceives of a sequence of populations, with both population size $N$ and sample size $n$ increasing. The theorem states that the standardized ratio estimator

$$\frac{\hat{\mu}_r - \mu}{\sqrt{\widehat{\text{var}}(\hat{\mu}_r)}}$$

has an asymptotic standard normal distribution as $n$ and $N - n$ tend to infinity, provided that certain conditions are satisfied. One of these conditions requires that the proportion of $\sigma_r^2$ due to outliers should not be too large, while another condition requires that the coefficient of variation of $\bar{x}$ should not be too large.

Of course, the central limit theorem cannot tell one exactly how large the sample size must be for a given population to ensure that the coverage probabilities for confidence intervals are adequate. A common rule of thumb prescribes a sample size of at least 30. Empirical studies by Royall and Cumberland (1985) with sample sizes of 32 showed that for some populations—particularly those in which the natural relationship between $y$ and $x$ was not a proportional ratio relationship—the coverage probabilities were much lower than the nominal level and tended to depend conditionally on the sample $\bar{x}$ values.

## 7.5  RATIO ESTIMATION WITH UNEQUAL PROBABILITY DESIGNS

For a design in which unit $i$ has probability $\pi_i$ of inclusion in the sample, used in sampling a population in which the variable of interest $y$ has a linear relationship through the origin with an auxiliary variable $x$, a generalization of the ratio estimator for estimating the population total has been suggested (Brewer 1963; Cassel et al. 1976, 1977, pp. 122, 151; Hájek 1971, 1981; Sukhatme and Sukhatme 1970). The generalized ratio estimator is

$$\hat{\tau}_G = \frac{\hat{\tau}_y}{\hat{\tau}_x} \tau_x$$

The components $\hat{\tau}_y$ and $\hat{\tau}_x$ are Horvitz–Thompson estimators of $\tau$ and $\tau_x$, respectively, that is,

$$\hat{\tau}_y = \sum_{i=1}^{\nu} \frac{y_i}{\pi_i}$$

and

$$\hat{\tau}_x = \sum_{i=1}^{\nu} \frac{x_i}{\pi_i}$$

Note that the usual ratio estimator is a special case of the generalized ratio estimator under simple random sampling, in which $\pi_i = n/N$ for all units.

Although $\hat{\tau}_y$ is an unbiased estimator of $\tau$ and $\hat{\tau}_x$ is unbiased for $\tau_x$, the generalized ratio estimator, as a ratio estimator, is not unbiased in the design sense for the population total. It is recommended in cases in which the $y$-values are roughly proportional to the $x$-values, so that the variance of the residuals $y_i - Rx_i$ is much smaller than the variance of the $y$-values themselves.

For an approximate formula for the mean square error or variance of $\hat{\tau}_G$, the Taylor series leads to using the variance formula for the Horvitz–Thompson estimator with $y_i - Rx_i$ as the variable of interest, where $R = \tau_y/\tau_x$ is the population ratio. For estimating the variance, the corresponding Horvitz–Thompson formula with the estimate $\hat{R} = \hat{\tau}_y/\hat{\tau}_x$ can be used (see Cochran 1977, p. 271).

The generalized ratio estimator may be written

$$\hat{\tau}_G = \hat{\tau}_y + \hat{R}(\tau_x - \hat{\tau}_x)$$

The first term in Taylor's formula, expanding about the point $(\tau_x, \tau)$, gives the approximation $\hat{\tau}_G \approx \hat{\tau}_y + R(\tau_x - \hat{\tau}_x)$.

The mean square error or variance of $\hat{\tau}_G$ may be approximated by substituting the population ratio $R$ for the estimate $\hat{R}$, giving

$$\hat{\tau} - \tau \approx \hat{\tau}_y + R(\tau_x - \hat{\tau}_x) - \tau = \hat{\tau}_y - R\hat{\tau}_x$$

since $R\tau_x = \tau$. Noting that $E(\hat{\tau}_y - R\hat{\tau}_x) = 0$, the approximation for the mean square error or variance is

$$\mathrm{var}(\hat{\tau}_G) \approx E(\hat{\tau}_y - R\hat{\tau}_x)^2 = \mathrm{var}(\hat{\tau}_y - R\hat{\tau}_x) = \mathrm{var}\left(\sum_{i=1}^{\nu} \frac{y_i - Rx_i}{\pi_i}\right)$$

The approximate variance is thus the variance of a Horvitz–Thompson estimator based on the variables $y_i - Rx_i$. Denoting $y_i - Rx_i$ by $y_i'$, the approximate formula is

$$\mathrm{var}(\hat{\tau}_G) \approx \sum_{i=1}^{N}\left(\frac{1 - \pi_i}{\pi_i}\right)y_i'^2 + \sum_{i=1}^{N}\sum_{j \neq i}\left(\frac{\pi_{ij} - \pi_i\pi_j}{\pi_i\pi_j}\right)y_i'y_j'$$

An estimator of this variance is obtained using $\hat{y}_i = y_i - \hat{R}x_i$ in the Horvitz–Thompson variance estimation formula:

$$
\begin{aligned}
\widehat{\text{var}}(\hat{\tau}_G) &= \sum_{i=1}^{\nu}\left(\frac{1-\pi_i}{\pi_i^2}\right)\hat{y}_i^2 + \sum_{i=1}^{\nu}\sum_{j\neq i}\left(\frac{\pi_{ij}-\pi_i\pi_j}{\pi_i\pi_j}\right)\frac{\hat{y}_i\hat{y}_j}{\pi_{ij}} \\
&= \sum_{i=1}^{\nu}\left(\frac{1}{\pi_i^2}-\frac{1}{\pi_i}\right)\hat{y}_i^2 + \sum_{i=1}^{\nu}\sum_{j\neq i}\left(\frac{1}{\pi_i\pi_j}-\frac{1}{\pi_{ij}}\right)\hat{y}_i\hat{y}_j
\end{aligned}
\tag{10}
$$

assuming all of the joint inclusion probabilities $\pi_{ij}$ are greater than zero.

Recall that the Horvitz–Thompson estimator has small variance when the $y$-values are approximately proportional to the inclusion probabilities, whereas if there is no such relationship, the variance of that estimator can be very large. An example in which the Horvitz–Thompson estimator performs very badly, because inclusion probabilities were poorly related to $y$-values, is the notorious "Circus Elephant Example" of Basu (1971, pp. 212–213). The circus owner in the example is planning to ship his 50 elephants and wishes to obtain a rough estimate of their total weight by weighing just one elephant. The average elephant ("Sambo" in Basu's example) is given probability $\pi = 99/100$ of being the one selected, while each of the other elephants is given inclusion probability of $1/4900$. Thus if Sambo is selected and found to weigh $y$, the total weight of the 50 elephants is estimated to be $\hat{\tau} = y/\pi = 100y/99$, or about the weight of one average elephant, while if the largest elephant ("Jumbo") is selected, the weight of the 50 elephants is estimated to be 4900 times Jumbo's weight! The generalized ratio estimator is recommended for avoiding this sort of problem (Hájek 1971, p. 236).

If unequal probability sampling has for one reason or another been used and there is no linear relationship between the $y$-values and either the inclusion probabilities or any auxiliary variable, the following estimator, which is the generalized ratio estimator with $x_i = 1$ for all units, can be used:

$$
\hat{\tau}_G = \frac{\sum_{i-1}^{\nu}(y_i/\pi_i)}{\sum_{i=1}^{\nu}(1/\pi_i)} N
\tag{11}
$$

The corresponding estimator of the population mean is

$$
\hat{\mu}_G = \frac{\sum_{i=1}^{\nu}(y_i/\pi_i)}{\sum_{i=1}^{\nu}(1/\pi_i)}
$$

For the estimator of variance $\hat{\tau}_G$ based on the Taylor approximation, one uses the Horvitz–Thompson variance estimation formula with $\hat{y}_i = y_i - \hat{\mu}_G$ as the variable of interest. The estimator of the mean square error or variance of $\hat{\mu}_G$ is obtained by dividing by $N^2$.

***Example 2.***   In the aerial survey (Example 1 of Chapter 6), a sample of $\nu = 3$ distinct observations was selected from a population of $N = 100$ units. The sample

$y$-values were 60, 14, and 1 with respective inclusion probabilities 0.1855, 0.0776, and 0.0394. The Horvitz–Thompson estimate was 529 of the animals in the study region (see Example 2 of Chapter 6). Using the same data, the generalized ratio estimate [using Equation (11)] is

$$\hat{\tau}_G = \frac{60/0.1855 + 14/0.0776 + 1/0.0394}{1/0.1855 + 1/0.0776 + 1/0.0394}(100)$$

$$= \frac{529.24}{43.66}(100) = 1212$$

The estimate of the mean per unit is $\hat{\mu}_G = 1212/100 = 12.12$.

For estimating the variance of $\hat{\tau}_G$ we will use $\hat{y}_1 = 60 - 12.12 = 47.88$, $\hat{y}_2 = 14 - 12.12 = 1.88$, and $\hat{y}_3 = 1 - 12.12 = -11.12$. The joint inclusion probabilities, computed in Example 2 of Chapter 6, are 0.0112, 0.0056, and 0.0023. The estimated variance [using Equation (10)] is

$$\widehat{\mathrm{var}}(\hat{\tau}_G) = \left(\frac{1}{0.1855^2} - \frac{1}{0.1855}\right)47.88^2 + \left(\frac{1}{0.0776^2} - \frac{1}{0.0776}\right)1.88^2$$

$$+ \left(\frac{1}{0.0394^2} - \frac{1}{0.0394}\right)(-11.12)^2$$

$$+ 2\left[\frac{1}{0.1855(0.0776)} - \frac{1}{0.0112}\right](47.88)(1.88)$$

$$+ 2\left[\frac{1}{0.1855(0.0394)} - \frac{1}{0.0056}\right](47.88)(-11.12)$$

$$+ 2\left[\frac{1}{0.0776(0.0394)} - \frac{1}{0.0023}\right](1.88)(-11.12)$$

$$= 176{,}714$$

with square root 420.

In this example, the ordinary Horvitz–Thompson is probably to be preferred, since there does seem to be a natural relationship between the $y$-value and the inclusion probability based on unit size. $\square$

## 7.6   MODELS IN RATIO ESTIMATION

Up to now, the population has been viewed as a collection of $N$ units with fixed $y$-values $y_i, \ldots, y_N$, with probability entering the situation only through the procedure by which the sample is selected. Another view is that the population $y$-values are a realization of a vector $\mathbf{Y} = (Y_1, \ldots, Y_N)$ of random variables. For example, if the population is a study region consisting of $N$ plots and the variable of interest $y_i$ for the $i$th plot is the volume of wood in the trees on that plot, one conceives of the amount of timber as a random variable depending on stochastic or unknown processes in nature, including such factors as rainfall and temperature.

If the population $Y$-values are random variables, so is any function of them such as their total $\tau = \sum_{i=1}^{N} Y_i$. Having observed the realized values of the $n$ random variables in the sample, one may wish to predict or estimate the value of another random variable such as $\tau$. The prediction approach to survey sampling under ratio and regression models has been described and developed in Cumberland and Royall (1981, 1988), Royall (1970, 1976a,b, 1988), Royall and Cumberland (1978, 1981a,b, 1985), Royall and Eberhardt (1975), Royall and Herson (1973a,b), and Royall and Pfeffermann (1982). Ericson (1969, 1988) describes some similar results attained through the related Bayesian approach.

The small-population example shows the sense in which the ratio estimator is biased under simple random sampling. The mean value of the estimator over all possible samples is not equal to the population parameter $\tau$. Hence, the ratio estimator is not design-unbiased under simple random sampling. Similarly, the associated variance estimator is not unbiased. However, under the additional assumption of an ordinary regression model for the relationship between the $x$'s and the $y$'s, with the regression line passing through the origin, the ratio estimator is *model-unbiased*. Consider the regression model

$$Y_i = \beta x_i + \varepsilon_i$$

in which $\beta$ is a fixed, though unknown parameter, the $x_i$'s are known, and the $\varepsilon_i$ are independent random variables with $\mathrm{E}(\varepsilon_i) = 0$. Under this model, the expected value of $Y_i$ given $x_i$ is

$$\mathrm{E}(Y_i) = \beta x_i$$

so the expected value of the estimator $\hat{\tau}_r$ is

$$\mathrm{E}(\hat{\tau}_r) = \frac{\sum_{i=1}^{n} \beta x_i}{\sum_{i=1}^{n} x_i} \tau_x = \beta \tau_x = \mathrm{E}\left(\sum_{i=1}^{N} Y_i\right) = \mathrm{E}(\tau)$$

no matter which sample $s$ is selected. The expectations above and throughout this section are with respect to the model assumed and are conditional on the sample selected.

If, in addition to the model assumptions above, the $Y_i$'s are assumed to be uncorrelated, with variance

$$\mathrm{var}(Y_i) = v_i \sigma_R^2$$

then it is well known from linear model results that the best linear unbiased estimator of $\beta$ is

$$\hat{\beta} = \frac{\sum_{i=1}^{n} (x_i Y_i / v_i)}{\sum_{i=1}^{n} (x_i^2 / v_i)}$$

The variance under the model of $\hat{\beta}$ is

$$\text{var}(\hat{\beta}) = \frac{\sigma_R^2}{\sum_{i=1}^{n}(x_i^2/v_i)}$$

An unbiased estimator of this variance is

$$\widehat{\text{var}}(\hat{\beta}) = \frac{\hat{\sigma}_R^2}{\sum_{i=1}^{n}(x_i^2/v_i)}$$

where

$$\hat{\sigma}_R^2 = \frac{1}{n-1}\sum_{i=1}^{n}\frac{(Y_i - \hat{\beta}x_i)^2}{v_i} \tag{12}$$

Since the $Y$-values in the population are random variables, so is their total $\tau = \sum_{i=1}^{N} Y_i$. Therefore, instead of estimating a fixed quantity, one needs to use the sample to predict the value of a random variable $\tau$. One would like a predictor $\hat{\tau}$ that is model unbiased for $\tau$, that is, $E(\hat{\tau}) = E(\tau)$, and having low mean square prediction error $E(\hat{\tau} - \tau)^2$ under the model. The unbiased linear predictor with the lowest mean square error can be shown to be

$$\hat{\tau} = \sum_s Y_i + \hat{\beta}\sum_{s'} x_i$$

where $s$ denotes summation over the sample and $s'$ denotes summation over the rest of the population. Thus, to predict the population total, one uses the known total in the sample and predicts the total for the rest of the population. Writing $Y_s$ and $x_s$ to denote the sum of the $Y$- and $x$-values, respectively, in the sample and $Y_{s'}$ and $x_{s'}$ for the sums in the rest of the population, the predictor may be written

$$\hat{\tau} = Y_s + \hat{\beta}x_{s'}$$

The mean square prediction error is

$$
\begin{aligned}
E(\hat{\tau} - \tau)^2 &= \text{var}(\hat{\tau} - \tau) \\
&= \text{var}(\hat{\beta}x_{s'} - Y_{s'}) \\
&= x_{s'}^2\,\text{var}(\hat{\beta}) + \text{var}(Y_{s'})
\end{aligned}
$$

since the two terms are independent under the assumed model. Thus,

$$E(\hat{\tau} - \tau)^2 = \left[\frac{x_{s'}^2}{\sum_s(x_i^2/v_i)} + \sum_{s'} v_i\right]\sigma_R^2$$

An unbiased estimator of the mean square error is obtained by substituting $\hat{\sigma}_R^2$ for $\sigma_R^2$.

If the variance of $Y_i$ is proportional to $x_i$, that is, $v_i = x_i$, the best linear estimator is

$$\hat{\beta} = \frac{\sum_{i=1}^n Y_i}{\sum_{i=1}^n x_i} = r$$

and the estimator of the population total is

$$\hat{\tau} = \sum_s Y_i + \frac{\sum_s Y_i}{\sum_s x_i} \sum_{s'} x_i = r\tau_x$$

the ratio estimator.

The mean square error of the ratio estimator under this model is

$$
\begin{aligned}
\mathrm{E}(\hat{\tau} - \tau)^2 &= \frac{\tau_x(\tau_x - x_s)}{x_s}\sigma_R^2 \\
&= \frac{N(N-n)}{n}\left(\frac{\mu_x \bar{x}_{s'}}{\bar{x}_s}\right)\sigma_R^2
\end{aligned}
\tag{13}
$$

An unbiased estimator of this mean square error is obtained by substituting for $\sigma_R^2$:

$$\hat{\sigma}_R^2 = \frac{1}{n-1}\sum_{i=1}^n \frac{(Y_i - \hat{\beta}x_i)^2}{x_i}$$

For estimating the population mean, the estimator of $\tau$ is divided by $N$ and the variance and variance estimate divided by $N^2$; that is, $\hat{\mu} = \hat{\tau}/N$, so that $\mathrm{E}(\hat{\mu} - \mu)^2 = \mathrm{E}(\hat{\tau} - \tau)^2/N^2$ and $\hat{E}(\hat{\mu} - \mu)^2 = \hat{E}(\hat{\tau} - \tau)^2/N^2$.

This estimator of mean square error, although unbiased when the model is true, has been found to be sensitive to departures from the assumed model (Royall 1988; Royall and Cumberland 1978, 1981a,b; Royall and Eberhardt 1975). Also, the mean square error estimator given earlier in Equation (2) tends to overestimate mean square error with samples having large $x$-values and underestimate with samples having small $x$-values. One robust alternative proposed is

$$\tilde{\mathrm{E}}(\hat{\tau} - \tau)^2 = N^2 \frac{\mu_x \bar{x}_{s'}}{\bar{x}_s^2(1 - V^2/n)}\widehat{\mathrm{var}}(\hat{\mu}_r)$$

where $V = \sum_{i=1}^n(x_i - \bar{x})^2/[(n-1)\bar{x}^2]$ and $\widehat{\mathrm{var}}(\hat{\mu}_r)$ is given by Equation (5). Another is the jackknife estimator

$$\hat{\mathrm{E}}_J(\hat{\tau} - \tau)^2 = \frac{(N-n)(n-1)}{Nn}\sum_{i=1}^n (t_{(i)} - \bar{t})^2$$

where $t_{(i)}$ is the ratio estimate of $\tau$ with the $i$th unit of the sample omitted and $\bar{t}$ is the mean of these estimates.

Under a model assuming the $Y$'s to be independent, the standard central limit theorem for independent random variables applies, although it is still required to conceive of both $n$ and $N - n$ becoming large. An approximate $100(1 - \alpha)\%$ confidence interval is given by

$$\hat{\tau} \pm t_{n-1}(\alpha/2)\sqrt{\hat{E}(\hat{\tau} - \tau)^2}$$

The confidence interval is exact under the model if the $Y_i$ are normally distributed.

**Example 3: Change Survey with Model.**   Suppose that the model $E(Y_i) = \beta x_i$, $\text{var}(Y_i) = x_i \sigma_R^2$, $\text{cov}(Y_i, Y_j) = 0$ for $i \neq j$ is assumed for the pocket change survey discussed earlier (Example 1). Then the best linear unbiased predictor of average change per person is the ratio estimate $\hat{\mu} = 3.25$. The term $\hat{\sigma}_R^2$ [using Equation (12)] is

$$\hat{\sigma}_R^2 = \frac{1}{9}\left\{ \frac{[\$8.75 - 0.975(\$8.35)]^2}{8.35} + \cdots + \frac{[\$1.25 - 0.975(\$0.50)]^2}{0.50} \right\}$$

$$= \$0.526$$

The sum of the $x$-values in the population is $\tau_x = 53(\$3.33) = \$176.5$, and the sum of the $x$-values in the sample is $x_s = \$42.75$. The model-unbiased estimate of mean square error [using Equation (13) divided by $N^2$] is

$$\hat{E}(\hat{\mu} - \mu)^2 = \frac{176.5(176.5 - 42.75)}{53^2(42.75)}\hat{\sigma}_R^2$$

$$= 0.1968(0.526) = 0.1034$$

A 95% confidence interval, which is exact under the assumed model if the $Y$-values are normally distributed, is

$$\$3.25 \pm 2.262\sqrt{0.1034} = \$3.25 \pm \$0.73 = (\$2.52, \$3.98) \qquad \square$$

## Types of Estimators for a Ratio

Consider the linear model

$$Y_i = \beta x_i + \epsilon_i$$

with

$$E(\epsilon_i) = 0$$

and

$$\mathrm{var}(\epsilon_i) = v_i \sigma_R^2$$

where $\beta$ and $\sigma_R^2$ are unknown parameters of the model. The best linear unbiased estimator of the ratio parameter $\beta$ is

$$\hat{\beta} = \frac{\sum_{i=1}^{n} (x_i Y_i / v_i)}{\sum_{i=1}^{n} (x_i^2 / v_i)}$$

Note that $\mathrm{E}(Y_i) = \beta x_i$ and $\mathrm{var}(Y_i) = v_i \sigma_R^2$.

When the variance of $\epsilon_i$ is proportional to $x_i$, that is, $v_i = x_i$, the best linear unbiased estimator is the *ratio estimator*

$$\hat{\beta} = \frac{\sum_{i=1}^{n} y_i}{\sum_{i=1}^{n} x_i}$$

When the standard deviation of $\epsilon_i$ is proportional to $x_i$, that is, $v_i = x_i^2$, the best linear estimator is the *mean-of-the-ratios estimator*

$$\hat{\beta} = \frac{1}{n} \sum_{i=1}^{n} \frac{y_i}{x_i}$$

When the variance of $\epsilon_i$ is constant, that is, $v_i = 1$, the best linear unbiased estimator is the *regression-through-the-origin estimator*

$$\hat{\beta} = \frac{\sum_{i=1}^{n} x_i y_i}{\sum_{i=1}^{n} x_i^2}$$

## 7.7 DESIGN IMPLICATIONS OF RATIO MODELS

If one believes the regression-through-the-origin model without question, one can question the necessity of random sampling. Regression theory asserts that under such a model the most precise estimates are obtained by deliberately selecting the $n$ observations with the largest $x$-values. If the assumed model is wrong, such a sample may lead to bias and inefficiency in the ratio estimator and does not offer the opportunity to examine the fit of the model.

A compromise approach, designed to provide *robustness* against departures from the assumed model, is to seek to obtain a sample that is *balanced* or *representative* in terms of the $x$'s. A sample balanced in $x$ is a sample for which the sample mean $\bar{x}$ of the $x$-values equals the population mean $\mu_x$ of the $x$-values. If the sample is balanced, the ratio estimator $\hat{\mu}_r$ equals the sample mean $\bar{y}$. The investigations by

Royall and his coauthors showed that with balanced samples, not only was the ratio estimator robust against departures from the assumed model, but estimators of its variance were also robust. Simple random sampling with a sufficiently large sample size tends to produce samples balanced in the auxiliary variable. Stratifying on the auxiliary variable and other, more complicated schemes may also be used to ensure samples that are at least approximately balanced. Purposively choosing a sample balanced in an auxiliary variable may seem temptingly ideal from a model-based viewpoint, but can lead to unexpected problems when other important auxiliary variables are unknown.

For the generalized ratio estimator, a representative sample is one for which $\hat{\mu}_x$, the Horvitz–Thompson estimator based on the $x$-values, equals $\mu_x$, the population mean of the $x$-values. With a representative sample, the generalized ratio estimator is identical to the Horvitz–Thompson estimator based on the $y$-values.

## EXERCISES

1. Consider the following data from a simple random sample of size $n = 2$ from a population of size $N = 8$, in which $y$ is the variable of interest and $x$ is an auxiliary variable: $y_1 = 50$, $x_1 = 10$; $y_2 = 22$, $x_2 = 2$. The population mean of the $x$'s is 5.

    (a) Give the ratio estimate of the mean of the $y$-values.

    (b) Estimate the variance of the ratio estimator above.

2. In a city of 72,500 people, a simple random sample of four households is selected from the 25,000 households in the population to estimate the average cost on food per household for a week. The first household in the sample had 4 people and spent a total of $150 in food that week. The second household had 2 people and spent $100. The third, with 4 people, spent $200. The fourth, with 3 people, spent $140.

    (a) Identify the sampling units, the variable of interest, and any auxiliary information associated with the units.

    (b) Describe two types of estimators for estimating the mean expenditure per household for a week's food in the city. Summarize some properties of each estimator.

    (c) Estimate mean expenditure using the first estimator, and estimate the variance of the estimator.

    (d) Estimate mean expenditure using the other estimator, and estimate the variance of the estimator.

    (e) Based on the data, which estimator appears preferable in this situation?

3. For a hypothetical survey to determine the number of pileated woodpecker nests, the study area is divided into $N = 4$ plots. For the $i$th plot in the population, $y_i$ is

the number of nests, while $x_i$ is the number of "snags" (old trees that provide nesting habitat). The values for each population unit follow: $y_1 = 2$, $x_1 = 20$; $y_2 = 3$, $x_2 = 25$; $y_3 = 0$, $x_3 = 0$; $y_4 = 1$, $x_4 = 15$. Consider a simple random sampling design with sample size $n = 2$.

(a) Make a table listing every possible sample of size 2, the probability of selecting each sample, the estimate $N\bar{y}$ of the population total for each sample, and the ratio estimate $\hat{\tau}_r$ for each sample.

(b) Give the value and variance expected for the estimator $N\bar{y}$.

(c) Compute the exact mean and mean square error of the ratio estimator $\hat{\tau}_r$.

4. Carry out a survey to estimate the mean or total of a population of your choice, in which an auxiliary variable is available along with the variable of interest. Examples include the average amount of pocket change carried by attendees to a lecture, as in Example 1, or eyeball estimates and careful measurements of any sort of object. In the process of carrying out the survey and making the estimate, think about or discuss with others the following:

(a) What practical problems arise in establishing a frame, such as a map or list of units, from which to select the sample?

(b) How is the sample selection actually carried out?

(c) What special problems arise in observing the units selected?

(d) Estimate the population mean or total using the sample mean.

(e) Give a 95% confidence interval for the population mean or total based on the sample mean.

(f) Estimate the population mean or total using the ratio estimator.

(g) Give a 95% confidence interval for the population mean or total based on the ratio estimator.

(h) How would you improve the survey procedure if you were to do it again?

5. Determine the first- and second-order terms of a Taylor series expansion of the ratio estimator. Use the approximation to obtain an approximate expression for the bias of the ratio estimator under simple random sampling.

6. Assume that the $Y$-values in the population are related to an auxiliary variable $x$ through the ratio model $Y_i = \beta x_i + \epsilon_i$, for $i = 1, 2, \ldots, N$, where the random errors $\epsilon_i$ are independent with $E(\epsilon_i) = 0$ and $\text{var}(\epsilon_i) = v_i \sigma_R^2$, in which the $v_i$ are known and $\beta$ and $\sigma_R^2$ are unknown constants. A simple random sample of size $n$ is selected. The best linear unbiased predictor of the population total $\tau$ is $\hat{\tau} = Y_s + \hat{\beta} x_{s'}$, where $Y_s$ is the total of the $Y$-values in the sample and $x_{s'}$ is the total of the $x$-values not in the sample and

$$\hat{\beta} = \frac{\sum_{i=1}^{n}(x_i Y_i / v_i)}{\sum_{i=1}^{n}(x_i^2 / v_i)}$$

When the variance of $Y_i$ is proportional to $x_i$, that is, $v_i = x_i$, the best linear unbiased estimator $\hat{\beta}$ is the *ratio* estimator $r = \bar{Y}/\bar{x}$.

**(a)** Show that when the variance of $Y_i$ is constant, that is, $v_i = 1$, $\hat{\beta}$ is the *regression-through-the-origin* estimator

$$\hat{\beta} = \frac{\sum_{i=1}^{n} x_i Y_i}{\sum_{i=1}^{n} x_i^2}$$

Compute the value of $\hat{\tau}$ with this estimator with the data in Example 1.

**(b)** Show that when the variance of $Y_i$ is proportional to $x_i^2$, that is, $v_i = x_i^2$, $\hat{\beta}$ is the *mean-of-the-ratios* estimator

$$\hat{\beta} = \frac{1}{n} \sum_{i=1}^{n} \frac{Y_i}{x_i}$$

Compute the value of $\hat{\tau}$ with this estimator with the data in Example 1.

CHAPTER 8

# Regression Estimation

In some sampling situations, there may be an auxiliary variable $x$ which is linearly related to the variable of interest $y$, at least approximately, but without $y$ being zero when $x$ is zero. For example, if $y$ were the yield per unit area of a plant crop and $x$ were the average concentration of an air pollutant in the vicinity of the plot, $y$ might tend to decrease with $x$ and be highest when $x$ was zero. In such a situation, a linear regression estimator rather than a ratio estimator might be appropriate.

In fact, there may be more than one auxiliary variable associated with each unit. In the example above, the auxiliary variable $x_1$ might be the level of atmospheric sulfur dioxide and $x_2$ the level of nitrous oxide, while another auxiliary variable $x_3$ might be an indicator variable equaling 1 if the plot's soil type had been categorized as favorable and zero otherwise. Multiple regression models or general linear statistical models can describe many such relationships between a variable of interest and a number of auxiliary variables.

The linear regression estimator with one auxiliary variable is described first, initially in the design-based or fixed-population context. Regression estimation with unequal probability designs and multiple regression models are covered in later sections of this chapter. Like the ratio estimator, the regression estimator is not design-unbiased under simple random sampling. Under usual regression model assumptions, however, the estimator is unbiased.

If a regression model describing a stochastic relationship between the auxiliary variables and the variable of interest is assumed, a natural objective of sampling is the "prediction" of some characteristic of the $y$-values of the population. The characteristic to be predicted may be the population mean or total or the $y$-value of a single unit not yet in the sample. The basic results of the linear prediction approach are summarized for the simple linear regression model with one auxiliary variable and then in general for multiple regression models with any number of auxiliary variables.

## 8.1  LINEAR REGRESSION ESTIMATOR

Suppose that for the $i$th unit in the population there is associated a value $y_i$ of the variable of interest and a value $x_i$ of an auxiliary variable, for $i = 1, \ldots, N$. The population mean and total of the $y$-values are denoted $\mu$ and $\tau$, respectively. The population mean and total of the $x$-values are denoted $\mu_x$ and $\tau_x$ and are assumed known. A simple random sample of $n$ units is selected.

The linear regression estimator for the population mean $\mu$ is

$$\hat{\mu}_L = a + b\mu_x$$

where

$$b = \frac{\sum_{i=1}^{n}(x_i - \bar{x})(y_i - \bar{y})}{\sum_{i=1}^{n}(x_i - \bar{x})^2}$$
$$a = \bar{y} - b\bar{x}$$

The value of $b$ gives the slope and $a$ gives the $y$-intercept of a straight line fitted to the data by least squares. Substituting for $a$, the estimator may be written

$$\hat{\mu}_L = \bar{y} + b(\mu_x - \bar{x})$$

Like the ratio estimator, the linear regression estimator is not design-unbiased under simple random sampling.

An approximate formula for the mean square error or variance of $\hat{\mu}_L$ is

$$\text{var}(\hat{\mu}_L) \approx \frac{N-n}{Nn(N-1)} \sum_{i=1}^{N}(y_i - A - Bx_i)^2$$

where

$$B = \frac{\sum_{i=1}^{N}(x_i - \mu_x)(y_i - \mu)}{\sum_{i=1}^{N}(x_i - \mu_x)^2}$$
$$A = \mu - B\mu_x$$

An estimator of this variance is provided by

$$\widehat{\text{var}}(\hat{\mu}_L) = \frac{N-n}{Nn(n-2)} \sum_{i=1}^{n}(y_i - a - bx_i)^2 \tag{1}$$

An approximate $(1 - \alpha)100\%$ confidence interval is given by

$$\hat{\tau}_L \pm t_{n-2}(\alpha/2)\sqrt{\widehat{\text{var}}(\hat{\mu}_L)}$$

where $t_{n-2}(\alpha/2)$ is the upper $\alpha/2$ point of the $t$-distribution with $n-2$ degrees of freedom.

The finite-population central limit theorem for the regression estimator, on which the confidence interval procedure above is based, is given in Scott and Wu (1981). However, empirical studies of the regression and ratio estimators for real populations with sample sizes of $n = 32$ by Royall and Cumberland (1985) show that the actual coverage probabilities, using the standard variance estimator as well as proposed alternatives, may be substantially lower than the nominal confidence level. In a subsequent theoretical study, Deng and Wu (1987) compare alternative variance estimators and propose that different estimators be used depending on whether the purpose is estimating mean square error or constructing a confidence interval. More research is needed on the topic.

The linear regression estimator of the population total $\tau$ is

$$\hat{\tau}_L = N\hat{\mu}_L = N(a + b\mu_x) = N\bar{y} + b(\tau_x - N\bar{x})$$

with variance formulas obtained by multiplying those for $\hat{\mu}_L$ by $N^2$.

***Example 1.*** To estimate the total yield of a crop in a field of $N = 100$ plots, $n = 4$ plots are selected by simple random sampling and the amount $y_i$ of the yield of each sample plot is measured. The yield of a plot depends on the amount $x_i$ of fertilizer applied to the plot, which is known for each plot in the population. The four sample $(x_i, y_i)$ pairs are $(50, 1410)$, $(100, 1690)$, $(150, 1680)$, and $(200, 1850)$.

The sample means are $\bar{y} = 1657.5$ and $\bar{x} = 125$. The other sample statistics are

$$b = \frac{(50 - 125)(1410 - 1657.5) + \cdots + (200 - 125)(1850 - 1657.5)}{(50 - 125)^2 + \cdots + (200 - 125)^2}$$

$$= \frac{32{,}750}{12{,}500} = 2.62$$

and $a = 1657.5 - 2.62(125) = 1330$.

The regression estimate of the total yield is

$$\hat{\tau}_L = 100[1657.5 + 2.62(100 - 125)] = 100(1592) = 159{,}200$$

To estimate the variance of $\hat{\tau}_L$, denote the value of the fitted regression line for the $i$th unit in the sample as $\hat{y}_i = a + bx_i$. The four fitted values are $\hat{y}_1 = 1330 + 2.62(50) = 1461$, $\hat{y}_2 = 1592$, $\hat{y}_3 = 1723$, and $\hat{y}_4 = 1854$. The estimated variance is

$$\widehat{\mathrm{var}}(\hat{\tau}_L) = N^2\widehat{\mathrm{var}}(\hat{\mu}_L)$$

$$= \frac{100(100 - 4)}{4(4 - 2)}[(1410 - 1461)^2 + \cdots + (1850 - 1854)^2]$$

$$= \frac{100(96)}{4}(7035) = 16{,}884{,}000$$

the square root of which is 4109.

The estimate based on the $y$-values alone would be $N\bar{y} = 165{,}750$. Because the units in the sample had overall higher-than-average fertilizer, the regression estimator has used the auxiliary information to adjust the estimate downward. The estimated variance of $N\bar{y}$ is $[100(96)/4](33{,}292) = 79{,}900{,}000$, the square root of which is 8939. The regression estimator appears to be more precise in this small example because of the low residual variation about the fitted regression line.                                                                                      □

## 8.2 REGRESSION ESTIMATION WITH UNEQUAL PROBABILITY DESIGNS

For a sampling design in which the inclusion probability for the $i$th unit is $\pi_i$, for $i = 1, \ldots, N$, the generalized ratio estimator of the population mean $\mu$ of the $y$-values is

$$\tilde{\mu}_y = \left( \sum_{i=1}^{\nu} \frac{y_i}{\pi_i} \right) \bigg/ \left( \sum_{i=1}^{\nu} \frac{1}{\pi_i} \right)$$

where the summation is over the $\nu$ distinct units in the sample. With an auxiliary variable $x$, the generalized ratio estimator based on the sample $x$-values is

$$\tilde{\mu}_x = \left( \sum_{i=1}^{\nu} \frac{x_i}{\pi_i} \right) \bigg/ \left( \sum_{i=1}^{\nu} \frac{1}{\pi_i} \right)$$

A generalized regression estimator, which is approximately (or asymptotically) unbiased for the population mean under the given design, is

$$\hat{\mu}_G = \tilde{\mu}_y + \hat{B}(\mu_x - \tilde{\mu}_x)$$

where $\hat{B}$ is a weighted regression slope estimator, based on the inclusion probabilities, given by

$$\hat{B} = \left[ \sum_{i=1}^{\nu} \frac{(x_i - \tilde{\mu}_x)(y_i - \tilde{\mu}_y)}{\pi_i} \right] \bigg/ \left[ \sum_{i=1}^{\nu} \frac{(x_i - \tilde{\mu}_x)^2}{\pi_i} \right]$$

An approximate expression for the mean square error or variance of $\hat{\mu}_G$ is obtained using

$$\operatorname{var}(\hat{\mu}_G) \approx \operatorname{var}\left[ \frac{1}{N} \sum_{i=1}^{\nu} \left( \frac{y_i - A - Bx_i}{\pi_i} \right) \right]$$

so that, defining the new variable $y_i' = y_i - A - Bx_i$, the Horvitz–Thompson variance formula may be used, giving

$$\text{var}(\hat{\mu}_G) \approx \frac{1}{N^2}\left[\sum_{i=1}^{N}\left(\frac{1-\pi_i}{\pi_i}\right)y_i'^2 + \sum_{i=1}^{N}\sum_{j\neq i}\left(\frac{\pi_{ij}-\pi_i\pi_j}{\pi_i\pi_j}\right)y_i'y_j'\right]$$

An estimator of this variance is obtained using $\hat{y}_i = y_i - \hat{A} - \hat{B}x_i$ in the Horvitz–Thompson variance estimation formula, where $\hat{A}$ is a weighted regression intercept estimator given by

$$\hat{A} = \frac{\sum_{i=1}^{\nu}\frac{y_i}{\pi_i} - \hat{B}\sum_{i=1}^{\nu}\frac{x_i}{\pi_i}}{\sum_{i=1}^{\nu}\frac{1}{\pi_i}}$$

The resulting estimator of variance is

$$\widehat{\text{var}}(\hat{\mu}_G) = \frac{1}{N^2}\left[\sum_{i=1}^{\nu}\left(\frac{1-\pi_i}{\pi_i^2}\right)\hat{y}_i^2 + \sum_{i=1}^{\nu}\sum_{j\neq i}\left(\frac{\pi_{ij}-\pi_i\pi_j}{\pi_i\pi_j}\right)\frac{\hat{y}_i\hat{y}_j}{\pi_{ij}}\right]$$

$$= \frac{1}{N^2}\left[\sum_{i=1}^{\nu}\left(\frac{1}{\pi_i^2}-\frac{1}{\pi_i}\right)\hat{y}_i^2 + \sum_{i=1}^{\nu}\sum_{j\neq i}\left(\frac{1}{\pi_i\pi_j}-\frac{1}{\pi_{ij}}\right)\hat{y}_i\hat{y}_j\right]$$

assuming that all of the joint inclusion probabilities $\pi_{ij}$ are greater than zero.

Generalized regression estimators are discussed in Brewer (1979), Brewer et al. (1988), Cassel et al. (1976, 1977), Hájek (1981), Isaki and Fuller (1982), Little (1983), Särndal (1980a,b), Särndal and Wright (1984), Tam (1988), and Wright (1983). General references include Särndal et al. (1992, pp. 225–229) and M. E. Thompson (1997, pp. 172–178).

## 8.3 REGRESSION MODEL

Like the ratio estimator, the regression estimator is not unbiased in the design sense under simple random sampling. That is, viewing the $y$ and $x$ values as fixed quantities, the expected value, over all possible samples, of the regression estimator of the population mean of the $y$'s does not exactly equal the true population mean. When an ordinary regression model is assumed, however, the regression estimator is unbiased under that model.

Suppose that a standard simple linear regression model is assumed to hold. That is, $Y_1, \ldots, Y_N$ are random variables whose distributions depend on the $x$-values, with

$$\text{E}(Y_i) = \alpha + \beta x_i$$

for $i = 1, \ldots, N$. The $Y$-values for different units are assumed to be uncorrelated and constant variance is assumed,

$$\text{var}(Y_i) = \sigma_L^2$$

for all $i$. The variance $\sigma_L^2$ can be thought of as the variation of the $Y$-values around the true regression line, which is determined by the parameters $\alpha$ and $\beta$.

Since the population $Y$-values are random variables, any function of them, such as their mean $\mu = (1/N) \sum_{i=1}^{N} Y_i$, is a random variable. We wish by observing the sample values to predict the value of the random variable $\mu$. It can be shown (see in particular the papers by Royall) that the linear predictor that minimizes the mean square error under the model is the linear regression estimator $\hat{\mu}_L$. Under the model, the linear regression estimator is model-unbiased, that is, $E(\hat{\mu}_L) = E(\mu)$, where the expectation is in terms of the model assumed, given the sample, and the unbiasedness holds for any given sample $s$.

The mean square prediction error, given the sample $s$, under the assumed model is

$$E(\hat{\mu}_L - \mu)^2 = \frac{N - n}{N} \left[ \frac{1}{n} + \frac{N}{N - n} \frac{(\bar{x} - \mu_x)^2}{\sum_{i=1}^{n} (x_i - \bar{x})^2} \right] \sigma_L^2 \qquad (2)$$

An unbiased estimator of mean square error is obtained by replacing $\sigma_L^2$ in formula (2) with the sample residual variance:

$$\hat{\sigma}_L^2 = \frac{1}{n - 2} \sum_{i=1}^{n} (Y_i - a - bx_i)^2$$

However, the simpler estimator [Equation (1)] is considered to be more robust to departures from the assumed model.

Regression models and the prediction approach in survey sampling are developed and discussed in Cumberland and Royall (1981, 1988), Hájek (1981), Pfeffermann and Nathan (1981), P. S. R. S. Rao (1988), Royall (1970, 1976b, 1988), Royall and Cumberland (1978, 1981b, 1985), Royall and Eberhardt (1975), Royall and Herson (1973a,b), and Royall and Pfeffermann (1982). Problems of estimating the regression parameters of an assumed model from survey data are discussed in Christensen (1987), Holt and Scott (1981), Nathan and Holt (1980), and Scott and Holt (1982), and are reviewed in Smith (1984).

## 8.4  MULTIPLE REGRESSION MODELS

Regression models readily extend to situations in which there is more than one auxiliary variable. Let $x_{i1}, x_{i2}, \ldots, x_{ip}$ be the values of $p$ auxiliary variables associated with the $i$th unit. Consider the multiple regression model

$$E(Y_i) = \sum_{k=1}^{p} \beta_k x_{ik}$$

$$\text{cov}(Y_i, Y_j) = \sigma_L^2 v_{ij}$$

for $i = 1, \ldots, N$ and $j = 1, \ldots, N$. Simple linear regression and ratio models are obtained as special cases.

Suppose that one wishes to predict some linear combination $\sum_{i=1}^{N} l_i Y_i$ of the population $y$-values. For example, with each coefficient $l_i = 1$, the function of interest is the population total; with each $l_i = 1/N$, it is the mean. If one wished only to predict the value on one unsampled unit, say unit $j$, one would have $l_j = 1$ and $l_i = 0$ for the other $N - 1$ units.

The linear prediction problem is to find a linear function $\sum_{i=1}^{n} a_i Y_i$ of the observations which is unbiased for the population quantity and has the lowest possible mean square prediction error. That is, find values $a_1, a_2, \ldots, a_n$ subject to

$$E\left( \sum_{i=1}^{n} a_i Y_i \right) = E\left( \sum_{i=1}^{N} l_i Y_i \right)$$

which minimize

$$E\left( \sum_{i=1}^{n} a_i Y_i - \sum_{i=1}^{N} l_i Y_i \right)^2$$

In matrix notation, the model is

$$E(\mathbf{Y}) = \mathbf{X}\beta$$

$$\text{var}(\mathbf{Y}) = \sigma_L^2 \mathbf{V}$$

Given a sample $s$ of $n$ of the units, the vector $\mathbf{Y}$ may be ordered so that the sample $y$-values hold the first $n$ places and may be written in partitioned form as

$$\mathbf{Y} = \begin{pmatrix} \mathbf{Y}_s \\ \mathbf{Y}_r \end{pmatrix}$$

where $\mathbf{Y}_s$ is the $n \times 1$ vector of sample values and $\mathbf{Y}_r$ the $(N - n) \times 1$ vector of the values in the rest of the population. Similarly, the $N \times p$ dimensional matrix $\mathbf{X}$ can be partitioned

$$\mathbf{X} = \begin{pmatrix} \mathbf{X}_s \\ \mathbf{X}_r \end{pmatrix}$$

Similarly, the variance–covariance matrix $\mathbf{V}$ may be partitioned

$$\mathbf{V} = \begin{pmatrix} \mathbf{V}_{ss} & \mathbf{V}_{sr} \\ \mathbf{V}_{rs} & \mathbf{V}_{rr} \end{pmatrix}$$

The predictor based on the sample may be written $\mathbf{a}'\mathbf{Y}_s$, where $\mathbf{a} = (a_1, \ldots, a_n)'$ is the vector of coefficients. The quantity to be predicted may be written $\mathbf{l}'\mathbf{Y} = \mathbf{l}'_s\mathbf{Y}_s + \mathbf{l}'_r\mathbf{Y}_r$, where $\mathbf{l} = (l_1, \ldots, l_N)$ and $\mathbf{l}_s$ and $\mathbf{l}_r$ contain the $l$-values for units in the sample and the rest of the population, respectively.

The mean square error for any such unbiased linear predictor is

$$E(\mathbf{a}'\mathbf{Y}_s - \mathbf{l}'\mathbf{Y})^2 = [(\mathbf{a} - \mathbf{l}_s)'\mathbf{V}_{ss}(\mathbf{a} - \mathbf{l}_s) - 2(\mathbf{h}_s - \mathbf{l})'\mathbf{V}_{sr}\mathbf{l}_r + \mathbf{l}'_r\mathbf{V}_{rr}\mathbf{l}_r]\sigma_L^2$$

Linear model theory (see, e.g., C. R. Rao 1965, Chap. 4; Royall and Herson 1973a; Seber 1977, pp. 60–64 and 84–86; also, the derivation of a somewhat more general prediction result is given in Chapter 20 of this book) gives the best linear unbiased estimator of the parameter vector $\beta$ as

$$\hat{\beta} = (\mathbf{X}'_s\mathbf{V}_{ss}^{-1}\mathbf{X}_s)^{-1}\mathbf{X}'_s\mathbf{V}_{ss}^{-1}\mathbf{Y}_s$$

The best linear unbiased predictor of the random quantity $\mathbf{l}'\mathbf{Y}$ is

$$\hat{\mathbf{a}}'\mathbf{Y}_s = \mathbf{l}'_s\mathbf{Y}_s + \mathbf{l}'_r[\mathbf{X}_r\hat{\beta} + \mathbf{V}_{rs}\mathbf{V}_{ss}^{-1}(\mathbf{Y}_s - \mathbf{X}_s\hat{\beta})] \tag{3}$$

The mean square error for this best predictor is

$$E(\hat{\mathbf{a}}'\mathbf{Y}_s - \mathbf{l}'\mathbf{Y})2 = [\mathbf{l}'_r(\mathbf{X}_r - \mathbf{V}_{rs}\mathbf{V}_{ss}^{-1}\mathbf{X}_s)(\mathbf{X}'_s\mathbf{V}_{ss}^{-1}\mathbf{X}_s)^{-1}(\mathbf{X}_r - \mathbf{V}_{rs}\mathbf{V}_{ss}^{-1}\mathbf{X}_s)'\mathbf{l}_r$$
$$+ \mathbf{l}'_r(\mathbf{V}_{rr} - \mathbf{V}_{rs}\mathbf{V}_{ss}^{-1}\mathbf{V}_{sr}\mathbf{l}_r)]\sigma_L^2 \tag{4}$$

An unbiased estimator $\hat{E}(\mathbf{a}'\mathbf{Y}_s - \mathbf{l}'\mathbf{Y})^2$ of the prediction mean square error is obtained by replacing $\sigma_L^2$ in expression (4) with

$$\hat{\sigma}_L^2 = \frac{\mathbf{r}'\mathbf{V}_{ss}^{-1}\mathbf{r}}{n - p}$$

where $\mathbf{r} = \mathbf{Y}_s - \mathbf{X}_s\hat{\beta}$ is the vector of residuals. However, a mean square error estimator of the form $\sum_{i=1}^\nu (a_i - l_i)^2(Y_i - \hat{y}_i)^2$, where $\hat{y}_i = \sum_{i=1}^\nu \hat{\beta}_k x_{ik}$, is considered more robust to departures from the assumed model [see Royall (1988) for a review].

It is often assumed in regression models that the $y$-values are uncorrelated, so that $\mathbf{V}$ is a diagonal matrix, the $i$th element on the diagonal being $v_i$, where $\text{var}(Y_i) = \sigma_L^2 v_i$. With this model, the predictor (3) simplifies to

$$\widehat{\mathbf{l}'\mathbf{Y}} = \mathbf{l}'_s\mathbf{Y}_s + \mathbf{l}'_r\mathbf{X}_r\hat{\beta}$$
$$= \mathbf{l}'_s\mathbf{Y}_s + (\mathbf{l}'\mathbf{X} - \mathbf{l}'_s\mathbf{X}_s)\hat{\beta}$$

For estimating a population total, this gives

$$\hat{\tau}_L = \sum_{i \in s} y_i + \sum_{i \in \bar{s}} (\hat{\beta}_1 x_{i1} + \hat{\beta}_2 x_{i2} + \cdots + \hat{\beta}_p x_{ip})$$

where $\bar{s}$ is the set of units not in the sample. That is, the total is estimated by adding the observed $y$-values in the sample to the estimated expected $y$-values outside the sample. This simple interpretation does not hold for the more general case (3) in which the covariances are not zero, since then a further adjustment is made based on the covariance structure.

For regression estimation of the population mean with unequal probability designs, an estimator that is approximately or asymptotically design-unbiased, whether the assumed model is true or not, is the generalized regression estimator:

$$\hat{\mu}_G = \hat{\mu}_y + \sum_{k=1}^{p} \hat{\beta}_k (\mu_k - \hat{\mu}_k)$$

where $\hat{\mu}_y$ is the Horvitz–Thompson estimator of $\mu$ and $\hat{\mu}_k$ is the Horvitz–Thompson estimator of the population mean of the $k$th auxiliary variable, that is, $\hat{\mu}_k = (1/N) \sum_{i=1}^{\nu} x_{ik}/\pi_i$. For the estimator $\hat{\beta}_k$ of the regression parameters, one may use either the usual best linear unbiased estimators or weighted estimators using the inclusion probabilities. For the case with zero covariances, in which $\mathbf{V}$ is a diagonal matrix with element $v_i$ for the variance of $y_i$, a weighted estimate is given by

$$\hat{\beta} = \left( \sum_{i \in s} \frac{\mathbf{x}_i \mathbf{x}'_k}{v_i \pi_i} \right)^{-1} \sum_{i \in s} \frac{\mathbf{x}_i y_i}{v_i \pi_i}$$

where $\mathbf{x}_i$ is the row of the $\mathbf{X}$ matrix corresponding to unit $i$ (cf. Sárndal et al. 1992, pp. 227–230; M. E. Thompson 1997, pp. 172–178).

Approximate variance formulas are obtained by approximating the parameter estimates with the true parameter values so that $E(\hat{\mu}_G - \mu)^2 \approx \mathrm{var}(\hat{\mu}_y - \sum_{k=1}^{p} \beta_k \hat{\mu}_k) = \mathrm{var}[(1/N) \sum_{i=1}^{\nu} Y'_i/\pi_i]$, where $Y'_i = Y_i - \sum_{k=1}^{p} \beta_k x_{ik}$. The approximate mean square error is then obtained from the Horvitz–Thompson formula with the variables $Y'_i$. An estimator of the mean square formula is obtained from the Horvitz–Thompson variance estimator formula with the variables $\hat{Y}_i = Y_i - \sum_{k=1}^{p} \hat{\beta}_k x_{ik}$.

## 8.5 DESIGN IMPLICATIONS OF REGRESSION MODELS

From Equation (2), the mean square error under an assumed simple linear regression model is minimized if a sample is chosen for which $\bar{x} = \mu_x$; that is, it is balanced in $x$. When the sample is balanced in $x$, the regression estimator is identical to the sample mean of the $y$-values, so balancing in $x$ gives the estimator robustness against departures from the assumed regression model. The studies of Cumberland and Royall (1988) and Royall and Cumberland (1981b) showed further that the usual estimators of mean square error behaved best when balanced samples were selected.

For more general linear models, it may be possible to determine a sample to minimize Equation (4). For robustness against departures from the assumed model, samples balanced in each $x$-variable, that is, samples for which $\bar{x}_k = \mu_k$ for $k = 1, \ldots, p_1$, are desired. With unequal probability sampling, a representative sample is defined as one for which $\hat{\mu}_k = \mu_k$ for each of the $p$ auxiliary variables. With such a sample, the generalized regression estimator is identical to the Horvitz–Thompson estimator based on the $y$-values.

Additionally, with the regression model $E(Y_i) = \sum_{k=1}^{p} \beta_k x_{ik}$ with $\text{var}(Y_i) = \sigma^2 v_i$ and the $Y$'s independent, a design with each inclusion probability $\pi_i$ proportional to the square root of $v_i$ is efficient in terms of minimizing the mean square error under the combination of the model and the design while having a generalized regression estimator which is approximately unbiased. Devising a selection procedure that produces balanced samples while having the desired individual selection probabilities may be no easy task. A stratified random sample with stratification based on the auxiliary variable and allocation based on the average value of $v_i$ within strata may come close.

One could go badly astray in purposively choosing, without any randomization, a sample balanced in an auxiliary variable $x$ while being unaware of the importance of one or more other auxiliary variables. Imagine, for example, a survey to estimate the abundance of beetles of a certain species in a study area. Beetle abundance is thought to be approximately linearly related to elevation $(x)$, so sample plots are systematically laid out alongside a road that cuts through the study area from low elevation to high elevation, so that $\bar{x} = \mu_x$. Unbeknown to the researchers, however, the truly important variable influencing beetle abundance in the study area is $x_2$, which is distance from roadways or other habitat disturbances. Thus, the carefully balanced sample is entirely unrepresentative and produces a misleading estimate. Of course, the researchers could redo the survey balancing on both known auxiliary variables, but what if there are further unknown and important auxiliary variables? For such reasons, many researchers would be more comfortable with a design incorporating randomization at some level, although stratification based on the known auxiliary variables could still be used.

## EXERCISES

1. In a survey to determine the amount of crop yield due to an air pollutant, a simple random sample of $n = 20$ plots were selected from $N = 1000$ in the population. The summary statistics on yield $y_i$ (in weight) and level of pollutant $x_i$ (in parts per million) were $\bar{y} = 10$, $\bar{x} = 6$, $\sum_{i=1}^{20}(x_i - \bar{x})(y_i - \bar{y}) = -60$, $\sum_{i=1}^{20}(x_i - \bar{x})^2 =$

30, and $\sum_{i=1}^{20} (y_i - a - bx_i)^2 = 80$. The mean pollutant level is $\mu_x = 5.0$.

(a) Estimate the mean yield for the population with a linear regression estimate.

(b) Estimate the variance of the linear regression estimate.

(c) Predict the yield on a plot in which the pollutant level is $x_i = 4$.

2. Carry out a survey to estimate the mean or total of a population of your choice, in which an auxiliary variable might be expected to have a linear relationship with the variable of interest. In the process of carrying out the survey and making the estimate, think about or discuss with others the following:

(a) Plot the $x$ and $y$ data.

(b) Estimate the population mean or total using the regression estimator.

(c) Give a 95% confidence interval for the population mean or total based on the regression estimator.

CHAPTER 9

# The Sufficient Statistic in Sampling

## 9.1 THE SET OF DISTINCT, LABELED OBSERVATIONS

The information obtained from a survey may include, in addition to the value of the variable of interest for each unit in the sample, the unit label associated with each value, the order in which the units were selected, and—for a with-replacement design—the number of times each unit was selected. How much of this information is relevant for estimating a population parameter? There is a simple answer: The *minimal sufficient statistic* for the finite population survey sampling situation is the unordered set of distinct, labeled observations (Basu 1969; Basu and Ghosh 1967; see also Cassel et al. 1977, pp. 35–39). Thus, the set of $y$-values alone is not enough—the identities of the associated units may be helpful in estimation. The order in which the values were selected is, on the other hand, more information than is necessary, at least in principle. The number of times a unit was selected is similarly not needed.

In practice, many widely used estimators in survey sampling are not functions of the minimal sufficient statistic. For example, in sampling with replacement, the sample mean of all $n$ observations depends on the number of times units are selected and hence is not a function of the minimal sufficient statistic. In unequal probability sampling without replacement, the Raj estimator, which uses conditional inclusion probabilities only, depends on order and hence is not a function of the minimal sufficient statistic. In unequal probability sampling with replacement, the Hansen–Hurwitz estimator depends on repeat selections and hence is not a function of the minimal sufficient statistic. Each of these estimators is used because it is simple to compute. In large surveys, it may be inconvenient to determine whether a unit in the sample has been selected previously. In unequal probability sampling without replacement, unconditional inclusion probabilities may be extremely difficult to compute.

In principle, however, one need consider only estimators that are functions of the minimal sufficient statistic. For any estimator that is not a function of the minimal sufficient statistic, one may obtain (using the Rao–Blackwell method) and estimator, depending on the minimal sufficient statistic, that is as good or better (see, e.g., Cassel et al. 1977). Results of minimal sufficiency in the sampling situation are illustrated below.

## 9.2 ESTIMATION IN RANDOM SAMPLING WITH REPLACEMENT

In Example 3 of Chapter 2 a nominal sample size of $n = 5$ led to the selection of $\nu = 4$ distinct units, since one was selected twice. The label of each unit in the sample, along with its $y$-value, is part of the data. From a population of $N = 100$ units, the selections, in order, were unit 37, with $y_{37} = 2$, unit 7, with $y_7 = 4$, unit 25, with $y_{25} = 0$, unit 7 again, with $y_7 = 4$ of course, and unit 15, with $y_{15} = 5$.

The minimal sufficient statistic $t$ consists of the set of distinct unit labels together with their $y$-values, that is,

$$t = \{(25,0), (37,2), (7,4), (15,5)\}$$

in which the order is immaterial.

The set of $y$-values obtained in the five selections were $\{0,2,4,4,5\}$. The sample mean of the five $y$-values was $\bar{y}_n = 3.0$. However, suppose instead that unit 37 had been selected twice, while units 7, 15, and 25 were each selected once. Then the minimal sufficient statistic would be exactly the same, but the sample mean of the five $y$-values $\{0,2,2,4,5\}$ would be $\bar{y}_n = 2.6$. Two other samples have the same value of the minimal sufficient statistic. One has $y$-values $\{0,0,2,4,5\}$ and sample mean $\bar{y}_n = 2.2$. The other has $y$-values $\{0,2,4,5,5\}$ and $\bar{y}_n = 3.2$.

Each of the four samples with the given value of $t$ has the same probability of being the one selected, because of the simple random sampling. Hence the conditional probability of each of the four samples, given $t$, is one-fourth. Thus, the conditional expectation of the estimator $\bar{y}_n$ given $t$ is

$$E(\bar{y}_n \mid t) = \frac{3.0 + 2.6 + 2.2 + 3.2}{4} = 2.75$$

Note that because of the equal probabilities, this estimator is in fact simply the sample mean $\bar{y}_\nu$ of the four distinct observations. But the Rao–Blackwell theorem, $\bar{y}_\nu$ has lower variance than $\bar{y}_n$. Both estimators are unbiased for the sample mean. The estimator $\bar{y}_\nu$ is a function of the sufficient statistic, while $\bar{y}_n$ is not.

The Horvitz–Thompson estimator is also a function of the minimal sufficient statistic and is also unbiased. Under the design simple random sampling with replacement, the inclusion probability for any unit $i$, since the draw-by-draw selection probability is $1/N$ and the $n$ selections are independent, is

$$\pi_i = 1 - (1 - 1/N)^n$$

The Horvitz–Thompson estimator under this design is

$$\hat{\mu}_\pi = \frac{1}{N}\sum_{i=1}^{\nu}\frac{y_i}{\pi_i} = \frac{\sum_{i=1}^{\nu}y_i}{N[1-(1-1/N)^n]}$$

For this example the Horvitz–Thompson estimator is

$$\hat{\mu}_\pi = \frac{11}{100[1-(1-0.01)^5]} = 2.24$$

## 9.3 ESTIMATION IN PROBABILITY-PROPORTIONAL-TO-SIZE SAMPLING

Next we consider Examples 1 and 2 of Chapter 6, in which selection was with probability proportional to size with replacement (PPS). In that example, a nominal sample size of $n=4$ had led to selection of $\nu=3$ distinct units, since one unit was selected twice. The Hansen–Hurwitz estimator, depending on the number of repeat observations and hence not a function of the minimal sufficient statistic, gave $\hat{\tau}_p = 800$ as an estimate of the number of animals in the study region (Example 1), while the Horvitz–Thompson estimator, which depended only on the distinct observations, gave $\hat{\tau}_\pi = 529$ (Example 2).

The Hansen–Hurwitz estimator may be improved using the Rao–Blackwell method. There are three possible samples of nominal size 4, giving the three distinct units obtained. Each of these three samples, however, has a different probability of being the one selected. Assigning sample unit labels 1, 2, and 3 to the three distinct units in the sample, we have $y_1 = 60$, $p_1 = 0.05$, $y_2 = 14$, $p_2 = 0.02$, and $y_3 = 1$, $p_3 = 0.01$.

For the sample obtained, unit 1 was selected twice and each of the other two once. The probability of selecting a sample in which unit 1 is selected twice and units 2 and 3 are selected once in $n=4$ independent selections is given by the multinomial probability

$$P(s_1) = \frac{4!}{2!\,1!\,1!}0.05^2(0.02)(0.01) = 6\times 10^{-6}$$

For the sample in which the second unit was selected twice, the probability is

$$P(s_2) = \frac{4!}{2!\,1!\,1!}0.05(0.02)^2(0.01) = (0.02)2.4\times 10^{-6}$$

and for the sample in which the third unit is selected twice,

$$P(s_3) = \frac{4!}{2!\,1!\,1!}0.05(0.02)(0.01)^2 = 1.2\times 10^{-6}$$

The sum of these probabilities gives the probability of obtaining the given value of the sufficient statistic, $P(t) = 9.6 \times 10^{-6}$. Thus, the conditional probabilities are $P(s_1 \mid t) = 6/9.6 = 0.625$, $P(s_2 \mid t) = 0.25$, and $P(s_3 \mid t) = 0.125$.

The Hansen–Hurwitz estimator for each of the three samples equals 800, 675, and 525, respectively. The improved estimate obtained as the conditional expectation of $\hat{\tau}_p$, given the minimal sufficient statistic, is thus

$$E(\hat{\tau}_p \mid t) = 800(0.625) + 675(0.25) + 525(0.125) = 734$$

This improved version of the Hansen–Hurwitz estimator has been known for some time (see Basu 1958; Pathak 1962) but has not been a popular estimator because of its computational complexity (see Cassel et al. 1977, p. 42; Chaudhuri and Vos 1988, p. 259).

## 9.4  COMMENTS ON THE IMPROVED ESTIMATES

By definition, a statistic $t$ is *sufficient* for a parameter $\theta$ if the conditional distribution of the data, given $t$, does not depend on $\theta$. In the sampling situation, the data are the sequence $\mathbf{D}$ of labels of the units selected, together with the associated $y$-values. The sequence not only retains order of selection, but may contain repeat selections. Writing the sequence of units selected as $\mathbf{s} = (i_1, i_2, \ldots, i_n)$, the data may be written $\mathbf{D} = \{(i, y_i) : i \in \mathbf{s}\}$. Implicit in the inclusion of the identity of a unit in the sample is any auxiliary information that may be associated with that unit, such as the values of auxiliary variables or the location of the unit in relation to other units.

In the sampling situation, the parameter of interest may be taken to be the vector $\mathbf{y} = y_1, y_2, \ldots, y_N$ of population $y$-values. Let $s$ denote the unordered set of distinct unit labels in the sample and let $t$ denote the unordered set of $y$-values, together with the associated unit labels [i.e., $t = \{(i, y_i) : i \in s\}$]. Then $t$ is not only sufficient for $\theta$, but minimally sufficient; that is, $t$ is a function of any other sufficient statistic for $\theta$.

With the two designs above, we obtained improved unbiased estimators, by the Rao–Blackwell method, which were functions of the minimal sufficient statistic. One would like to be able to say that such an estimator was the unique unbiased estimator of lowest variance. However, in each example, a different function of the minimal sufficient statistic—the Horvitz–Thompson estimator—was also unbiased for the population mean (or total). When two different estimators based on the minimal sufficient statistic can have the same expectation, the minimal sufficient statistic is said to be *not complete*. In such a situation, one cannot usually make sweeping statements about one estimator being best. The lack of completeness of the minimal sufficient statistic in sampling is basically due to the presence of the unit labels in the data. On the one hand, the lack of completeness caused by this extra information in the data has been the underlying source of much of the difficulty and lack of optimality results in the theory of sampling. On the other hand, the label information is often useful in obtaining better estimators than could be obtained otherwise.

For model-based predictive methods in sampling, a related concept called *predictive sufficiency* applies (Bjørnstad 1990; Bolfarine and Zacks 1992; Lauritzen 1974; Skinner 1983; Thompson and Seber 1996, Chap. 3). To be sufficient for predicting an unknown but random population quantity $Z$ such as the population total, a statistic $T$ must first have the property that the conditional distribution of the data $D$ given $T$ does not depend on any unknown parameters $\phi$ of the population model. Second, $D$ and $Z$ must be conditionally independent given $T$. The first property just says that $T$ must be sufficient in the ordinary sense for inference about the parameter $\phi$. For the i.i.d. normal model presented in Chapter 2, with a fixed sample size design the minimal predictive sufficient statistic consists of the sample mean and variance. With a random sample size design, the realized sample size $\nu$ is part of the minimal sufficient statistic. For the general linear regression model with known covariance matrix, the minimal predictive sufficient statistic consists of the usual sufficient statistic for estimating the parameters, supplemented by sample statistics related to the covariances between the observed units and the quantity, such as the population total, to be predicted.

CHAPTER 10

# Design and Model

## 10.1 USES OF DESIGN AND MODEL IN SAMPLING

In the design-based approach to survey sampling, the values of a variable of interest ($y$-values) of the population are viewed as fixed quantities and the selection probabilities introduced with the design are used in determining the expectations, variances, biases, and other properties of estimators. In the model-based approach, on the other hand, the values of the variables of interest in the population are viewed as random variables, and the properties of estimators depend on the joint distribution of these random variables.

One reason for the historical reliance on design-based methods in sampling, in addition to the elimination of personal biases in selecting the sample, is that in many cases—and especially with natural populations—very little may be known about the population. Most researchers find it reassuring in such a situation to know that the estimation method used is unbiased no matter what the nature of the population itself. Such a method is called *design-unbiased*: The expected value of the estimator, taken over all samples which might be selected, is the correct population value. Design-unbiased estimators of the variance, used for constructing confidence intervals, are also available for most such designs.

One area of sampling in which the model-based approach has received considerable attention is in connection with ratio and regression estimation. In many sampling situations involving auxiliary variables, it seems natural to researchers to postulate a theoretical model for the relationship between the auxiliary variables and the variable of interest. A model can, of course, also be assumed for populations without auxiliary variables. For example, if the $N$ variables $Y_1, \ldots, Y_N$ can be assumed to be independent and identically distributed, many standard statistical results apply without reference to how the sample is selected. However, it is difficult to cite examples of survey situations in which a model of independent, identically distributed $y$-values can be assumed with confidence. In fact, a pervasive problem with the model approach to sampling is that for many real populations, attempts to specify models have been far from adequate. Typically, the models become mathematically complex while still not being suitably realistic. In particular, any model assuming that the $y$-values are independent (or have an exchangeable distribution) ignores the tendency in many populations for nearby or related units to be correlated.

Moreover, many survey programs need to produce results that will be used by people of widely different viewpoints and often conflicting preferences regarding whether an estimate should be higher or lower. For example, a demographic survey may be used to allocate governmental resources from one district to another; a fishery survey may be used to determine the amount of commercial catch allowed. It would be hard in such a siuation to propose a model that would seem acceptable or realistic to all interested parties. In such a situation, the elimination of ordinary human selection biases through some sort of random selection procedure can be a powerful pragmatic argument in favor of an approach that is at least partially design-based.

With some populations, however, experience may have established convincingly that certain types of patterns are typical of the $y$-values of that type of population. For example, in spatially distributed geological and ecological populations, the $y$-values of nearby units may be positively correlated, with the strength of the relationship decreasing with distance. If such tendencies are known to exist, they can be used in obtaining efficient predictors of unknown values and in devising efficient sampling procedures. This model-based approach has been prevalent in sampling for mining and geological studies, in which the cost of sampling is particularly high and the economic incentive is strong for obtaining the most precise possible estimates for a given amount of sampling effort.

Sources of nonsampling error must be modeled if they are to be taken into account. Problems of differential response, missing data, measurement errors, and detectability must be modeled in some way in order to adjust for biases and to assess the uncertainty of estimates.

## 10.2 CONNECTIONS BETWEEN THE DESIGN AND MODEL APPROACHES

Let $\mathbf{y} = (y_1, y_2, \ldots, y_N)$ denote the vector of $y$-values associated with the $N$ units of the population. From the model viewpoint, these $y$-values are random variables with some joint distribution $F$. Let $P(\mathbf{s})$ denote the probability under the design of selecting sample $\mathbf{s}$, where $\mathbf{s}$ is a sequence or subset of the units in the population.

From the sample of $n$ units, one wishes to estimate or predict the value of some quantity $y_0$, where $y_0$ may, for example, be the population mean, the population total, or the $y$-value at a unit not in the sample. The predictor or estimator $\hat{y}_0$ is a function of the $y$-values of the sample.

An estimator or predictor $\hat{y}_0$ is said to be *design-unbiased* for $y_0$ if its conditional expectation, given the realization of the $N$ population $y$-values, is the realized value of $y_0$, that is, if

$$\mathrm{E}(\hat{y}_0 \mid \mathbf{y}) = y_0$$

Notice that, although $y_0$ may be viewed as a random variable, with a distribution determined by $F$, the design-unbiased estimator $\hat{y}_0$ is unbiased under the design for the realized value of $y_0$—the actual value that $y_0$ has taken on at the time of the survey. The distribution $F$, which produced the population $y$-values, is thus irrelevant to this unbiasedness.

An estimator or predictor $\hat{y}_0$ is said to be *model-unbiased* for $y_0$ if, given any sample **s**, the conditional expectation of $\hat{y}_0$ equals the expectation of $y_0$, that is, if

$$E(\hat{y}_0 \mid s) - E(y_0)$$

No matter what sampling design gave rise to the sample **s**, the model-unbiased predictor $\hat{y}_0$ is unbiased under the population distribution $F$ for $y_0$ given the sample **s** obtained. The design that produced the sample **s** is thus irrelevant to this unbiasedness.

An estimator or predictor $\hat{y}_0$ is *unbiased* (i.e., unconditionally unbiased) for $y_0$ if the expectation of $\hat{y}_0$ equals the expectation of $y_0$, that is, if

$$E(\hat{y}_0) - E(y_0)$$

Any estimator that is *either* design-unbiased or model-unbiased for $y_0$ will be (unconditionally) unbiased for $y_0$, by a well-known property of expectation.

Thus, if the desired end is simply unbiasedness, it can be achieved through either the design or the model approach. However, some authors philosophically demand one or the other types of unbiasedness—design unbiasedness, so that assumptions about the population are not relied upon, or model unbiasedness, so that the particular sample obtained is taken into account.

The mean square error associated with predicting $y_0$ with $\hat{y}_0$ is

$$E(y_0 - \hat{y}_0)^2$$

the expectation being taken with respect to both the distribution of the population values and the design. If $\hat{y}_0$ is unbiased for $y_0$, the mean square error is the variance of the difference:

$$E(y_0 - \hat{y}_0)^2 = \text{var}(y_0 - \hat{y}_0)$$

From the model viewpoint, interest focuses on the conditional mean square error, given the sample **s**. If $\hat{y}_0$ is model-unbiased for $y_0$, this mean square error is a conditional variance:

$$E[(y_0 - \hat{y}_0)^2 \mid s] = \text{var}(y_0 - \hat{y}_0 \mid s)$$

From the design viewpoint, the concern is with the conditional mean square error given the realized population $y$-values. When $\hat{y}_0$ is design-unbiased for $y_0$, this conditional mean square error is

$$E[(y_0 - \hat{y}_0)^2 \mid \mathbf{y}] = \text{var}(\hat{y}_0 \mid \mathbf{y})$$

If $\hat{y}_0$ is design-unbiased, the unconditional mean square error may be written

$$E(y_0 - \hat{y}_0)^2 = E[\text{var}(\hat{y}_0 \mid \mathbf{y})]$$

If $\hat{y}_0$ is model-unbiased, the unconditional mean square error may be written

$$E(y_0 - \hat{y}_0)^2 = E[\text{var}(y_0 - \hat{y}_0 \mid \mathbf{s})]$$

Estimators of variance may in similar fashion be design- or model-unbiased. A variance estimator that is either design-unbiased or model-unbiased will be unconditionally unbiased. Thus, with the design simple random sampling, the usual estimator

$$\widehat{\text{var}}(\hat{y}) = \left(\frac{N - n}{N}\right)\frac{s^2}{n}$$

which is design-unbiased for $\text{var}(\bar{y})$, is unbiased for the true mean square error no matter what distribution may give rise to the population.

## 10.3   SOME COMMENTS

A main result of the preceding section is that a sampling strategy is unconditionally unbiased if it is either design-unbiased or model-unbiased. Even so, the two approaches may lead to conflicting recommendations. An assumed-ratio model may suggest purposive selection of the units with the highest $x$-values; such a procedure is certainly not design-unbiased. The sample mean may be design-unbiased under simple random sampling; but under an assumed model the sample mean for the particular sample selected may not be model-unbiased.

Some advantages of a design-based approach include obtaining unbiased or approximately unbiased estimators (and estimators of variance) that do not depend on any assumptions about the population—a sort of nonparametric approach—obtaining estimates acceptable (if grudgingly) by users with differing and conflicting interests, avoiding ordinary human biases in selection, obtaining fairly representative or balanced samples with high probability, and avoiding the potentially disastrous effects of important but unknown auxiliary variables. Some benefits of a model-based approach include assessing the efficiency of standard designs and estimators under different assumptions about the population, suggesting good designs to use—or good samples to obtain—for certain populations, deriving estimators that make the most efficient use of the sample data, making good use of auxiliary information, dealing with observational data obtained without any proper sampling design, and dealing with missing data and other nonsampling errors.

For a real population, however, even the best model is something one not so much believes as tentatively entertains. Under the assumption of the model, one can outline an efficient course of action in carrying out a survey. It is also nice to be able to say that if that assumption is wrong, the strategy still has certain desirable properties—for example, the estimator is still unbiased, if less efficient. One approach combining design and model considerations uses the best available model to suggest an efficient design and form of estimator of the population mean or total while seeking unbiasedness or approximate unbiasedness under the design, and using estimators of variance that are robust against departures from the model. With this approach, one looks for a strategy with low unconditional mean square error, subject to the required (exact or approximate) design unbiasedness. Such an approach has been useful in the development of such survey methods as the generalized ratio and regression estimators under probability designs. "Model-assisted" strategies such as these, using models to suggest good sampling designs and inference procedures but seeking to have good design-based properties that provide robustness against any possible departures from the assumed model, are described in depth in Särndal et al. (1992).

Reviews of the ideas and issues involved in the relationship of design and model in sampling are found in Cassel et al. (1977), Godambe (1982), Hansen et al. (1983), Hedayat and Sinha (1991), Särndal (1978), Smith (1976, 1984), Sugden and Smith (1984), M. E. Thompson (1997), and Thompson and Seber (1996).

## 10.4 LIKELIHOOD FUNCTION IN SAMPLING

In the design-based, fixed-population approach to sampling, the values $(y_i, \ldots, y_N)$ of the variable of interest are viewed as fixed or given for all units in the population. With this approach, the unknown values $y_i$ of the variable of interest in the population are the unknown parameters. For designs that do not depend on any unobserved $y$-values the likelihood function is constant, equal to the probability of selecting the sample obtained, for every potential value $\mathbf{y}$ of the population consistent with the sample data (Basu 1969).

In the model-based approach, the population values $\mathbf{y}$ are viewed as realizations from a stochastic distribution. Suppose that there is a population model $f(\mathbf{y}; \theta)$, giving the probability that the $y$-values in the population take on the specific set of values $\mathbf{y} = (y_1, y_2, \ldots, y_N)$. This probability may depend on an unknown parameter $\theta$ as well as on the auxiliary variables. The distribution may also depend on auxiliary variables. However, the dependence of the data, sampling design, and model on auxiliary variables will be left implicit in this section for notational simplicity. Also for ease of notation, assume that the variable of interest is a discrete random variable, so that sums rather than integrals are involved in the likelihood function.

The likelihood function is the probability of obtaining the observed data as a function of the unknown parameters. The data in sampling consist of the units in the sample together with their associated values of the variable of interest and any auxiliary variables recorded. For simplicity, the data can be written $d = (s, \mathbf{y}_s)$, where $s$ is the set or sequence of units selected and $\mathbf{y}_s$ represents the $y$-values in the sample. Let $p$ denote the sampling design giving for every possible sample the probability that it is the one selected. Now in general, the design can depend on auxiliary variables $\mathbf{x}$ that are known for the whole population and even on the variable of interest $\mathbf{y}$. For example, in surveys that rely on volunteers or that involve nonresponse, the probability of volunteering or of responding, and hence being in the sample, is often related to the variable of interest. The adaptive sampling designs in the last part of this book also depend on the variable of interest. Thus, the sampling design can be written $p(s \mid \mathbf{y})$.

The likelihood function is thus the probability that the sample $s$ is selected and the values $\mathbf{y}_s$ are observed and can be written

$$L_d(\theta) = \sum p(s \mid \mathbf{y}) f(\mathbf{y}; \theta)$$

where the sum is over possible realizations of the population $\mathbf{y}$ that are consistent with the observed data $d$. Since the $y$-values in the sample are fixed by the data, the sum is over all possible values $\mathbf{y}_{\bar{s}}$ for the units not in the sample.

An important point to note is that in general the likelihood function depends on both the design and the model. A prevalent mistake in statistics and other fields is to analyze data through careful modeling but without considering the procedure by which the sample is selected. The "likelihood" based on the model only, without consideration of the design, was termed the *face-value likelihood* by Dawid and Dickey (1977) because inference based on it alone takes the data at face value without considering how the data were selected.

There are certain conditions, however, under which the design can be ignored for inference. For any design in which the selection of the sample depends on $y$-values only through those values $\mathbf{y}_s$ included in the data, the design probability can be moved out of the sum and forms a separate factor in the likelihood. Then the likelihood can be written

$$L_d(\theta) = p(s \mid \mathbf{y}_s) \sum_{\mathbf{y}_{\bar{s}}} f(\mathbf{y}; \theta)$$

The design then does not affect the value of estimators or predictors based on direct likelihood methods such as maximum likelihood or Bayes estimators. For any such "ignorable" design, the sum in the likelihood above, over all values of $\mathbf{y}$ leading to the given data value, is simply the marginal probability of the $y$ and values associated with the sample data. This marginal distribution depends on what sample was selected but does not depend on how that sample was selected. For likelihood-based inference with a design ignorable in this sense, the face-value likelihood gives the correct inference.

Likelihood-based inference, such as maximum likelihood estimation or prediction and Bayes methods, is simplified if the design can be ignored at the inference stage. The fact that the sampling design does not affect the value of a Bayes or likelihood-based estimator in survey sampling was noted by Godambe (1966) for designs that do not depend on any values of the variable of interest and by Basu (1969) for designs that do not depend on values of the variable of interest outside the sample. Scott and Smith (1973) showed that the design could become relevant to inference when the data lacked information about the labels of the units in the sample. Rubin (1976) gave exact conditions for a missing data mechanism—of which a sampling design can be viewed as an example—to be relevant in frequentist and likelihood-based inference. For likelihood-based methods such as maximum likelihood and Bayes methods, the design is "ignorable" if the design or mechanism does not depend on values of the variable of interest outside the sample or on any parameters in the distribution of those values. For frequency-based inference such as design- or model-unbiased estimation, however, the design is relevant if it depends on any values of the variable of interest, even in the sample. Scott (1977) showed that the design is relevant to Bayes estimation if auxiliary information used in the design is not available at the inference stage. Sugden and Smith (1984) gave general and detailed results on when the design is relevant in survey sampling situations. Thompson and Seber (1996) discuss the underlying inference issues for adaptive designs, in which the selection procedure deliberately takes advantage of observed values of the variable of interest (and see the descriptions of these designs in later chapters of this book).

The concept of design ignorability thus depends on the model assumed, the design used, and the data collected. It is important to underscore that a design said to be "ignorable" for likelihood-based inference might not be ignorable for a frequentist-based inference, such as model-unbiased estimation, and that even though a design may be ignorable at the inference stage, in that, for example, the way an estimator is calculated does not depend on the design used—the design is still relevant a priori to the properties of the estimator. Ironically, in the real world, it is quite possible that the only data sets for which the designs are truly "ignorable" for inference purposes are those that were obtained through deliberately planned and carefully implemented sampling designs.

# PART III

# Some Useful Designs

# CHAPTER 11

# Stratified Sampling

In stratified sampling, the population is partitioned into regions or strata, and a sample is selected by some design within each stratum. Because the selections in different strata are made independently, the variances of estimators for individual strata can be added together to obtain variances of estimators for the whole population. Since only the within-stratum variances enter into the variances of estimators, the *principle of stratification* is to partition the population in such a way that the units within a stratum are as similar as possible. Then, even though one stratum may differ markedly from another, a stratified sample with the desired number of units from each stratum in the population will tend to be "representative" of the population as a whole.

A geographical region may be stratified into similar areas by means of some known variable such as habitat type, elevation, or soil type. Even if a large geographic study area appears to be homogeneous, stratification into blocks can help ensure that the sample is spread out over the whole study area. Human populations may be stratified on the basis of geographic region, city size, sex, or socioeconomic factors.

In the following, it is assumed that a sample is selected by some probability design from each of the $L$ strata in the population, with selections in different strata independent of each other. The variable of interest associated with the $i$th unit of stratum $h$ will be denoted $y_{hi}$. Let $N_h$ represent the number of units in stratum $h$ and $n_h$ the number of units in the sample from that stratum. The total number of units in the population is $N = \sum_{h=1}^{L} N_h$ and the total sample size is $n = \sum_{h=1}^{L} n_h$. The total of the $Y$-values in stratum $h$ is $\tau_h = \sum_{i=1}^{N_h} y_{hi}$ and the mean for that stratum is $\mu_h = \tau_h / N_h$. The total for the whole population is $\tau = \sum_{h=1}^{L} \tau_h$. The overall population mean is $\mu = \tau / N$.

The design is called *stratified random sampling* if the design within each stratum is simple random sampling. Figure 11.1 shows a stratified random sample from a population of $N = 400$ units. The sizes of the $L = 4$ strata are $N_1 = 200$, $N_2 = 100$, and $N_3 = N_4 = 50$. Within each stratum, a random sample without replacement has been selected independently. The total sample size of $n = 40$ has been allocated proportional to stratum size, so that $n_1 = 20$, $n_2 = 10$, and $n_3 = n_4 = 5$.

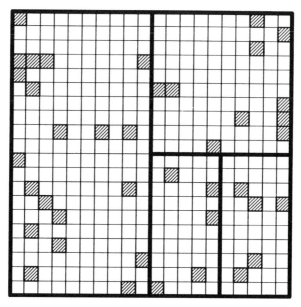

**Figure 11.1.** Stratified random sample.

The results in the next section are written to allow for the possibility of any design within a given stratum, provided that the selections are independent between strata; then specific results for stratified random sampling are given.

## 11.1   ESTIMATING THE POPULATION TOTAL

### With Any Stratified Design

Suppose that within stratum $h$ any specified design is used to select the sample $s_h$ of $n_h$ units, and suppose that one has an estimator $\hat{\tau}_h$ which is unbiased for $\tau_h$ with respect to that design. Let $\mathrm{var}(\hat{\tau}_h)$ denote the variance of $\hat{\tau}_h$, and suppose that one has an unbiased estimator $\widehat{\mathrm{var}}(\hat{\tau}_h)$ of that variance.

Then an unbiased estimator of the overall population total $\tau$ is obtained by adding together the stratum estimators:

$$\hat{\tau}_{\mathrm{st}} = \sum_{h=1}^{L} \hat{\tau}_h$$

Because of the independence of the selections in different strata, the variance of the stratified estimator is the sum of the individual stratum variances:

$$\mathrm{var}(\hat{\tau}_{\mathrm{st}}) = \sum_{h=1}^{L} \mathrm{var}(\hat{\tau}_h)$$

An unbiased estimator of that variance is the sum of individual stratum estimators:

$$\widehat{\mathrm{var}}(\hat{\tau}_{\mathrm{st}}) = \sum_{h=1}^{L} \widehat{\mathrm{var}}(\hat{\tau}_h)$$

## With Stratified Random Sampling

If the sample is selected by simple random sampling without replacement in each stratum, then

$$\hat{\tau}_h = N_h \bar{y}_h$$

is an unbiased estimator of $\tau_h$, where

$$\bar{y}_h = \frac{1}{n_h} \sum_{i=1}^{n_h} y_{hi}$$

is the sample mean for stratum $h$.

An unbiased estimator for the population total $\tau$ is

$$\hat{\tau}_{\mathrm{st}} = \sum_{h=1}^{L} N_h \bar{y}_h$$

having variance

$$\mathrm{var}(\hat{\tau}_{\mathrm{st}}) = \sum_{h=1}^{L} N_h (N_h - n_h) \frac{\sigma_h^2}{n_h}$$

where

$$\sigma_h^2 = \frac{1}{N_h - 1} \sum_{i=1}^{N_h} (y_{hi} - \mu_h)^2$$

is the finite population variance from stratum $h$.

An unbiased estimator of the variance of $\hat{\tau}_{\mathrm{st}}$ is

$$\widehat{\mathrm{var}}(\hat{\tau}_{\mathrm{st}}) = \sum_{h=1}^{L} N_h (N_h - n_h) \frac{s_h^2}{n_h}$$

where

$$s_h^2 = \frac{1}{n_h - 1} \sum_{i=1}^{n_h} (y_{hi} - \bar{y}_h)^2$$

is the sample variance from stratum $h$.

## 11.2   ESTIMATING THE POPULATION MEAN

### With Any Stratified Design

Since $\mu = \tau/N$, the stratified estimator for $\mu$ is

$$\hat{\mu}_{st} = \frac{\hat{\tau}_{st}}{N}$$

Assuming that the selections in different strata have been made independently, the variance of the estimator is

$$\text{var}(\hat{\mu}_{st}) = \frac{1}{N^2} \text{var}(\hat{\tau}_{st})$$

with unbiased estimator of variance

$$\widehat{\text{var}}(\hat{\mu}_{st}) = \frac{1}{N^2} \widehat{\text{var}}(\hat{\tau}_{st})$$

### With Stratified Random Sampling

With stratified random sampling, an unbiased estimator of the population mean $\mu$ is the *stratified sample mean*:

$$\bar{y}_{st} = \frac{1}{N} \sum_{h=1}^{L} N_h \bar{y}_h \tag{1}$$

Its variance is

$$\text{var}(\bar{y}_{st}) = \sum_{h=1}^{L} \left(\frac{N_h}{N}\right)^2 \left(\frac{N_h - n_h}{N_h}\right) \frac{\sigma_h^2}{n_h} \tag{2}$$

An unbiased estimator of this variance is

$$\widehat{\text{var}}(\bar{y}_{st}) = \sum_{h=1}^{L} \left(\frac{N_h}{N}\right)^2 \left(\frac{N_h - n_h}{N_h}\right) \frac{s_h^2}{n_h} \tag{3}$$

*Example 1.*   The results of a stratified random sample are summarized in Table 11.1. Substituting in Equation (1), the estimate of the population mean is

$$\bar{y}_{st} = \frac{1}{41}[20(1.6) + 9(2.8) + 12(0.6)]$$

$$= \frac{1}{41}(32 + 25.2 + 7.2)$$

$$= \frac{64.4}{41} = 1.57$$

**Table 11.1. Results of a Stratified Random Sample**

| Stratum $h$ | $N_h$ | $n_h$ | $\bar{y}_h$ | $s_h^2$ |
|---|---|---|---|---|
| 1 | 20 | 5 | 1.6 | 3.3 |
| 2 | 9 | 3 | 2.8 | 4.0 |
| 3 | 12 | 4 | 0.6 | 2.2 |

The estimator $\hat{\tau}$ of the population total, obtained by multiplying by 41, is 64.4. The estimated variance of $\bar{y}_{\text{st}}$ [from Equation (3)] is

$$\widehat{\text{var}}(\bar{y}_{\text{st}}) = \frac{1}{41^2}\left[20(20-5)\frac{3.3}{5} + 9(9-3)\frac{4.0}{3} + 12(12-4)\frac{2.2}{4}\right]$$
$$= \frac{322.8}{41^2} = 0.192$$

The estimated variance for the estimator of the population total, obtained by multiplying by $41^2$, is 322.8. □

## 11.3 CONFIDENCE INTERVALS

When all the stratum sample sizes are sufficiently large, an approximate $100(1-\alpha)\%$ confidence interval for the population total is provided by

$$\hat{\tau}_{\text{st}} \pm t\sqrt{\widehat{\text{var}}(\hat{\tau}_{\text{st}})}$$

where $t$ is the upper $\alpha/2$ point of the normal distribution. For the mean, the confidence interval is $\hat{\mu}_{\text{st}} \pm t\sqrt{\widehat{\text{var}}(\hat{\mu}_{\text{st}})}$. As a rule of thumb, the normal approximation may be used if all the sample sizes are at least 30. With small sample sizes, the $t$-distribution with an approximate degrees of freedom may be used. The Satterthwaite (1946) approximation for the degrees of freedom $d$ to be used is

$$d = \left(\sum_{h=1}^{L} a_h s_h^2\right)^2 \bigg/ \left[\sum_{h=1}^{L}(a_h s_h^2)^2/(n_h-1)\right] \tag{4}$$

where $a_h = N_h(N_h - n_h)/n_h$. If all the stratum sizes $N_h$ are equal and all the sample sizes $n_h$ are equal and all the sample variances $s_h^2$ are equal then the degrees of freedom are $n - L$.

Satterthwaite's formula is based on approximating the distribution of a linear combination of sample variances with a chi-square distribution. Some possible refinements to Satterthwaite's formula are discussed in Ames and Webster (1991).

***Example 2: Confidence Interval.*** For Example 1, the variance coefficients are $a_1 = 20(20 - 5)/5 = 60$, $a_2 = 9(9 - 3)/3 = 18$, and $a_3 = 12(12 - 4)/4 = 24$. The estimated degrees of freedom [from Equation (4)] are

$$d = \frac{[60(3.3) + 18(4.0) + 24(2.2)]^2}{[60(3.3)]^2/4 + [18(4.0)]^2/2 + [24(2.2)]^2/3}$$

$$= \frac{322.8^2}{13,322.28} = 7.82$$

or about 8 degrees of freedom. The approximate 95% confidence interval for the mean is $1.57 \pm 2.306\sqrt{0.192} = 1.57 \pm 1.01 = (0.56, 2.58)$. ☐

## 11.4 THE STRATIFICATION PRINCIPLE

Since the formula for the variance of the estimator of the population mean or total with stratified sampling contains only within-stratum population variance terms, the estimators will be more precise the smaller the $\sigma_h^2$. Equivalently, estimation of the population mean or total will be most precise if the population is partitioned into strata in such a way that *within each stratum, the units are as similar as possible.* Thus, in a survey of a plant or animal population, the study area might be stratified into regions of similar habitat or elevation, with the idea that within strata, abundances will be more similar than between strata. In a survey of a human population, stratification may be based on socioeconomic factors or geographic region.

***Example 3: Comparison with Simple Random Sampling.*** Consider a small population of $N = 6$ units, divided into two strata of $N_h = 3$ units each, in order to examine the effectiveness of stratified sampling in comparison to simple random sampling. The population $y$-values in stratum 1 are 2, 0, and 1. In stratum 2, the $y$-values are 5, 9, and 4. Thus, the overall population mean is $\mu = 3.5$, and the overall population variance is $\sigma^2 = 10.7$. Within stratum 1, the population mean is $\mu_1 = 1$ and the population variance is $\sigma_1^2 = 1.0$. Within stratum 2, $\mu_2 = 6$ and $\sigma_2^2 = 7.0$.

For simple random sampling with sample size $n = 4$, the sample mean is an unbiased estimator of the population mean and has variance $\text{var}(\bar{y}) = [(6 - 4)/6] (10.7/4) = 0.89$. For stratified random sampling with sample sizes $n_1 = n_2 = 2$, so that the total sample size is still 4, the stratified sample mean $\bar{y}_{st}$ is an unbiased estimator of the population mean having variance $\text{var}(\bar{y}_{st}) = (3/6)^2 [(3 - 2)/3](1/2) + (3/6)^2[(3 - 2)/3](7/2) = 0.33$. For this population, stratification has been effective because the units within each stratum are relatively similar. ☐

## 11.5 ALLOCATION IN STRATIFIED RANDOM SAMPLING

Given a total sample size $n$, one may choose how to allocate it among the $L$ strata. If each stratum is the same size and one has no prior information about the population, a reasonable choice would be to assume equal sample sizes for the strata, so that for stratum $h$ the sample size would be

$$n_h = \frac{n}{L}$$

If the strata differ in size, *proportional allocation* could be used to maintain a steady sampling fraction throughout the population. If stratum $h$ has $N_h$ units, the sample size allocated to it would be

$$n_h = \frac{nN_h}{N}$$

The allocation scheme that estimates the population mean or total with the lowest variance for a fixed total sample size $n$ under stratified random sampling is *optimum allocation*:

$$n_h = \frac{nN_h\sigma_h}{\sum_{k=1}^{L} N_k\sigma_k}$$

The stratum population standard deviations $\sigma_h$ may in practice be estimated with sample standard deviations from past data.

In some sampling situations, the cost of sampling, measured in terms of time or money, differs from stratum to stratum, and total cost may be described by the linear relationship

$$c = c_0 + c_1 n_1 + c_2 n_2 + \cdots + c_L n_L$$

where $c$ is the total cost of the survey, $c_0$ is an "overhead" cost, and $c_h$ is the cost per unit observed in stratum $h$. Then for a fixed total cost $c$, the lowest variance is achieved with sample size in stratum $h$ proportional to $N_h\sigma_h/\sqrt{c_h}$, that is,

$$n_h = \frac{(c - c_0)N_h\sigma_h/\sqrt{c_h}}{\sum_{k=1}^{L} N_k\sigma_k\sqrt{c_k}}$$

Thus, the optimum scheme allocates larger sample size to the larger or more variable strata and smaller sample size to the more expensive or difficult-to-sample strata.

***Example 4: Allocation.***  A population consists of three strata of sizes $N_1 = 150$, $N_2 = 90$, and $N_3 = 120$, so that the total population size is $N = 360$. Based on sample standard deviations from previous surveys, the standard deviations within each stratum are estimated to be approximately $\sigma_1 \approx 100$, $\sigma_2 \approx 200$, and $\sigma_3 \approx 300$.

Proportional allocation of a sample of total size $n = 12$ is given by $n_1 = 12(150)/360 = 5$, $n_2 = 12(90)/360 = 3$, and $n_3 = 12(120)/360 = 4$.

Assuming equal cost per unit of sampling in each stratum, the optimal allocation of a total sample size of $n = 12$ between the three strata is

$$n_1 = \frac{12(150)(100)}{150(100) + 90(200) + 120(300)} = 2.6$$

$$n_2 = \frac{12(90)(200)}{150(100) + 90(200) + 120(300)} = 3.1$$

$$n_3 = \frac{12(120)(300)}{150(100) + 90(200) + 120(300)} = 6.3$$

Rounding to whole numbers gives $n_1 = 3$, $n_2 = 3$, and $n_3 = 6$. $\square$

## 11.6 POSTSTRATIFICATION

In some situations it may be desired to classify the units of a sample into strata and to use a stratified estimate, even though the sample was selected by simple random, rather than stratified, sampling. For example, a simple random sample of a human population may be stratified by sex after selection of the sample, or a simple random sample of sites in a fishery survey may be poststratified on depth. In contrast to conventional stratified sampling, with poststratification, the stratum sample sizes $n_1, n_2 \ldots, n_L$ are random variables.

With proportional allocation in conventional stratified random sampling, the sample size in stratum $h$ is fixed at $n_h = nN_h/N$ and the variance [Equation (2)] simplifies to $\mathrm{var}(\bar{y}_{st}) = [(N - n)/nN] \sum_{h=1}^{L} (N_h/N)\sigma_h^2$. With poststratification of a simple random sample of $n$ units from the whole population, the sample size $n_h$ in stratum $h$ has expected value $nN_h/N$, so that the resulting sample tends to approximate proportional allocation. With poststratification the variance of the stratified estimator $\bar{y}_{st} = \sum_{k=1}^{L} (N_h/N)\bar{y}_h$ is approximately

$$\mathrm{var}(\bar{y}_{st}) \approx \frac{N - n}{nN} \sum_{h=1}^{L} \left(\frac{N_h}{N}\right)\sigma_h^2 + \frac{1}{n^2} \left(\frac{N - n}{N - 1}\right) \sum_{h=1}^{L} \frac{N - N_h}{N} \sigma_h^2 \qquad (5)$$

and the variance of $\hat{\tau}_{st} = N\bar{y}_{st}$ is $\mathrm{var}(\hat{\tau}_{st}) = N^2 \mathrm{var}(\bar{y}_{st})$. The first term is the variance that would be obtained using a stratified random sampling design with proportional allocation. An additional term is added to the variance with poststratification, due to the random sample sizes.

For a variance estimate with which to construct a confidence interval for the population mean with poststratified data from a simple random sample, it is recommended to use the standard stratified sampling method [Equation (3)] rather than substituting the sample variances directly into Equation (5). With poststratification, the standard formula [Equation (3)] estimates the conditional variance [given by

Equation (2)] of $\bar{y}_{st}$ given the sample sizes $n_1, \ldots, n_L$, while Equation (5) is the unconditional variance [and see the comments of J. N. K. Rao (1988, p. 440)].

To use poststratification, the relative size $N_h/N$ of each stratum must be known. If the relative stratum sizes are not known, they may be estimated using double sampling (see Chapter 14). Further discussion of poststratification may be found in Cochran (1977), Hansen et al. (1953), Hedayat and Sinha (1991), Kish (1965), Levy and Lemeshow (1991), Singh and Chaudhary (1986), and Sukhatme and Sukhatme (1970). Variance approximations for poststratification vary among the sampling texts. The derivation for the expression given here is given in Section 11.8 under the heading "Poststratification Variance."

## 11.7 POPULATION MODEL FOR A STRATIFIED POPULATION

A simple model for a stratified population assumes that the population $Y$-values are independent random variables, each having a normal distribution, and with means and variances depending on stratum membership. Under this model, the value $Y_{hi}$ for the $i$th unit in stratum $h$ has a normal distribution with mean $\mu_h$ and variance $\sigma_h^2$, for $h = 1, \ldots, L$, $i = 1, \ldots, N_h$, and the $Y_{hi}$ are independent. A stratified sample $s$ is selected using any conventional design within each stratum.

Since for each unit $Y_{hi}$ is a random variable, the population total $T = \sum_{h=1}^{L} \sum_{i=1}^{N_h} Y_{hi}$ is also a random variable. Since the $Y$-values are observed only for units in the sample, we wish to predict $T$ using a predictor $\hat{T}$ computed from the sample data. Desirable properties to have in a predictor $\hat{T}$ include model unbiasedness,

$$E(\hat{T} \mid s) = E(T)$$

where expectation is taken with respect to the model. In addition, we would like the mean square prediction error $E(\hat{T} - T)^2$ to be as low as possible.

For a given sample the best unbiased predictor of the population total $T$ is

$$\hat{T} = \sum_{h=1}^{L} N_h \bar{y}_h$$

which is the standard stratified sampling estimator. Without the assumption of normality, the predictor $\hat{T}$ is best linear unbiased. This result is a special case of prediction results about the general linear regression model.

In addition, a model-unbiased estimator of the mean square prediction error is the standard stratified variance estimator

$$\hat{E}(\hat{T} - T)^2 = \sum_{h=1}^{L} \left(\frac{N_h}{N}\right)^2 \left(\frac{N_h - n_h}{N_h}\right) \frac{S_h^2}{n_h}$$

in which $S_h^2$ is the sample variance within stratum $h$.

## 11.8  DERIVATIONS FOR STRATIFIED SAMPLING

**Optimum Allocation**

Consider the variance of the estimator $\hat{\tau}_{st}$ as a function $f$ of the sample sizes, with the total sample sizes given. The object is to choose $n_1, n_2, \ldots, n_L$ to minimize

$$f(n_1, \ldots, n_L) = \text{var}(\hat{\tau}_{st}) = \sum_{h=1}^{L} N_h \sigma_h^2 \left( \frac{N_h}{n_h} - 1 \right)$$

subject to the constraint

$$\sum_{h=1}^{L} n_h = n$$

The Lagrange multiplier method may be used to solve such a problem. Write $g(n_1, \ldots, n_L) = \sum n_h - n$. The solution is obtained by differentiating the function $H = f - \lambda g$ with respect to each $n_h$ and $\lambda$, where $\lambda$ is the Lagrange multiplier, and setting the partial derivatives equal to zero. The partial derivatives are

$$\frac{\partial H}{\partial n_h} = -\left( \frac{N_h^2 \sigma_h^2}{n_h^2} \right) - \lambda = 0$$

for $h = 1, \ldots, L$. Differentiating with respect to $\lambda$ reproduces the constraint $\sum n_h - n = 0$. Solving for $n_h$ gives

$$n_h = \frac{n N_h \sigma_h}{\sum_{k=1}^{L} N_k \sigma_k}$$

To verify that the solution gives a minimum of the variance function, as opposed to a maximum or saddle point, the second derivatives are examined. Writing $H_{hk}$ for the second partial derivative $\partial^2 H / \partial n_h \partial n_k$ gives

$$H_{hh} = \frac{N^2 \sigma_h^2}{n_h^3} \qquad h = 1, \ldots, L$$

$$H_{hk} = 0 \qquad h \neq k$$

$$H_{\lambda h} = -1$$

$$H_{\lambda \lambda} = 0$$

A sufficient condition for the solution to be a minimum is that for any set of numbers $a_1, \ldots, a_L$ satisfying $\sum_{h=1}^{L} H_{\lambda h} a_h = 0$, the double sum $\sum_{h=1}^{L} \sum_{K=1}^{L} H_{hk} a_h a_k$ is invariably positive (Hancock 1960, p. 115). Since $H_{hk} = 0$ for $h \neq k$, $\sum_{h=1}^{L} \sum_{K=1}^{L} H_{hk} a_h a_k = \sum_{h=1}^{L} N_h^2 \sigma_h^2 / n_h^3$, which is invariably positive.

The derivation proceeds similarly when the constraint depends on cost. An alternative derivation uses the Cauchy–Schwartz inequality (see, e.g., Cochran 1977, p. 97).

### Poststratification Variance

With simple random sampling, the number $n_h$ of sample units in stratum $h$ has a hypergeometric distribution with $E(n_h) = nN_h/N$ and

$$\text{var}(n_h) = n(N_h/N)(1 - N_h/N)[(N - n)/(N - 1)]$$

The poststratification estimator $\bar{y}_{st}$ is unbiased for the population mean $\mu$ provided that samples in which any of the $n_h$ are zero are excluded. Then

$$\text{var}(\bar{y}_{st}) = E[\text{var}(\bar{y}_{st} \mid n_1, \ldots, n_L)]$$

$$= E\left[\sum_{h=1}^{L} \left(\frac{N_h}{N}\right)^2 \left(\frac{N_h - n_h}{N_h}\right) \frac{\sigma_h^2}{n_h}\right]$$

$$= \sum_{h=1}^{L} \left(\frac{N_h}{N}\right)^2 \sigma_h^2 \left[E\left(\frac{1}{n_h}\right) - \frac{1}{N_h}\right]$$

Using a Taylor series approximation for $1/n_h$, whose first derivative is $-n_h^{-2}$ and a second derivation $2n_h^{-3}$, and taking expectation gives the approximation

$$E\left(\frac{1}{n_h}\right) \approx \left[\frac{1}{E(n_h)}\right] + \left[\frac{1}{(E(n_h))^3}\right] \text{var}(n_h)$$

$$= \frac{N}{nN_h} + \left(\frac{N}{nN_h}\right)^2 \left(\frac{N - N_h}{N}\right) \left(\frac{N - n}{N - 1}\right)$$

Substituting this approximation into the variance expression gives

$$\text{var}(\bar{y}_{st}) \approx \sum_{h=1}^{L} \left(\frac{N_h}{N}\right)^2 \sigma_h^2 \left[\frac{N}{nN_h} + \left(\frac{N}{nN_h}\right)^2 \left(\frac{N - N_h}{N}\right) \left(\frac{N - n}{N - 1}\right) - \frac{1}{N_h}\right]$$

$$= \frac{N - n}{nN} \sum_{h=1}^{L} \left(\frac{N_h}{N}\right) \sigma_h^2 + \frac{1}{n^2} \left(\frac{N - n}{N - 1}\right) \sum_{h=1}^{L} \left(\frac{N - N_h}{N}\right) \sigma_h^2$$

which completes the derivation.

**EXERCISES**

1. The following results were obtained from a stratified random sample:
   Stratum 1: $N_1 = 100$, $n_1 = 50$, $\bar{y}_1 = 10$, $s_1^2 = 2800$
   Stratum 2: $N_2 = 50$, $n_2 = 50$, $\bar{y}_2 = 20$, $s_2^2 = 700$
   Stratum 3: $N_3 = 300$, $n_3 = 50$, $\bar{y}_3 = 30$, $s_3^2 = 600$
   (a) Estimate the mean for the whole population.
   (b) Give a 95% confidence interval for the mean.

2. Allocate a total sample size of $n = 100$ between two strata having sizes $N_1 = 200$
   and $N_2 = 300$ and variances $\sigma_1^2 = 81$ and $\sigma_2^2 = 16$ (a) using proportional
   allocation and (b) using optimal allocation (assume equal costs).

3. Use stratified random sampling to estimate the mean or total of a population of
   your choice. In the process of carrying out the survey and making the estimate,
   think about or discuss with others the following:
   (a) What practical problems arise in establishing a frame, such as a map or list of
       units, from which to select the sample?
   (b) How is the sample selection actually carried out?
   (c) What special problems arise in observing the units selected?
   (d) Estimate the population mean or total.
   (e) Estimate the variance of the estimator above.
   (f) Give a 95% confidence interval for the population mean or total.
   (g) Using the stratum sample variances from your data, give the proportional and
       the optimum allocations of a sample of size 200 in a future survey.
   (h) How would you improve the survey procedure if you were to do it again?

CHAPTER 12

# Cluster and Systematic Sampling

Although systematic sampling and cluster sampling seem on the surface to be opposites—the one spacing out the units of a sample and the other bunching them together—the two designs share the same structure. The population is partitioned into *primary units*, each primary unit being composed of *secondary units*. Whenever a primary unit is included in the sample, the $y$-values of every secondary unit within it are observed.

In systematic sampling, a single primary unit consists of secondary units spaced in some systematic fashion throughout the population. In cluster sampling, a primary unit consists of a cluster of secondary units, usually in close proximity to each other. In the spatial setting, a systematic sample primary unit may be composed of a collection of plots in a grid pattern over the study area. Cluster primary units include such spatial arrangements as square collections of adjacent plots or long, narrow strips of adjacent units. A cluster sample consisting of a simple random sample of 10 primary units, each consisting of eight secondary units, is shown in Figure 12.1. A systematic sample with two randomly selected starting points is shown in Figure 12.2. The systematic sample consists of two primary units (distinguished by their shading), each with 16 secondary units.

The key point in any of the systematic or clustered arrangements is that whenever any secondary unit of a primary unit is included in the sample, all the secondary units of that primary unit are included. Even though the actual measurements may be made on secondary units, it is the primary units that are selected. In principle, one could dispense with the concept of the secondary units, regarding the primary units as the sampling units and using, as the variable of interest for any primary unit, the total of the $y$-values of the secondary units within it. Then all properties of estimators may be obtained based on the design by which the sample of primary units is selected. However, several common features of systematic and cluster sampling make these designs worth considering as special cases:

1. In systematic sampling, it is not uncommon to have a sample size of 1, that is, a single primary unit.

**129**

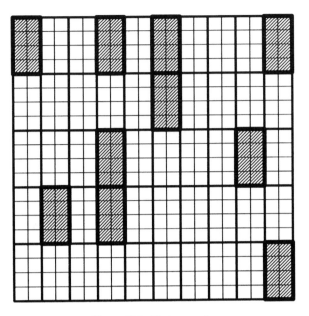

**Figure 12.1.** Cluster sample.

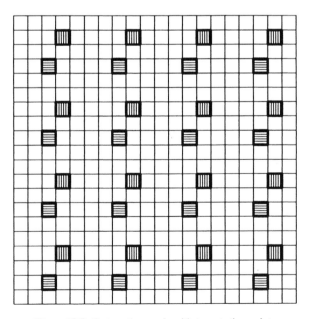

**Figure 12.2.** Systematic sample with two starting points.

2. In cluster sampling, the size of the cluster may serve as auxiliary information that may be used either in selecting clusters with unequal probabilities or in forming ratio estimators.
3. The size and shape of clusters may affect efficiency.

Let $N$ denote the number of primary units in the population and $n$ the number of primary units in the sample. Let $M_i$ be the number of secondary units in the $i$th primary unit. The total number of secondary units in the population is $M = \sum_{i=1}^{N} M_i$. Let $y_{ij}$ denote the value of the variable of interest of the $j$th secondary unit in the $i$th primary unit. The total of the $y$-values in the $i$th primary unit will be denoted simply $y_i$, that is, $y_i = \sum_{j=1}^{M_i} y_{ij}$. The population total is $\tau = \sum_{i=1}^{N} \sum_{j=1}^{M_i} y_{ij} = \sum_{i=1}^{N} y_i$. The population mean per primary unit is $\mu_1 = \tau/N$. The population mean per secondary unit is $\mu = \tau/M$.

## 12.1 PRIMARY UNITS SELECTED BY SIMPLE RANDOM SAMPLING

### Unbiased Estimator

When primary units are selected by simple random sampling without replacement, an unbiased estimator of the population total $\tau$ is

$$\hat{\tau} = \frac{N}{n}\sum_{i=1}^{n} y_i = N\bar{y} \tag{1}$$

where $\bar{y} = (1/n)\sum_{i=1}^{n} y_i$, the sample mean of the primary unit totals.
The variance of this estimator is

$$\text{var}(\hat{\tau}) = N(N-n)\frac{\sigma_u^2}{n} \tag{2}$$

where $\sigma_u^2$ is the finite population variance of the primary unit totals,

$$\sigma_u^2 = \frac{1}{N-1}\sum_{i=1}^{N}(y_i - \mu_1)^2$$

An unbiased estimate of the variance of $\hat{\tau}$ is

$$\widehat{\text{var}}(\hat{\tau}) = N(N-n)\frac{s_u^2}{n}$$

where $s_u^2$ is the sample variance of the primary unit totals,

$$s_u^2 = \frac{1}{n-1} \sum_{i=1}^{n} (y_i - \bar{y})^2$$

These results are familiar from simple random sampling. An unbiased estimator of the mean per primary unit $\mu_1$ is $\bar{y} = \hat{\tau}/N$, and an unbiased estimator of the mean per secondary unit $\mu$ is $\hat{\mu} = \hat{\tau}/M$. The variance of $\bar{y}$ is $\mathrm{var}(\bar{y}) = (1/N^2)\mathrm{var}(\hat{\tau})$, and the variance of $\hat{\mu}$ is $\mathrm{var}(\hat{\mu}) = (1/M^2)\mathrm{var}(\hat{\tau})$. The estimated variances are obtained similarly by dividing the estimated variance of $\hat{\tau}$ by $N^2$ or $M^2$.

### Ratio Estimator

If primary unit total $y_i$ is highly correlated with primary unit size $M_i$, a ratio estimator based on size may be efficient. The ratio estimator of the population total is

$$\hat{\tau}_r = rM$$

were the sample ratio $r$ is

$$r = \frac{\sum_{i=1}^{n} y_i}{\sum_{i=1}^{n} M_i}$$

The population ratio is the mean per secondary unit $\mu$. As a ratio estimator, $\hat{\tau}_r$ is not unbiased, but the bias tends to be small with large sample sizes, and the mean square error may be considerably less than that of the unbiased estimator when the $y_i$ and the $M_i$ tend to be proportionally related.

An approximate formula for the mean square error or variance of the ratio estimator is

$$\mathrm{var}(\hat{\tau}_r) \approx \frac{N(N-n)}{n(N-1)} \sum_{i-1}^{N} (y_i - M_i\mu)^2$$

An estimator of this variance is given by

$$\widehat{\mathrm{var}}(\hat{\tau}_r) = \frac{N(N-n)}{n(n-1)} \sum_{i=1}^{n} (y_i - rM_i)^2$$

or the adjusted estimator

$$\widetilde{\mathrm{var}}(\hat{\tau}_r) = \left(\frac{nM}{N\sum_{i=1}^{n} M_i}\right)^2 \widehat{\mathrm{var}}(\hat{\tau}_r)$$

Alternative variance estimators for ratio estimators are discussed in the chapter on auxiliary data and ratio estimation (Chapter 7) and are reviewed in J. N. K. Rao (1988, pp. 402–403), P. S. R. S. Rao (1988, pp. 454–456), and Royall (1988, pp. 402–403).

To estimate the population mean $\mu_1$ per primary unit, the ratio estimator would be $\hat{\mu}_{1r} = \hat{\tau}_r/N$, for which the mean square error formulas above would be divided by $N^2$. To estimate the population mean $\mu$ per secondary unit, the ratio estimator is $\hat{\mu}_r = \hat{\tau}_r/M = r$, for which one would divide the mean square error expressions above by $M^2$.

## 12.2  PRIMARY UNITS SELECTED WITH PROBABILITIES PROPORTIONAL TO SIZE

Suppose that the primary units are selected with replacement with draw-by-draw selection probabilities proportional to the sizes of the primary units, that is, $p_i = M_i/M$. One way to carry out such a design is to select $n$ secondary units from the $M$ in the population, using simple random sampling with replacement: A primary unit is selected every time any of its secondary units is selected.

### Hansen–Hurwitz (PPS) Estimator

An unbiased estimator of the population total under sampling with replacement with probabilities proportional to size, based on the Hansen–Hurwitz estimator, is

$$\hat{\tau}_p = \frac{M}{n}\sum_{i=1}^{n} \frac{y_i}{M_i}$$

with each observation utilized in the sum as many times as its primary unit is selected. The variance of this estimator is

$$\mathrm{var}(\hat{\tau}_p) = \frac{M}{n}\sum_{i=1}^{N} M_i(\bar{y}_i - \mu)^2$$

where $\bar{y}_i = y_i/M_i$. An unbiased estimator of this variance is

$$\widehat{\mathrm{var}}(\hat{\tau}_p) = \frac{M^2}{n(n-1)} \sum_{i=1}^{n} (\bar{y}_i - \hat{\mu}_p)^2$$

where $\hat{\mu}_p = \hat{\tau}_p/M$.

The estimator $\hat{\mu}_p$ is unbiased for the population mean per secondary unit $\mu$ under the probability-proportional-to-size selection, while $\hat{\mu}_{1p} = \hat{\tau}_p/N$ is unbiased for the population mean per primary unit. Variance formulas for these estimators are obtained by dividing the variance expressions for $\hat{\tau}_p$ by $M^2$ or $N^2$.

### Horvitz–Thompson Estimator

A Horvitz–Thompson estimator can also be computed for this design, using the inclusion probabilities

$$\pi_i = 1 - (1 - p_i)^n$$

and joint inclusion probabilities

$$\pi_{ij} = \pi_i + \pi_j - [1 - (1 - p_i - p_j)^n]$$

based on the selection probabilities $p_i = M_i/M$.

The Horvitz–Thompson estimator for the population total is

$$\hat{\tau}_\pi = \sum_{i=1}^{\nu} \frac{y_i}{\pi_i}$$

where $\nu$ is the number of distinct primary units in the sample. Variance formulas for this estimator were given in Section 6.2.

### 12.3 THE BASIC PRINCIPLE

Since every secondary unit is observed within a selected primary unit, the within-primary-unit variance does not enter into the variances of the estimators. Thus, the basic *systematic and cluster sampling principle* is that to obtain estimators of low variance or mean square error, the population should be partitioned into clusters in such a way that one cluster is similar to another. Equivalently, the within-primary-unit variance should be as great as possible in order to obtain the most precise estimators of the population mean or total. The ideal primary unit contains the full diversity of the population and hence is "representative."

With natural populations of spatially distributed plants, animals, or minerals, and with many human populations, the condition above is typically satisfied by systematic primary units, in which the secondary units are spaced apart, but not by clusters of geographically adjacent units. Cluster sampling is more often than not carried out for reasons of convenience or practicality rather than to obtain lowest variance for a given number of secondary units observed.

## 12.4 SINGLE SYSTEMATIC SAMPLE

Many surveys utilizing a systematic design select a single starting unit at random and then observe every secondary unit at the appropriate spacing from there. Thus the sample consists of a single primary unit selected at random. From a sample of size 1 it is possible to obtain an unbiased estimator of the population mean or total, but it is not possible to obtain an unbiased estimator of its variance.

Naively proceeding as if the $M_1$ secondary units in the single systematic primary unit were a simple random sample from the $M$ secondary units in the population and using the variance formula from simple random sampling leads to good variance estimates only if the units of the population can reasonably be conceived as being in random order. With many natural populations, in which nearby units tend to be similar to each other, this procedure tends to overestimate the variance of the estimator of the population mean or total.

A variety of procedures for estimating variance from a single systematic sample are discussed in Bellhouse (1988a), Murthy and Rao (1988), and Wolter (1984). One of the simplest is to combine pairs (or larger groups) of adjacent units into "strata" and estimate variance as if stratified random sampling had been used.

***Example 1.*** The distinctions between the estimators can be illustrated with a systematic sample selected from a population in which the number $N$ of possible systematic samples does not divide evenly into the number $M$ of (secondary) units in the population. In a survey of bald eagle nests, a coastline 1300 km in length is divided into units of 100-km length, so that there are $M = 13$ of these units in the population. A "one-in-three" systematic sample is selected by choosing at random one of the first three units and then including that unit and every third unit thereafter in the sample. For every unit included in the sample, all eagle nests are counted using research vessels and aircraft. Thus, a single systematic sample is selected, but the size $M_i$ of the various possible samples differs. The number of primary units (the number of possible systematic samples) is $N = 3$, of which $n = 1$ will be selected. If the first unit is chosen as a starting point, $M_1 = 5$ units will be observed, whereas if either of the other starting points is chosen, $M_i = 4$ units will be observed.

Suppose that unit 3 is selected at random from the first three, and the $y$-values observed on the survey are 5, 1, 10, and 18 nests. Then the unbiased estimate of the total number of nests on the coastline is

$$\hat{\tau} = \frac{3}{1}(5 + 1 + 10 + 18) = 102$$

However, presented with data from such a survey, many people would choose to take the average of the four units observed and multiply by the number of units in the population, obtaining the estimate

$$\hat{\tau}_r = \frac{5 + 1 + 10 + 18}{4}(13) = 110.5$$

As a ratio estimator, this estimator is not unbiased with the design used.

With a single starting point ($n = 1$), the PPS estimator is identical to the ratio estimator. Hence the second method above would give an unbiased estimate if the systematic sample were selected with probability proportional to size. This could be accomplished by selecting one unit out of the 13 in the population at random and then including in the sample that unit and every third unit to the right and to the left. □

## 12.5 VARIANCE AND COST IN CLUSTER AND SYSTEMATIC SAMPLING

The effectiveness of cluster or systematic sampling depends both on the variance resulting from using primary units of a given size and shape and the cost of sampling such units. As a starting point, the variance of selecting $n$ primary units may be compared with a simple random sample of an equivalent number of secondary units. The average size of clusters in the population is $\bar{M} = M/N$, so the expected number of secondary units in a simple random sample of $n$ primary units is $n\bar{M}$.

For the unbiased estimate of the population total based on a simple random sample of $n\bar{M}$ secondary units, write $\hat{\tau}_{\text{srs}} = M\bar{y}$. The variance of this design–estimator combination is

$$\text{var}(\hat{\tau}_{\text{srs}}) = M^2\left(\frac{(N\bar{M} - n\bar{M})}{nN\bar{M}^2}\right)\sigma^2$$

$$= N^2\left[\frac{\bar{M}(N - n)}{nN}\right]\sigma^2$$

where $\sigma^2$ is the finite population variance for secondary units,

$$\sigma^2 = \sum_{i=1}^{N}\sum_{j=1}^{\bar{M}}\frac{(y_{ij} - \mu)^2}{N\bar{M} - 1}$$

and $\mu = \tau/N\bar{M}$.

For a cluster or repeated systematic sample, with a simple random sample of $n$ primary units, the unbiased estimator [see Equation (1)] will be denoted $\hat{\tau}_u$, with the subscript $u$ indicating that the design with which the estimator is used is a random sample of primary units of type $u$. The label $u$ identifies the size, shape, or arrangement of primary units, which could be, for example, square clusters, rectangular clusters, or systematic samples. The variance of this design–estimator combination is

$$\text{var}(\hat{\tau}_u) = N^2 \left( \frac{N-n}{nN} \right) \sigma_u^2$$

where $\sigma_u^2 = \sum_{i=1}^{N} (y_i - \mu_1)^2 / (N-1)$ and $\mu_1 = \tau/N$.

The relative efficiency of the cluster (or systematic) sample to the simple random sample of equivalent sample size, defined as the ratio of the variances, is

$$\frac{\text{var}(\hat{\tau}_{\text{srs}})}{\text{var}(\hat{\tau}_u)} = \frac{\bar{M}\sigma^2}{\sigma_u^2}$$

Thus cluster (systematic) sampling is efficient if the variance $\sigma_u^2$ between primary units is small relative to the overall population variance $\sigma^2$.

To estimate this relative efficiency using data from a cluster or systematic sampling design, the usual sample variance $s^2$ cannot be used as an estimate of $\sigma^2$, because the data were not obtained with simple random sampling. Instead, $\sigma^2$ can be estimated using analysis of variance of the cluster (systematic) sample data as follows.

For simplicity, suppose that each of $N$ primary units has an equal number $\bar{M}$ of secondary units. The total sum of squares in the population can be partitioned as

$$\sum_{i=1}^{N} \sum_{j=1}^{\bar{M}} (y_{ij} - \mu)^2 = \sum_{i=1}^{N} \sum_{j=1}^{\bar{M}} (y_{ij} - \bar{y}_i)^2 + \bar{M} \sum_{i=1}^{N} (\bar{y}_i - \mu)^2$$

where $\bar{y}_i = \sum_{j=1}^{\bar{M}} y_{ij}/\bar{M}$. The first term on the right contains the within-primary-unit sum of squares and the second term the between-primary-unit sum of squares.

Write $\sigma_w^2 = \sum_{i=1}^{N} \sum_{j=1}^{\bar{M}} (y_{ij} - \bar{y}_i)^2 / [N(\bar{M}-1)]$ for the within-primary-unit variance and $\sigma_b^2 = \sum_{i=1}^{N} (\bar{y}_i - \mu)^2 / (N-1)$ for the variance between primary unit means. (Note that $\sigma_u^2 = \bar{M}^2 \sigma_b^2$ is the variance between primary unit totals.) An unbiased estimate of $\sigma_w^2$ using the random sample of clusters is $s_w^2 = \sum_{i=1}^{n} \sum_{j=1}^{\bar{M}} (y_{ij} - \bar{y}_i)^2 / [n(\bar{M}-1)]$, and an unbiased estimate of $\sigma_b^2$ is $s_b^2 = \sum_{i=1}^{n} (\bar{y}_i - \hat{\mu})^2 / (n-1)$, where $\hat{\mu} = \hat{\tau}_u/\bar{M}$. The sum-of-squares equality may be written

$$(N\bar{M} - 1)\sigma^2 = N(\bar{M} - 1)\sigma_w^2 + (N-1)\bar{M}\sigma_b^2$$

An unbiased estimate of $\sigma^2$ from the simple random cluster sample is

$$\hat{\sigma}^2 = \frac{N(\bar{M} - 1)s_w^2 + (N-1)\bar{M}s_b^2}{N\bar{M} - 1}$$

The estimated relative efficiency of cluster sampling (simple random sample of $n$ clusters) based on the data from the cluster sample is $\hat{\sigma}^2 / \bar{M} s_b^2 = \bar{M} \hat{\sigma}^2 / s_u^2$. Given a cluster sample with equal-sized clusters, one can thereby compare the efficiency of a variety of smaller units.

The variance of cluster or systematic sampling can alternatively be examined in terms of the correlation within primary units. The within-primary-unit correlation coefficient is defined as

$$\rho = \frac{\sum_{i=1}^{N} \sum_{j=1}^{\bar{M}} \sum_{j' \neq j} (y_{ij} - \mu)(y_{ij'} - \mu)}{(\bar{M} - 1)(N\bar{M} - 1)\sigma^2}$$

The sum of squares in the primary unit variance $\sigma_u^2$ may be written

$$\sum_{i=1}^{N} (y_i - \mu_1)^2 = \sum_{i=1}^{N} \left( \sum_{j=1}^{\bar{M}} y_{ij} - \bar{M}\mu \right)^2$$

$$= \sum_{i=1}^{N} \left[ \sum_{j=1}^{\bar{M}} (y_{ij} - \mu) \right]^2$$

$$= \sum_{i=1}^{N} \left[ \sum_{j=1}^{\bar{M}} \sum_{j'=1}^{\bar{M}} (y_{ij} - \mu)(y_{ij'} - \mu) \right]$$

$$= \sum_{i=1}^{N} \sum_{j=1}^{\bar{M}} (y_{ij} - \mu)^2 + \sum_{i=1}^{N} \sum_{j=1}^{\bar{M}} \sum_{j' \neq j} (y_{ij} - \mu)(y_{ij'} - \mu)$$

Substituting into Equation (2), the variance with cluster sampling may be written

$$\text{var}(\hat{\tau}) = \frac{N^2(N-n)}{nN} \left( \frac{M-1}{N-1} \right) \sigma^2 [1 + (\bar{M} - 1)\rho]$$

$$\approx \frac{N^2 \bar{M}(N-n)}{nN} \sigma^2 [1 + (\bar{M} - 1)\rho]$$

If $\rho$ is zero, the variance with cluster sampling will be approximately the same as the variance of a simple random sample of an equal number ($n\bar{M}$) of secondary units. If $\rho$ is greater than zero, the simple random sample will give lower variance. If $\rho$ is less than zero, the cluster sample gives lower variance.

With many natural populations, units near each other tend to be similar, so with compact clusters, $\rho$ is greater than zero. For such populations, the value of $\rho$, and hence the variance of $\hat{\tau}$, will tend to be larger with square clusters, in which the secondary units are close together, than with long, thin clusters, in which at least some of the secondary units are far apart. With systematic sampling, the secondary units of each primary unit are spaced relatively far apart, so that $\rho$ may well be less than zero. For this reason, systematic sampling is inherently efficient with many real populations.

The advantage of cluster sampling is that it is often less costly to sample a collection of units in a cluster than to sample an equal number of secondary units selected at random from the population. Considering the case with equal-sized clusters, let $\bar{M}_u$ be the number of secondary units in a primary unit of type $u$. Let $\sigma_u^2$ be the population variance for that type of unit. Let $c_u$ be the cost of measuring a randomly selected unit of that type, so that the cost of a sample of $n_u$ units is $c_u n_u$. Ignoring the finite population correction factor in Equation (2), the variance of an estimator $\hat{\tau}_u = N\bar{y}_u$ of the population total is approximately $\mathrm{var}(\hat{\tau}_u) \approx N_u^2 \sigma_u^2/n_u$. For a fixed cost $C$, $n_u = C/c_u$ and the variance is $\mathrm{var}(\hat{\tau}_u) \approx N_u^2 \sigma_u^2 c_u/C = c_u \sigma_u^2/(\bar{M}_u^2 C)$. The choice of primary unit giving the lowest variance is the one giving the smallest value of $c_u \sigma_u^2/\bar{M}_u^2$. For specified variance, the primary unit giving the lowest cost is again the one giving the smallest value of $c_u \sigma_u^2/\bar{M}_u^2$ (see Cochran 1977, p. 234).

In principle, the ideal size and shape of primary unit can be determined by a variance function and a cost function, each depending on the size and shape of primary unit. Such functions are not necessarily simple in real sampling situations. Examples of such functions are discussed in Cochran (1977), Hansen et al. (1953), Jessen (1978), and Kish (1965).

## EXERCISES

1. Assume that the following are data from cluster sampling with simple random sampling of clusters. There are 10 clusters (primary units) and a total of 100 secondary units in the population. For each of the $n = 3$ selected clusters, $y_i$ is the cluster total for the variable of interest and $M_i$ is cluster size: $y_1 = 4$, $M_1 = 5$; $y_2 = 12$, $M_2 = 20$; $y_3 = 7$, $M_3 = 10$.

   (a) Give an unbiased estimate of the population total.

   (b) Estimate the variance of that estimator.

2. Using the data of Exercise 1 and assuming simple random sampling, (a) give the ratio-to-size estimate of the population total and (b) estimate the variance of that estimator.

3. Using the data of Exercise 1, but assuming that the sample was obtained with selection probabilities proportional to cluster size (PPS), with replacement, (a) give an unbiased estimate of the population total and (b) estimate the variance of that estimator.

4. Use random sampling of clusters to estimate the mean or total of a population of your choice. In the process of carrying out the survey and making the estimate, think about or discuss with others the following:

(a) What practical problems arise in establishing a frame, such as a map or list of units, from which to select the sample?

(b) How is the sample selection actually carried out?

(c) What special problems arise in observing the units selected?

(d) Estimate the population mean or total.

(e) Estimate the variance of the estimator above.

(f) Give a 95% confidence interval for the population mean or total.

(g) How would you improve the survey procedure if you were to do it again?

5. To estimate the number of typographical errors in a 65-page manuscript, a systematic sample of pages is selected by first selecting a random number between 1 and 10 and including in the sample that numbered page and every tenth page thereafter. The random number selected was 6. The number of typographical errors on the sample pages were 1, 0, 2, 3, 0, and 1. Assume that no errors on sample pages were missed.

(a) Give an unbiased estimate, under the design used, of the total number of errors in the manuscript. What design was used?

(b) The person doing the survey estimated the total number of errors in the manuscript by $65(1 + 0 + 2 + 3 + 0 + 1)/6 = 75.83$. Which estimator was used? Is it unbiased with the design used?

(c) The variance of the estimator was estimated by $65(65 - 6)(1.37)/6$, where 1.37 is the sample variance of the six error counts. Is this unbiased for the actual variance of the estimator of the total number of errors? Discuss.

6. Use repeated systematic sampling to estimate the mean or total of a population of your choice. In the process of carrying out the survey and making the estimate, think about or discuss with others the following:

(a) What practical problems arise in establishing a frame, such as a map or list of units, from which to select the sample?

(b) How is the sample selection actually carried out?

(c) What special problems arise in observing the units selected?

(d) Estimate the population mean or total.

(e) Estimate the variance of the estimator above.

(f) Give a 95% confidence interval for the population mean or total.

(g) How would you improve the survey procedure if you were to do it again?

7. With a systematic sample having a single randomly selected starting point, the expansion estimator $\hat{\tau} = N\bar{y}$ is design unbiased for the population total $\tau$, but no design-unbiased estimator of $\text{var}(\hat{\tau})$ exists. Assume for simplicity that each possible sample (primary unit) has the same number of secondary units, that is, $M_i = m$ for all $i$. With the single starting point the sample has only $n = 1$ primary units, and $\bar{y}$, the sample mean of the primary unit totals, is the sum of the $m$

secondary units in the primary unit selected. $N$ is the number of possible starting points (the number of possible systematic samples), so the total number of secondary units in the population is $M = Nm$.

Now consider a model-based approach, in which the population $Y$-values are assumed to be independent, identically distributed random variables, each with mean $\beta$ and variance $\gamma$. Under the model, show that $\hat{\tau}$ is (model) unbiased for the population total, find the mean square prediction error $E(\hat{\tau} - \tau)^2$, and find an unbiased estimator of $E(\hat{\tau} - \tau)^2$. [*Hint*: The notation may be simplified to avoid double subscripts by relabeling the $M = Nm$ secondary units in the population from 1 to $M$, with $Y_j$ representing the $Y$-value of the $j$th secondary unit. The population total can be partitioned into sample and nonsample parts as $\tau = \sum_{j=1}^{M} Y_j = \sum_{j \in s} Y_j + \sum_{j \notin s} Y_j.$]

# CHAPTER 13

# Multistage Designs

If, after selecting a sample of primary units, a sample of secondary units is selected from each of the primary units selected, the design is referred to as *two-stage sampling*. If in turn a sample of tertiary units is selected from each selected secondary unit, the design is *three-stage sampling*. Higher-order *multistage designs* are also possible.

Multistage designs are used for a variety of practical reasons. To obtain a sample of fish caught in a commercial fishery, it may be necessary first to take a sample of boats and then a sample of fish from each selected boat. To obtain a sample of plants of some species, it may be convenient to take a sample of plots first and then a sample of plants from each selected plot.

In a single-stage design such as cluster sampling, the variability of the estimator occurs because different samples of primary units will give different values of the estimate. With two-stage designs, the estimator has variability even for a given selection of primary units, because different subsamples of secondary units will give rise to different values of the estimator.

Let $N$ denote the number of primary units in the population and $M_i$ the number of secondary units in the $i$th primary unit. Let $y_{ij}$ denote the value of the variable of interest for the $j$th secondary unit in the $i$th primary unit. The total of the $y$-values in the $i$th primary unit is $y_i = \sum_{j=1}^{M_i} y_{ij}$. The mean per secondary unit in the $i$th primary unit is $\mu_i = y_i/M_i$. The population total is $\tau = \sum_{i=1}^{N} \sum_{j=1}^{M_i} y_{ij}$. The population mean per primary unit is $\mu_1 = \tau/N$, while the population mean per secondary unit is $\mu = \tau/M$, where $M = \sum_{i=1}^{N} M_i$ is the total number of secondary units in the population.

Figure 13.1 shows a two-stage sample in which a simple random sample of $n = 10$ primary units was selected at the first stage and, at the second stage, a simple random sample of $m_i = 4$ of the $M_i = 8$ of the secondary units in each selected primary unit was selected. Figure 13.2 shows a two-stage sample with the same total number of secondary units, but with $n = 20$ primary units selected and $m_i = 2$ secondary units selected from each primary unit.

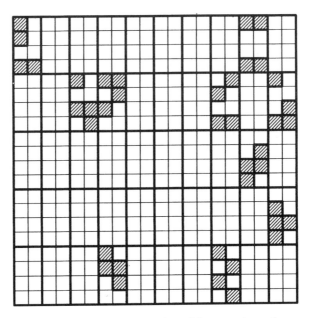

**Figure 13.1.** Two-stage sample of 10 primary units and four secondary units per primary unit.

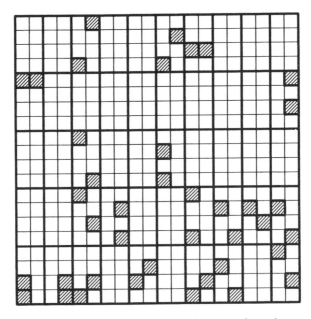

**Figure 13.2.** Two-stage sample of 20 primary units and two secondary units per primary unit.

## 13.1  SIMPLE RANDOM SAMPLING AT EACH STAGE

Consider first a two-stage design with simple random sampling at each stage. At the first stage, a simple random sample without replacement of $n$ primary units is selected. From the $i$th selected primary unit, a simple random sample without replacement of $m_i$ secondary units is selected, for $i = 1, \ldots, n$.

### Unbiased Estimator

Since simple random sampling is used at the second stage, an unbiased estimator of the total $y$-value for the $i$th primary unit in the sample is

$$\hat{y}_i = \frac{M_i}{m_i} \sum_{j=i}^{m_i} y_{ij} = M_i \bar{y}_i \tag{1}$$

where $\bar{y}_i = (1/m_i) \sum_{j=1}^{m_i} y_{ij} = \hat{y}_i / M_i$. Then, since simple random sampling is used at the first stage, an unbiased estimator of the population total is

$$\hat{\tau} = \frac{N}{n} \sum_{i=1}^{n} \hat{y}_i \tag{2}$$

The variance of $\hat{\tau}$ is

$$\mathrm{var}(\hat{\tau}) = N(N-n) \frac{\sigma_u^2}{n} + \frac{N}{n} \sum_{i=1}^{N} M_i(M_i - m_i) \frac{\sigma_i^2}{m_i} \tag{3}$$

where

$$\sigma_u^2 = \frac{1}{N-1} \sum_{i=1}^{N} (y_i - \mu_1)^2 \tag{4}$$

and for $i = 1, \ldots, N$,

$$\sigma_i^2 = \left( \frac{1}{M_i - 1} \right) \sum_{j=1}^{M_i} (y_{ij} - \mu_i)^2 \tag{5}$$

The first term of the variance expression above is the variance that would be obtained if every secondary unit in a selected primary unit were observed, while the second term contains the variance due to estimating the primary unit values $y_i$ from subsamples of secondary units. The quantity $\sigma_u^2$ in the first term is the population variance among primary unit totals, while $\sigma_i^2$ in the second term is the population variance within the $i$th primary unit.

An unbiased estimator of the variance of $\hat{\tau}$ is obtained by replacing the population variances with sample variances:

$$\widehat{\text{var}}(\hat{\tau}) = N(N-n)\frac{s_u^2}{n} + \frac{N}{n}\sum_{i=1}^{n} M_i(M_i - m_i)\frac{s_i^2}{m_i} \tag{6}$$

where

$$s_u^2 = \frac{1}{n-1}\sum_{i=1}^{n}(\hat{y}_i - \hat{\mu}_1)^2 \tag{7}$$

and for $i = 1, \ldots, n$,

$$s_i^2 = \left(\frac{1}{m_i - 1}\right)\sum_{j=1}^{m_i}(y_{ij} - \bar{y}_i)^2 \tag{8}$$

and $\hat{\mu}_1 = (1/n)\sum_{i=1}^{n}\hat{y}_i$.

To estimate population means, $\hat{\mu}_1 = \hat{\tau}/N$ is an unbiased estimator of the population mean per primary unit $\mu_1$, for which the variance expressions above would be divided by $N^2$, and $\hat{\mu} = \hat{\tau}/M$ is unbiased for the mean per secondary unit $\mu$, with the variance expressions divided by $M^2$.

***Example 1.*** A simple random sample of $n = 3$ primary units is selected from a population of $N = 100$ primary units. From each of the primary units selected, a random sample of $m_i = 2$ secondary units is selected. The sizes of the three primary units selected are $M_1 = 24$, $M_2 = 20$, and $M_3 = 15$ secondary units. The y-values for the first primary unit in the sample are 8 and 12. For the second primary unit in the sample, the y-values are 0 and 0. For the third primary unit, the y-values are 1 and 3.

The estimates for the total of each sample primary unit [from Equation (1)] are

$$\hat{y}_1 = \frac{24}{2}(8 + 12) = 240$$

$$\hat{y}_2 = \frac{20}{2}(0 + 0) = 0$$

$$\hat{y}_3 = \frac{15}{2}(1 + 3) = 30$$

The estimate of the population total [from Equation (2)] is

$$\hat{\tau} = \frac{100}{3}(240 + 0 + 30) = 100(90) = 9000$$

The estimate of the mean per primary unit is $\hat{\mu}_1 = 90$.

The sample variance between primary unit totals [from Equation (7)] is

$$s_u^2 = \frac{1}{3-1}\left[(240-90)^2 + (0-90)^2 + (30-90)^2\right] = 17{,}100$$

The sample means within primary units are $\bar{y}_1 = (8+12)/2 = 10$, $\bar{y}_2 = 0$, and $\bar{y}_3 = 2$. The sample variances within primary units are

$$s_1^2 = \frac{1}{2-1}\left[(8-10)^2 + (12-10)^2\right] = 8$$

and similarly, $s_2^2 = 0$ and $s_3^2 = 2$.

The estimated variance of $\hat{\tau}$ [from Equation (6)] is

$$\widehat{\mathrm{var}}(\hat{\tau}) = 100(100-3)\frac{17{,}100}{3}$$
$$+ \frac{100}{3}\left[24(24-2)\frac{8}{2} + 20(20-2)\frac{0}{2} + 15(15-2)\frac{2}{2}\right]$$
$$= 55{,}290{,}000 + 76{,}900 = 55{,}366{,}900$$

giving an estimated standard error of 7441. □

**Ratio Estimator**

A ratio estimator of the population total based on the sizes of the primary units is

$$\hat{\tau}_r = \hat{r}M$$

where

$$\hat{r} = \frac{\sum_{i=1}^n \hat{y}_i}{\sum_{i=1}^n M_i}$$

An approximate mean square error or variance formula for this estimator is

$$\mathrm{var}(\hat{\tau}_r) \approx \frac{N(N-n)}{n(N-1)}\sum_{i=1}^N (y_i - M_i\mu)^2 + \frac{N}{n}\sum_{i=1}^n M_i(M_i - m_i)\frac{\sigma_i^2}{m_i}$$

An estimator of the variance of $\hat{\tau}$ is

$$\widehat{\mathrm{var}}(\hat{\tau}_r) = \frac{N(N-n)}{n(n-1)}\sum_{i=1}^n (\hat{y}_i - M_i\hat{r})^2 + \frac{N}{n}\sum_{i=1}^n M_i(M_i - m_i)\frac{s_i^2}{m_i}$$

Estimators for the population means $\mu_1$ and $\mu$ are $\hat{\mu}_{1r} = \hat{\tau}_r/N$ and $\hat{\mu}_r = \hat{\tau}_r/M = \hat{r}$, for which the variance expressions are divided by $N^2$ and $M^2$, respectively.

***Example 2.*** The ratio estimate of the population total for Example 1, with $M = 2500$ secondary units in the population, is

$$\hat{\tau}_r = \frac{240 + 0 + 30}{24 + 20 + 15}(2500) = 11{,}441$$

For the mean square error estimate the sum of the squared residuals is

$$\sum_{i=1}^{n}(\hat{y}_i - M_i\hat{r})^2 = [240 - 4.58(24)]^2 + [0 - 4.58(20)]^2 + [30 - 4.58(15)]^2 = 26{,}814$$

The estimated mean square error of $\hat{\tau}_r$ is

$$\widehat{\mathrm{var}}(\hat{\tau}_r) = 100(100 - 3)\frac{26{,}814}{3(3 - 1)} + 76{,}900 = 43{,}426{,}200$$

the square root of which is 6590. $\qquad\qquad\qquad\qquad\qquad\qquad\qquad\qquad$ □

## 13.2  PRIMARY UNITS SELECTED WITH PROBABILITY PROPORTIONAL TO SIZE

Next consider a two-stage design in which primary units are selected with replacement, with probabilities proportional to size, and a sample of secondary units is selected independently using simple random sampling without replacement each time a primary unit is selected. Since the second-stage samples are selected independently, any secondary unit could appear in the sample more than once because of the with-replacement sampling at the first stage.

With this design, an unbiased estimator of the population total is

$$\hat{\tau}_p = \frac{M}{n}\sum_{i=1}^{n}\frac{\hat{y}_i}{M_i} = \frac{M}{n}\sum_{i=1}^{n} = \bar{y}_i$$

where $\bar{y}_i = (1/m_i)\sum_{j=1}^{m_i} y_{ij}$, the sample mean within the $i$th primary unit of the sample, and $\hat{y}_i = M_i\bar{y}_i$.

The variance is

$$\mathrm{var}(\hat{\tau}_p) = \frac{M}{n}\sum_{i=1}^{N}M_i(\mu_i - \mu)^2 + \frac{M}{n}\sum_{i=1}^{N}\left[\frac{M_i - m_i}{m_i(M_i - 1)}\sum_{j=1}^{M_i}(y_{ij} - \mu_i)^2\right]$$

An unbiased estimate of this variance is

$$\widehat{\text{var}}(\hat{\tau}_p) = \frac{M^2}{n(n-1)} \sum_{i=1}^{n} (\bar{y}_i - \hat{\mu}_p)^2$$

where $\hat{\mu}_p = \hat{\tau}_p / M$.

**Example 3.**  The unbiased estimate of the population total with the probability-proportional-to-size design with the data from Example 1, with $M = 2500$ secondary units in the population, is

$$\hat{\tau}_p = \frac{2500}{3} \left( \frac{240}{24} + \frac{0}{20} + \frac{30}{15} \right) = \frac{2500}{3}(10 + 0 + 2) = 10,000$$

The sample variance of the three numbers 10, 0, and 2 is 28. An unbiased estimate of the variance of $\hat{\tau}_p$ is

$$\widehat{\text{var}}(\hat{\tau}_p) \frac{2500^2}{3}(28) = 58,333,333$$

giving a standard error of 7638.                                              □

## 13.3   ANY MULTISTAGE DESIGN WITH REPLACEMENT

Notice that the estimate of variance is particularly simple for the design with probability-proportional-to-size sampling with replacement. In fact, variance estimation is equally simple for *any* multistage design in which primary units are selected with replacement with known draw-by-draw selection probabilities $P_i$, subsampling is done independently between different primary units, and an unbiased estimator $\hat{y}_i$ is available for the total in any selected primary unit $i$. Then an unbiased estimator of the population total $\tau$ is

$$\hat{\tau}_p = \frac{1}{n} \sum_{i=1}^{n} \frac{\hat{y}_i}{p_i}$$

In the probability-proportional-to-size design of the preceding section, $\hat{y}_i = M_i \bar{y}_i$ and $p_i = M_i / M$.

An unbiased estimator of the variance of this estimator is

$$\widehat{\text{var}}(\hat{\tau}_p) = \frac{1}{n(n-1)} \sum_{i=1}^{n} \left( \frac{\hat{y}_i}{p_i} - \hat{\tau} \right)^2$$

This result follows from the independence of the $n$ selections, due to the with-replacement primary selection and the independent selections in different primary units. Thus, the $\hat{y}_i/p_i$ are a random sample from some distribution, and hence their sample variance estimates the variance of that distribution. The simple result holds no matter how many stages in the design (see J. N. K. Rao 1975, 1988, pp. 431–432).

Even though the *estimator* of variance is simple with such a design, the actual variance depends on the specific designs used at each stage and has terms for each stage of sampling, as exemplified by the design in Section 13.2.

## 13.4 COST AND SAMPLE SIZES

The practical advantage of two-stage sampling, relative to a simple random sample of the same number of secondary units, is that it is often easier or less expensive to observe many secondary units in a cluster than to observe the same number of secondary units randomly spread over the population. Consider the unbiased estimator $\hat{\tau}$ with simple random sampling of $n$ primary units and simple random sampling of $m_i$ secondary units from the $i$th selected primary unit. For simplicity, consider the case in which the primary units are all the same size, with $M_i = \bar{M}$ for all $i$ and in which the subsample size in each selected primary unit is $m$ secondary units.

Suppose that the average cost of sampling is described by the cost function

$$C = c_0 + c_1 n + c_2 nm$$

where $C$ is the total cost of the survey, $c_0$ a fixed overhead cost, $c_1$ the cost per primary unit selected, $c_2$ the cost per secondary unit. The number of secondary units in the sample is $nm$. For a fixed cost $C$, the minimum value of $\text{var}(\hat{\tau})$ is obtained with subsample sizes

$$m_{\text{opt}} = \sqrt{\frac{c_1 \sigma_w^2}{c_2 \left(\sigma_b^2 - \sigma_w^2/\bar{M}\right)}}$$

where $\sigma_b^2$ is the variance between primary unit means,

$$\sigma_b^2 = \frac{\sum_{i=1}^{N}(\mu_i - \mu)^2}{N - 1}$$

and $\sigma_w^2$ is the average within-primary-unit variance,

$$\sigma_w^2 = \frac{1}{N}\sum_{i=1}^{N}\sigma_i^2$$

If $\sigma_b^2$ is not greater than $\sigma_w^2/\bar{M}$, however, use $m_{\mathrm{opt}} = \bar{M}$. Using the optimal choice of $m$, one can then solve the cost equation for $n$, giving $n = (C - c_0)/(c_1 + c_2 m_{\mathrm{opt}})$.

The derivation of $m_{\mathrm{opt}}$ can be carried out using Lagrange's method. With equal primary unit and subsample sizes, and using the identity $\sigma_b^2 = \sigma_u^2/\bar{M}^2$, the variance of $\hat{\tau}$ under simple random sampling can be written

$$\mathrm{var}(\hat{\tau}) = \frac{N(N-n)\bar{M}^2\sigma_b^2}{n} + \frac{N^2\bar{M}(\bar{M}-m)\sigma_w^2}{nm}$$

Write $V(n,m) = \mathrm{var}(\hat{\tau})$ and define the function $F = V - \lambda(c_0 + c_1 n + c_2 nm)$, in which $\lambda$ is the Lagrange multiplier. The function $F$ is differentiated with respect to $n$, $m$, and $\lambda$, the derivatives set to zero, and the equations solved for $m$ and $n$.

To achieve a specified value of $\mathrm{var}(\hat{\tau})$ at minimum cost, the optimal choice of subsample size in each selected primary unit is again $m_{\mathrm{opt}}$. The variance equation with $m = m_{\mathrm{opt}}$ is then solved for $n$.

More complicated formulas for optimal subsample sizes are available for random sampling of unequal-sized primary units and for probability-proportional-to-size sampling of primary units (see Singh and Chaudhary 1986). A more exact solution to the optimization problem, taking account of the discrete nature of the sample sizes $n$ and $m$, is described in Hedayat and Sinha (1991).

## 13.5   DERIVATIONS FOR MULTISTAGE DESIGNS

Whatever the sampling design used at each stage, two basic results about conditional expectations and variances underlie the properties of estimators from two-stage designs. Let $T$ represent some estimator, such as an estimator of the population total, and let $s_1$ be the set of primary units in the sample. Both $T$ and $s_1$ are random, depending on the sample selected. The first result, regarding the expected value of the estimator, is

$$\mathrm{E}\left[\mathrm{E}(T \mid s_1)\right] = \mathrm{E}(T) \tag{9}$$

The conditional expectation of $T$ given the selection of primary units is taken over all possible subselections of secondary units from those primary units. The unconditional expected value is then obtained by considering all possible samples of primary units. The second result, regarding the variance of $T$, is

$$\mathrm{var}(T) = \mathrm{var}[\mathrm{E}(T \mid s_1)] + \mathrm{E}[\mathrm{var}(T \mid s_1)] \tag{10}$$

The first term on the right-hand side contains the between-primary-unit variance, while the second term contains the within-primary-unit variance. The conditional variance, like the conditional expectation, is obtained by considering all possible subsamples of secondary units given the primary units selected. The unconditional variance and expectation are taken over all possible samples of primary units.

## Unbiased Estimator

Consider the two-stage design, with simple random sampling at each stage and the unbiased estimator $\hat{\tau}$. Given the sample $s_1$ of primary units, the conditional expectation of the estimator $\hat{y}_i$ for the $i$th primary unit in the sample is the actual total $y_i$, because of the simple random sampling within primary unit $i$ at the second stage; that is,

$$E(\hat{y}_i \mid s_1) = y_i$$

The conditional expectation of $\hat{\tau}$ is

$$E(\hat{\tau} \mid s_1) = E\left(\frac{N}{n}\sum_{i=1}^{n}\hat{y}_i\right) = \frac{N}{n}\sum_{i=1}^{n}y_i$$

The unconditional expectation, using Equation (9), is

$$E(\hat{\tau}) = E[E(\hat{\tau} \mid s_1)] = E\left(\frac{N}{n}\sum_{i=1}^{n}y_i\right) = \tau$$

because of the simple random sampling at the first stage.
    The variance of the conditional expectation of $\hat{\tau}$ is

$$\text{var}[E(\hat{\tau} \mid s_1)] = \text{var}\left(\frac{N}{n}\sum_{i=1}^{n}y_i\right) = N(N-n)\frac{\sigma_u^2}{n}$$

because of the simple random sampling of primary units at the first stage.
    The conditional variance of $\hat{y}_i$, given the first stage selection $s_1$, is

$$\text{var}(\hat{y}_i \mid s_1) = M_i(M_i - m_i)\frac{\sigma_i^2}{m_i}$$

since $\hat{y}_i$ is the estimate of the total $y_i$ under simple random sampling within primary unit $i$. Thus the conditional variance of $\hat{\tau}$, since the second-stage selections for different primary units are made independently, is

$$
\begin{aligned}
\text{var}(\hat{\tau} \mid s_1) &= \text{var}\left(\frac{N}{n}\sum_{i=1}^{n}\hat{y}_i\right) \\
&= \left(\frac{N}{n}\right)^2\sum_{i=1}^{n}\text{var}(\hat{y}_i \mid s_1) \\
&= \frac{N^2}{n^2}\sum_{i=1}^{n}M_i(M_i - m_i)\frac{\sigma_i^2}{m_i}
\end{aligned}
$$

To obtain the expected value of this variance, over all possible samples of primary units, let the indicator variable $z_i$ equal 1 if the $i$th primary unit is in the sample and zero otherwise, and write the conditional variance as

$$\text{var}(\hat{\tau} \mid s_1) = \frac{N^2}{n^2} \sum_{i=1}^{N} M_i(M_i - m_i) z_i \frac{\sigma_i^2}{m_i}$$

Since $E(z_i) = n/N$, the inclusion probability for a primary unit under simple random sampling, the expected conditional variance is

$$E[\text{var}(\hat{\tau} \mid s_1)] = \frac{N}{n} \sum_{i=1}^{N} M_i(M_i - m_i) \frac{\sigma_i^2}{m_i}$$

Combining the two terms according to Equation (10), yields

$$\text{var}(\hat{\tau}) = N(N - n) \frac{\sigma_u^2}{n} + \frac{N}{n} \sum_{i=1}^{N} M_i(M_i - m_i) \frac{\sigma_i^2}{m_i}$$

## Ratio Estimator

For the ratio estimator with simple random sampling at each of the two stages, the conditional expectation of $\hat{y}_i$ is $y_i$, because of the simple random sampling at the second stage. Hence, the conditional expectation of $\hat{\tau}_r$ given $s_1$ is the ordinary ratio estimator. Thus, the approximate variance of the ratio estimator gives the first component of variance, while the second component, based on the simple random sampling at the second stage, remains unchanged.

## Probability-Proportional-to-Size Sampling

When the primary units are selected with probabilities proportional to size, with replacement, the conditional expectation of $\hat{\tau}_p$ is

$$E(\hat{\tau}_p \mid s_1) = \frac{M}{m} \sum_{i=1}^{n} \frac{1}{M_i} E(y_i \mid S_1) = \frac{M}{n} \sum_{i=1}^{n} \frac{Y_i}{M_i} = \frac{1}{n} \sum_{i=1}^{n} \frac{y_i}{p_i}$$

where $p_i = M_i/M$ is the selection probability for primary unit $i$. Thus, the conditional expectation has the form of a Hansen–Hurwitz estimator, from which follow unbiasedness for $\tau$ and the first term of the variance. For the second component of the variance, the conditional variance is

$$\text{var}(\hat{\tau}_p) = \text{var}\left( m_n \sum_{i=1}^{n} \bar{y}_i \mid s_1 \right) = \frac{M^2}{n^2} \sum_{i=1}^{n} \left( \frac{M_i - m_i}{M_i} \right) \frac{\sigma_i^2}{m_i}$$

since, given primary unit $i$ in the sample, $\bar{y}_i$ is the sample mean of a simple random sample of size $m_i$. To obtain the expected value of this conditional variance, over the probability-proportional-to size selection of primary units, let $v_i$ be the number of times primary unit $i$ is selected. The random variable $v_i$ is a binomial random variable with expected value $E(v_i) = np_i = mN_i/M$. Writing

$$\text{var}(\hat{\tau}_p) = \frac{M^2}{n^2} \sum_{i=1}^{N} \left( \frac{M_i - m_i}{M_i} \right) \frac{v_i \sigma_i^2}{m_i}$$

the expectation is

$$E[\text{var}(\hat{\tau}_p)] = \frac{M}{n} \sum_{i=1}^{N} \left[ \frac{M_i - m_i}{m_i(M_i - 1)} \sum_{j=1}^{M_i} (y_{ij} - \mu_i)^2 \right]$$

the second-stage component of variance.

## More Than Two Stages

For designs with more than two stages, the expected values and variances of estimators can be broken down further. For the third stage of sampling, the conditional variance in Equation (10) may be decomposed into

$$\text{var}(\hat{\tau} \mid s_2) = \text{var}[E(\hat{\tau} \mid s_1, s_2) \mid s_1] + E[\text{var}(\hat{\tau} \mid s_1, s_2) \mid s_1]$$

where $s_2$ is the given sample of secondary units. Letting $E_1$ denote expectation conditional on $s_1$ and $E_{12}$ denote expectation conditional on $s_1$ and $s_2$, with similar subscripting denoting the conditional variances, the decomposition may be written

$$\text{var}(\hat{\tau}) = \text{var}[E_1(\hat{\tau})] + E\{\text{var}_1[E_{12}(t)]\} + E\{E_1[\text{var}_{12}(\hat{\tau})]\}$$

For each additional stage of sampling, the last term is further decomposed, adding one more component to the variance.

## EXERCISES

1. A population consists of $N = 10$ primary units, each of which consists of $M_i = 6$ secondary units. A two-stage sampling design selects 2 primary units by simple random sampling (without replacement) and 3 secondary units from each selected primary unit, also by simple random sampling. The observed values of the variable of interest are 7, 5, 3 from the first primary unit selected and 4, 2, 3 from the second primary unit selected.

(a) Estimate the population mean $\mu$ per secondary unit.

(b) Estimate the variance of the estimator above.

2. Use two-stage sampling, with simple random sampling at each stage, to estimate the mean or total of a population of your choice. In the process of carrying out the survey and making the estimate, think about or discuss with others the following:

(a) What practical problems arise in establishing a frame, such as a map or list of units, from which to select the sample?

(b) How is the sample selection actually carried out?

(c) What special problems arise in observing the units selected?

(d) Estimate the population mean or total.

(e) Estimate the variance of the estimator above.

(f) Give a 95% confidence interval for the population mean or total.

(g) How would you improve the survey procedure if you were to do it again?

CHAPTER 14

# Double or Two-Phase Sampling

In the earlier discussions of ratio and regression estimation, the values of the auxiliary variables were assumed known for the entire population. In conventional stratified sampling, the auxiliary information necessary for classifying units into strata is assumed known for every unit in the population. In some situations, however, the values of the auxiliary variables, like those of the variable of interest, can be known only through sampling. If the auxiliary values are easier or cheaper to obtain than the values of the variable of interest, it may be convenient to observe the auxiliary variables on a larger sample than that on which the variable of interest is recorded.

*Double sampling* refers to designs in which initially a sample of units is selected for obtaining auxiliary information only, and then a second sample is selected in which the variable of interest is observed in addition to the auxiliary information. The second sample is often selected as a subsample of the first. The purpose of double sampling is to obtain better estimators by using the relationship between auxiliary variables and the variable of interest. Double sampling is also known as *two-phase sampling*.

In surveys to estimate the volume of trees in a stand, "eyeball" estimates of volume by trained observers may conveniently be obtained for a large sample of standing trees, while accurate volume measurements requiring felling may be limited to a small subsample of these trees. In surveys of abundance of some animal or plant species, accurate counts may be made by ground crews, while less accurate counts covering a much larger area may be made from the air. In such cases, the variable of interest is the accurate measurement, while the less accurate but easier to obtain measurements serve as auxiliary data.

In double sampling for stratification, the units in the initial sample are classified into strata. The second sample is then selected by stratified sampling from the initial sample. An important special use of this method is adjustment for nonresponse in surveys. Figure 14.1 depicts a double sample, with a simple random sample of $n' = 60$ units on which $x_i$ is observed and a subsample (doubly shaded) of $n = 20$ units on which $y_i$ is observed in addition.

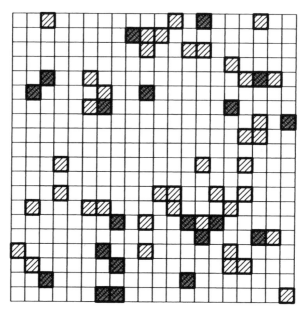

**Figure 14.1.** Double sample. The variable of interest and an auxiliary variable are both recorded on the double-shaded (dark) units. On single-shaded units, only the auxiliary variable is observed.

## 14.1   RATIO ESTIMATION WITH DOUBLE SAMPLING

The variable of interest for the $i$th unit will be denoted $y_i$ as usual, while the auxiliary variable will be denoted $x_i$. Let $n'$ be the number of units in the first sample and $n$ the number of units in the second sample. For each unit in the second sample, both $y_i$ and $x_i$ are observed, while for the rest of the units in the first sample, only $x_i$ is observed.

If the $y_i$ and $x_i$ are highly correlated, with a linear relationship in which $y_i = 0$ when $x_i = 0$, ratio estimation with double sampling may lead to improved estimates of the population mean or total.

Suppose that from a population of $N$ units, a sample of $n'$ units on which the $x$-values are observed is selected by random sampling without replacement. From the $n'$ units selected, a subsample of $n$ units, on which the $y$-values are observed in addition to the $x$-values, is selected, also by random sampling without replacement.

From the small sample containing both $x$ and $y$ values, one obtains the sample ratio

$$r = \frac{\sum_{i=1}^{n} y_i}{\sum_{i=1}^{n} x_i}$$

where the summations are over the units in the second sample.

From the full sample for which $x$-values were obtained, one can estimate the population total $\tau_x$ of the $x$-values:

$$\hat{\tau}_x = \frac{N}{n'}\sum_{i=1}^{n'} x_i$$

in which the summation is over the entire initial sample.

The ratio estimator of the population total $\tau$ of the $y$-values is

$$\hat{\tau}_r = r\hat{\tau}_x$$

An approximate formula for the mean square error or variance of this estimator is

$$\text{var}(\hat{\tau}_r) \approx N(N - n')\frac{\sigma^2}{n'} + N^2\left(\frac{n' - n}{n'}\right)\frac{\sigma_r^2}{n}$$

where $\sigma^2$ is the usual finite population variance of the $y$-values and $\sigma_r^2$ is the population variance about the ratio line given by

$$\sigma_r^2 = \frac{1}{N - 1}\sum_{i=1}^{N}(y_i - Rx_i)^2$$

in which the population ratio $R$ is $R = \tau/\tau_x$.

An estimate of this variance is provided by

$$\widehat{\text{var}}(\hat{\tau}_r) = N(N - n')\frac{s^2}{n'} + N^2\left[\frac{n' - n}{n'n(n - 1)}\right]\sum_{i=1}^{n}(y_i - rx_i)^2$$

where $s^2$ is the sample variance of the $y$-values of the subsample.

**Example 1.**   In a survey of moose abundance, a total of 240 moose were seen in 20 plots in an aerial survey. When ground crews were sent in to thoroughly examine five of these plots—in which 56 moose had been seen from the air—a total of 70 moose were discovered. The study area consists of 100 plots of equal size. Estimate the number of moose in the study area.

The variable of interest $y$ is the actual number of moose in a plot, presumably determined by the ground crews. The number observed from the air—less accurate but enabling a larger number of plots to be covered—is an auxiliary variable $x$. From the subsample of $n = 5$ units observed from both air and ground, the sample ratio is

$$r = \frac{70}{56} = 1.25$$

From the full sample of $n' = 20$ units observed from the air, an estimate of the total in the population is

$$\hat{\tau}_x = \frac{100}{20}(240) = 1200$$

The ratio estimate of the number of moose in the study area, based on the double-sampling design, is

$$\hat{\tau}_r = r\hat{\tau}_x = 1.25(1200) = 1500$$

Thus, the total number of moose in the study area is estimated to be 1500.

In sampling problems of this type, the reciprocal of the sample ratio is an estimate of *detectability*. Thus, $56/70 = 1/1.25 = 0.8$ estimates the probability that any given moose in a selected plot is detected from the air.    □

## 14.2 ALLOCATION IN DOUBLE SAMPLING FOR RATIO ESTIMATION

Double sampling for ratio estimation is effective in situations in which the variable of interest $y$ tends to be linearly related to an auxiliary variable $x$, with $y$ tending to be zero when $x$ is zero, and it is easier or cheaper to measure $x$ than $y$. The ideal proportion of the sample to subsample in double sampling depends on the relative costs of observing the two variables and on the strength of the ratio relationship between them.

Suppose that the cost of observing the $x$-variable on one unit is $c'$ and the cost of observing the $y$-variable is $c$, so that the total cost $C$ is given by

$$C = c'n' + cn$$

Then for a fixed cost $C$ the lowest variance of the estimator $\hat{\tau}_r$ is obtained by using the following subsampling fraction:

$$\frac{n}{n'} = \sqrt{\frac{c'}{c}\left(\frac{\sigma_r^2}{\sigma^2 - \sigma_r^2}\right)}$$

## 14.3 DOUBLE SAMPLING FOR STRATIFICATION

In some sampling situations, the units can be assigned to strata only after the sample is selected. For example, a sample of people selected by random dialing can be stratified by sex, age, or occupation only after the phone calls are made. A random sample of fish in a survey to estimate mean age may be stratified into size classes after selection. The methods of poststratification, described in Chapter 11, were useful in such situations only if the relative proportion $W_h = N_h/N$ of population units in stratum $h$ is known for each stratum. If the population proportions are not known, double sampling may be used, with an initial (large) sample used to classify the units into strata and then a stratified sample selected from the initial sample. Double sampling can be advantageous if the units tend to be similar within strata and if the auxiliary characteristic on which classification into strata is based is easier or more inexpensive to measure than the variable of interest.

An initial simple random sample of size $n'$ units is selected from the population of $N$ units. These units are classified into strata, with $n'_h$ observed to be in stratum $h$, for $h = 1, \ldots, L$. The population proportion $W_h$ of units in stratum $h$ is estimated with the sample proportion $w_h = n'_h/n'$. A second sample is then selected by stratified random sampling from the first sample, with $n_h$ units selected from the $n'_h$ sample units in stratum $h$, and the variable of interest $y_{hi}$ recorded for each unit in this second sample. The sample mean in stratum $h$ in the second sample is $\bar{y}_h = \sum_{i=1}^{n_h} y_{hi}/n_h$. The population mean is estimated using

$$\bar{y}_d = \sum_{h=1}^{L} w_h \bar{y}_h$$

The estimator $\bar{y}_d$ is unbiased for the population mean $\mu$ and has variance

$$\mathrm{var}(\bar{y}_d) = \frac{N - n'}{N} \frac{\sigma^2}{n'} + \mathrm{E} \sum_{h=1}^{L} \left[ \left(\frac{n'_h}{n'}\right)^2 \left(\frac{n'_h - n_h}{n'_h}\right) \frac{\sigma^2_{h(s_1)}}{n_h} \right] \tag{1}$$

where $\sigma^2$ is the overall population variance and $\sigma^2_{h(s_1)}$ is the population variance within stratum $h$ for the particular first-phase sample $s_1$. The first term in Equation (1) is the variance of the expected value of $\bar{y}_d$ given the initial sample. It is the variance that would be obtained if the $y$-values were recorded for the entire initial sample and the sample mean used as an estimator. The second term is the expected value, over all possible initial samples, of the conditional variance of the stratified sample mean $\bar{y}_d$ given the initial sample. The expectation is left in the second term because $n'_h$, $n_h$, and $\sigma^2_{h(s_1)}$ are random variables depending on the first-phase sample.

An unbiased estimate of the variance of $\bar{y}_d$ with double sampling, using the stratum sample variances $s_h^2$ from the second sample, is

$$\widehat{\mathrm{var}}(\bar{y}_d) = \frac{N - 1}{N} \sum_{h=1}^{L} \left(\frac{n'_h - 1}{n' - 1} - \frac{n_h - 1}{N - 1}\right) \frac{w_h s_h^2}{n_h} + \frac{N - n'}{N(n' - 1)} \sum_{h=1}^{L} w_h (\bar{y}_h - \bar{y}_d)^2 \tag{2}$$

If the second-stage sampling fractions $n_h/n_h'$ are fixed in advance, and samples for which any $n_h' = 0$ are excluded, the variance is approximately

$$\mathrm{var}(\bar{y}_d) = \frac{N - n'}{Nn'}\sigma^2 + \sum_{h=1}^{L}\frac{W_h\sigma_h^2}{n'}\left(\frac{n_h'}{n_h} - 1\right)$$

The estimator of variance remains the same for this case.

## 14.4   DERIVATIONS FOR DOUBLE SAMPLING

### Approximate Mean and Variance: Ratio Estimation

As in two-stage sampling, the estimators of double sampling are evaluated by conditioning on the first sample. Let $s_1$ denote the initial selection of $n'$ units. Given $s_1$, the estimator $\hat{\tau}_r$ is a ratio estimator based on a simple random sample of $n$ of the $n'$ units in $s_1$, so that its conditional expectation is approximately

$$\mathrm{E}(\hat{\tau}_r \mid s_1) \approx N\bar{y}_{s1}$$

where $\bar{y}_{s1}$ is the mean of the $y$-values of the $n'$ units in $s_1$. Since $\bar{y}_{s1}$ is the sample mean of a simple random sample, the unconditional expectation is thus approximately

$$\mathrm{E}(\hat{\tau}_r) \approx \mathrm{E}(N\bar{y}_{s1}) = \tau$$

and

$$\mathrm{var}[\mathrm{E}(\hat{\tau}_r \mid s_1)] \approx \mathrm{var}(N\bar{y}_{s1}) = N^2\left(\frac{N - n'}{Nn'}\right)\sigma^2$$

The conditional variance given $s_1$ of the ratio estimator $\hat{\tau}_r$ is approximately

$$\mathrm{var}(\hat{\tau}_r \mid s_1) \approx \frac{N^2(n' - n)}{n'n(n' - 1)}\sum_{i=1}^{n'}(y_i - r_1 x_i)^2$$

where $r_1$ is the total of the $y$-values divided by the total of the $x$-values in the sample $s_1$ of $n'$ units. Its expected value over all possible selections of the $n'$ units is, as a variance estimator in ratio estimation under simple random sampling, approximately

$$\mathrm{E}[\mathrm{var}(\hat{\tau}_r \mid s_1)] \approx \frac{N^2(n' - n)}{n'n}\sigma_r^2$$

The variance of the estimator $\hat{\tau}_r$ is then obtained from the decomposition

$$\text{var}(\hat{\tau}_r) = \text{var}[E(\hat{\tau}_r \mid s_1)] + E[\text{var}(\hat{\tau}_r \mid s_1)]$$

giving

$$\text{var}(\hat{\tau}_r) \approx N(N - n')\frac{\sigma^2}{n'} + N^2\left(\frac{n' - n}{n'}\right)\frac{\sigma_r^2}{n}$$

The estimator of this variance uses the unbiased estimator $s^2$ for $\sigma^2$ and a ratio variance estimator for $\sigma_r^2$.

**Optimum Allocation for Ratio Estimation**

The problem is to find $n'$ and $n$ which minimize $\text{var}(\hat{\tau}_r)$ subject to $c'n' + cn = C$. The problem can be solved using Lagrange's method. Let $f(n', n) = \text{var}(\hat{\tau}_r)$, which can be written

$$f(n', n) = N^2[n'^{-1}(\sigma^2 - \sigma_r^2) + n^{-1}\sigma_r^2 - N^{-1}\sigma^2]$$

Let $g(n', n) = c'n' + cn - C$. Define $H(n', n, \lambda) = f - \lambda g$. The solution is obtained by setting the partial derivatives of $H$ equal to zero:

$$\frac{\partial H}{\partial n'} = \frac{-N^2(\sigma^2 - \sigma_r^2)}{n'^2} - \lambda c' = 0$$

$$\frac{\partial H}{\partial n} = \frac{-N^2(\sigma_r^2)}{n^2} - \lambda c = 0$$

Eliminating the Lagrange multiplier $\lambda$ by dividing the first equation by $N^2 c'$ and the second by $-N^2 c$ and adding the two equations together gives

$$\frac{\sigma^2 - \sigma_r^2}{c'n'^2} = \frac{\sigma_r^2}{cn^2}$$

or

$$\frac{n}{n'} = \sqrt{\frac{c'}{c}\left(\frac{\sigma_r^2}{\sigma^2 - \sigma_r^2}\right)}$$

### Expected Value and Variance: Stratification

The conditional expected value of $\bar{y}_d$ given the first-phase sample $s_1$ is

$$E(\bar{y}_d \mid s_1) = E\left(\sum_{h=1}^{L} \frac{n'_h}{n'} \bar{y}_h \mid s_1\right) = \sum_{h=1}^{L} \frac{n'_h}{n'} E(\bar{y}_h \mid s_1)$$

since given $s_1$, the stratum sizes $n'_h$ are fixed. Because of the simple random sampling within strata at the second phase, the expected value of $\bar{y}_h$ given $s_1$ is the actual mean $\bar{y}'_h$ of the $y$-values for the $n'_h$ units in that stratum. Thus the conditional expectation of the estimator is

$$E(\bar{y}_d \mid s_1) = \left(\sum_{h=1}^{L} \frac{n'_h}{n'} \bar{y}'_h \mid s_1\right) = \frac{1}{n'} \sum_{i \in s_1} y_i$$

The last term on the right is the mean of the $y$-values for the entire first-phase sample, which may be denoted $\bar{y}_{s_1}$.

With the simple random sampling at the first phase, the unconditional expectation is

$$E(\bar{y}_d) = E[E(\bar{y}_d \mid s_1)] = E(\bar{y}_{s_1}) = \mu$$

The variance of $\bar{y}_d$ given $s_1$, with the stratified random sampling at the second phase, is

$$\text{var}(\bar{y}_d \mid s_1) = \sum_{h=1}^{L} \text{var}\left(\frac{n'_h}{n'} \bar{y}_h \mid s_1\right) = \sum_{h=1}^{L} \left(\frac{n'_h}{n'}\right)^2 \left(\frac{n'_h - n_h}{n'_h}\right) \frac{\sigma_h^2(s_1)}{n_h}$$

where

$$\sigma_h^2(s_1) = \frac{\sum_{i \in s_{1h}} (y_i - \bar{y}'_h)^2}{n'_h - 1}$$

is the finite population variance for the set of units in stratum $h$ of the first-phase sample, denoted by $s_{1h}$.

The variance of the conditional expectation is $\text{var}[E(\bar{y}_d)] = \text{var}(\bar{y}_{s_1})$, which is based on a simple random sample of size $n'$. Using the decomposition $\text{var}(\bar{y}_d) = \text{var}[E(\bar{y}_d) \mid s_1] + E[\text{var}(\bar{y}_d \mid s_1)]$, the variance expression (1) is obtained.

The unbiased estimator of variance and other results on double sampling for stratification, including optimal sampling fractions, are given in J. N. K. Rao (1973). Additional discussions of double sampling for stratification are given in Cochran (1977) and Singh and Chaudhary (1986). Results for designs with unequal inclusion probabilities at each phase are given in Särndal and Swensson (1987) and Särndal et al. (1992).

## 14.5 NONSAMPLING ERRORS AND DOUBLE SAMPLING

*Nonsampling errors* refers to differences between estimates and population quantities that do not arise solely from the fact that only a sample, instead of the whole population, is observed. For example, in surveys asking questions on sensitive topics, such as illicit drug use or sexual behaviors, respondents may tend to understate behaviors perceived as socially stigmatized and overstate behaviors perceived to be held in high esteem. Giving the questionnaire to every member of the population will not eliminate the difference between the recorded values and the actual quantities of interest. Similarly, in a survey of wintering waterfowl in which biologists count birds in selected plots from an aircraft, the errors in counting and detection will not be eliminated by flying over every plot in the study region.

In the idealized sampling situation there is a *frame*, such as a telephone book or a map, which lists or indexes every unit in the population. A sample of these units is selected using a known design, and the variables of interest and any relevant auxiliary variables are observed without error for every unit in the sample. Nonsampling errors can occur when the frame does not correspond exactly to the target population, when auxiliary information associated with units in the frame is not accurate, when some of the selected units cannot be observed, when variables associated with units in the sample are observed with measurement error or recorded inaccurately, and when the actual probabilities of selection differ from those of the presumed design. Nonsampling errors can also be introduced when the process of observing units in the sample affects the values of variables, as when sample subjects required to monitor their diets unconsciously select more healthful food items during the study period or when destructive sampling methods are used.

Sampling errors tend to become smaller the larger the sample size used. With nonsampling errors, in contrast, the errors tend to persist as sample size increases. The first line of defense against nonsampling errors is to make every effort to keep them as small as possible. For example, in telephone surveys frame error is reduced, although not eliminated, by using random number dialers rather than phone directories. The random digit dialers eliminate frame errors associated with unlisted telephone numbers and out-of-date directories but not those associated with people in the target population who do not have telephones or who have mobile phones not reachable by the random dialing scheme. Efforts to reduce or eliminate nonresponse in surveys are usually imperative because the tendency toward nonresponse may be associated, in unknown ways, with variables of interest, so that with a high percentage of nonresponse, the remaining sample of respondents my not be representative of the overall population. Measurement errors are kept to a minimum with careful design of survey questionnaires, unobtrusive interview methods, observational methods that increase the detectability of animals, and precise instruments for physical measurements.

A variety of methods are available for adjusting for nonresponse and measurement errors. A wide literature exists on reducing and adjusting for nonsampling errors, with many of the methodologies having developed in response to the particular realities of the substantive field of study.

A number of the methods for dealing with nonsampling errors are based on double sampling, either as a deliberately imposed design method for overcoming the errors or as a way of modeling and adjusting for the errors. For example, in the moose survey of Example 1, double sampling with ratio estimation provided a way to adjust for the measurement error caused by imperfect detectability, calibrating the imperfect observations using the ratio of the accurate intensive observations to the imperfect observations in the second phase sample. The following sections describe how double sampling is used to overcome problems in nonresponse. The first method described involves an implemented double sampling design utilizing "callbacks" or intensive effort to obtain a response from a subsample of nonrespondents. Other methods model the response mechanism and view it as a natural sampling design selecting the survey respondents as a subsample in the second phase, using double-sampling estimators to adjust for the nonresponse.

### Nonresponse, Selection Bias, or Volunteer Bias

The self-selection of volunteers for studies on socially sensitive behaviors produces a form of nonresponse or selection bias. Some methods assume that the population is divided into responders and nonresponders—or volunteers and refusers—with the characteristic of each person fixed. Methods for dealing with this situation include double sampling with a second, more intensive effort to sample the nonrespondents. In many situations it is considered more realistic to model the response as a stochastic phenomenon, so that each person has a certain probability of responding given selection in the sample, or of volunteering given the opportunity to do so.

### Double Sampling to Adjust for Nonresponse: Callbacks

An important use of double sampling for stratification involves the use of callbacks to adjust for nonresponse in surveys. In surveys of human populations, nonrespondents—people who were selected to be in the sample but are either unavailable or unwilling to respond—are often not representative with respect to the variable of interest. In such a situation, the sample mean of the respondents would be biased as an estimate of the population mean. Similarly, in surveys of natural populations, selected sites which for one reason or another remain unobserved are often not representative of the study region as a whole. For example, in a trawl survey of a fish population, some selected locations may not be observed when the ocean floor terrain is found to be too rocky or otherwise obstructed for the net. The habitat and hence the species composition at such sites may be not typical of the population as a whole, so that the sample mean of the trawled sites would give a biased estimate of the population mean.

In such cases the nonresponding units may be considered a separate stratum—but a unit is not known to be in that stratum until after the sample is selected. A subsample of the nonresponse stratum is then taken, usually requiring more intensive effort or different methods than that used in the initial sample. In the case of the telephone survey, the subsample consists of callbacks or visits by interviewers. In the case of the fishery survey, a different type of net or other sampling method may be needed for difficult sites.

Let stratum 1 consist of the responding units and stratum 2 the nonresponding units. An initial simple sample of $n'$ units is selected. Let $n'_1$ be the number of responding units and $n'_2$ the number of nonresponding units in the initial part of the survey. In the callback or second effort, responses are obtained for a simple random sample of $n_2$ of the initial nonrespondents. Conceptually, the stratified sample at the second stage includes all $n'_1$ of the initial respondents and $n_2$ of the initial $n'_2$ nonrespondents.

An unbiased estimate of the population mean is provided by the double-sampling stratified estimator

$$\bar{y}_d = \frac{n'_1}{n'}\bar{y}_1 + \frac{n'_2}{n'}\bar{y}_2$$

An expression for the variance is obtained from Equation (1), with $n_1 = n'_1$:

$$\mathrm{var}(\bar{y}_d) = \frac{N - n'}{N}\frac{\sigma^2}{n'} + \mathrm{E}\sum_{h=1}^{L}\left(\frac{n'_h}{n'}\right)^2\left(\frac{n'_h - n_h}{n'_h}\right)\frac{\sigma^2_{h(s_1)}}{n_h} \tag{3}$$

where $\sigma^2$ is the overall population variance and $\sigma^2_{h(s_1)}$ is the population variance within stratum $h$ for the particular first-phase sample $s_1$. The first term in Equation (3) is the variance of the expected value of $\bar{y}_d$ given the initial sample. It is the variance that would be obtained if the $y$-values were recorded for the entire initial sample and the sample mean used as an estimator. The second term is the expected value, over all possible initial samples, of the conditional variance of the stratified sample mean $\bar{y}_d$ given the initial sample. The expectation is left in the second term because $n'_h$, $n_h$, and $\sigma^2_{h(s_1)}$ are random variables depending on the first-phase sample.

An unbiased estimate of the variance of $\bar{y}_d$ with double sampling, using the stratum sample variances $s^2_h$ from the second sample, is

$$\widehat{\mathrm{var}}(\bar{y}_d) = \frac{N-1}{N}\sum_{h=1}^{L}\left(\frac{n'_h - 1}{n' - 1} - \frac{n_h - 1}{N - 1}\right)\frac{w_h s^2_h}{n_h}$$
$$+ \frac{N - n'}{N(n' - 1)}\sum_{h=1}^{L} w_h(\bar{y}_h - \bar{y}_d)^2 \tag{4}$$

## Response Modeling and Nonresponse Adjustments

Consider a population of $N$ units—people, for example—in which we are interested in a behavioral characteristic $y$. Let $s$ denote the set of people who are selected (asked) to be in the study and $r$ be the subset of those who respond or volunteer. Let $\pi_i$ denote the probability that the $i$th person (unit $i$) is selected to be in the study. Given the sample selected $s$ with person $i$ in it, let $\pi_{i|s}$ denote the probability that this person responds given that he or she has been selected, that is, the probability that the person agrees to volunteer. [Rosenbaum and Rubin (1983) termed this probability the *response propensity score*.] Then an unbiased estimator of the population total of the $y$-values is

$$\hat{\tau} = \frac{1}{N} \sum_{i \in r} \frac{y_i}{\pi_i \pi_{i|s}}$$

If, on the other hand, the sample mean $\bar{y}$ of the respondents is used to summarize the results of the study, its expected value is $E(\bar{y}) = \sum_{i=1}^{N} y_i \pi_i \pi_{i|s}$, which in general equals the population mean only if the selection and response probabilities are all equal. The problem in real studies is either to control or estimate the selection and response probabilities.

The nonresponse situation can be viewed as sampling in two phases, with the natural response mechanism doing the selection at the second phase. The sample $s$ is selected by a probability design with known inclusion probabilities $\pi_i$ and joint inclusion probabilities $\pi_{ij}$. Because the response mechanism that selects the sub-sample $r$ of respondents is not known, it must be modeled.

A common assumption is that units within identifiable groups have equal response probabilities and that their responses are independent. With this assumption, the assumed natural sampling design is stratified Bernoulli sampling. That is, from the $n_h$ units in group $h$ in the first-phase sample, the second-phase units are assumed selected by $n_h$ independent trials, each with the unknown response probability $\theta_{hs}$. The numbers $n_h$ and $m_h$ that are selected and respond in group $h$ are thus random variables. Conditional on the sample sizes $n_h$ and $m_h$, however, the Bernoulli sample has the property of a simple random sample, so that any subset of $m_h$ units is equally likely to be the responding subsample. The conditional response probability for person $i$ in group $h$, given the first-phase sample $s$ and given the second-phase group sample sizes $m_1, \ldots, m_{Hs}$, is

$$\pi_{i|s,\mathrm{m}} = \frac{m_h}{n_h}$$

and the joint response probability for two units in group $h$ is

$$\pi_{ij|s,\mathrm{m}} = \frac{m_h}{n_h} \left( \frac{m_h - 1}{n_h - 1} \right)$$

where $n_h$ is the number of people in group $h$ and $m_h$ is the number of people who respond. Särndal et al. (1992, pp. 577–582) term this the *response homogeneity group model*. Since the double-sampling properties are derived first conditionally on the first-phase sample, the definition of the groups, including their number $H_s$ and which group a given unit belongs to, can depend on the first-phase sample.

Provided that $m_h \geq 1$ for all groups, an estimator of the population total under this model is

$$\hat{\tau} = \sum_{i \in r} \frac{y_i}{\pi_i \pi_{i|s,\mathbf{m}}}$$

The variance of this estimator is

$$\text{var}(\hat{\tau}) = \sum_{i=1}^{N} \sum_{j=1}^{N} \left( \frac{\pi_{ij} - \pi_i \pi_j}{\pi_i \pi_j} \right) y_i y_j + \text{E}\left[ \sum_{h=1}^{H_s} n_h^2 \left( \frac{n_h - m_h}{n_h} \right) \frac{\sigma_{hs}^2}{m_h} \right]$$

where $\sigma_{hs}^2$ is the finite-population variance of $y_i/\pi_i$ in the group $h$ of the first-phase sample $s$.

An estimator of this variance, provided that $m_h \geq 2$ for all groups, is

$$\widehat{\text{var}}(\hat{\tau}) = \sum_{i \in s} \sum_{j \in s} \left( \frac{\pi_{ij} - \pi_i \pi_j}{\pi_i \pi_j \pi_{ij}} \right) y_i y_j + \sum_{h=1}^{H_s} n_h^2 \left( \frac{n_h - m_h}{n_h} \right) \frac{s_{hs}^2}{m_h}$$

where $s_{hs}^2$ is the sample variance for the respondents in group $h$.

Additional methods for nonresponse based on two-phase sampling models use ratio and regression estimation utilizing auxiliary information from the entire sample, the subsample of respondents, or both (cf. Särndal and Swensson 1987, Särndal et al. 1992). Models for nonresponse and model-based methods for estimation in the presence of nonresponse are described in Little (1982). Pfeffermann (1988) describes a regression-based approach with adjustments based on inclusion probabilities for the sampling design and response mechanism. References to further methods for dealing with missing data include Little and Rubin (1987), Rubin (1987), and Schafer (1997). General references to nonsampling errors in surveys include Biemer et al. (1991), Groves and Couper (1998), and Lessler and Kalsbeek (1992).

## EXERCISES

1. In an aerial survey of four plots selected by simple random sampling, the numbers of ducks detected were 44, 55, 4, and 16. Careful examination of photoimagery of the first and third of these plots (selected as a simple random subsample) revealed the actual presence of 56 and 6 ducks, respectively. The study area consists of 10 plots.

    (a) Estimate the total number of ducks in the study area by using a ratio estimator. Estimate the variance of the estimator.

**(b)** Suppose that the photo analysis doubles the cost of observing a plot. Estimate the optimal subsampling fraction.

2. Use double sampling to estimate the mean or total of a population of your choice in which an auxiliary variable exists that is easier to observe than the variable of interest. In the process of carrying out the survey and making the estimate, think about or discuss with others the following:

   **(a)** What practical problems arise in establishing a frame, such as a map or list of units, from which to select the sample?

   **(b)** How is the sample selection actually carried out?

   **(c)** What special problems arise in observing the units selected?

   **(d)** Estimate the population mean or total using ratio estimation.

   **(e)** Estimate the variance of the estimator above.

   **(f)** Give a 95% confidence interval for the population mean or total.

   **(g)** How would you improve the survey procedure if you were to do it again?

3. In a survey to estimate average household monthly medical expenses, 500 households were selected at random from a population of 5000 households. Of the selected households, 336 had children in the household and 164 had no children. A stratified subsample of 112 households with children and 41 households without children was then selected, and monthly medical expenditure data were collected from households in the subsample. For the households with children, the sample mean expenditure was $280 with a sample standard deviation of 160; for the households without children, the respective figures were $110 and 60. Estimate mean monthly medical expenditure for households in the population, and estimate the variance of the estimate.

# PART IV

# Methods for Elusive and Hard-to-Detect Populations

CHAPTER 15

# Network Sampling and
# Link-Tracing Designs

In a survey to estimate the prevalence of a rare disease, a random sample of medical centers is selected. From the records of each medical center in the sample, records of patients treated for that disease are obtained. However, a given patient may have been treated at more than one medical center. The more medical centers at which a given patient has been treated, the higher is the probability that that patient's records will be obtained in the sample.

In another survey, also with the purpose of estimating the prevalence of a rare characteristic in a population, a simple random sample of households is selected. At a selected household, the adult occupants are asked to report on the occurrence of the characteristic not only in themselves but also in their siblings. Thus, a person with several siblings who are living in different households has a higher inclusion probability than one with no siblings living in separate households. Even within a single household, the inclusion probabilities for different occupants are not necessarily equal.

Designs of the type described above are referred to as *network sampling* or *multiplicity sampling*. In network sampling, a simple random sample or stratified random sample of units (selection units) is selected, and all observation units (people in the examples) linked to any of the units selected are included or observed. The *multiplicity* of a person is the number of selection units—medical centers or households—to which a person is linked. Defining a network to be a set of observation units with a given linkage pattern, a network may be linked with more than one selection unit (siblings living in more than one household), and a single selection unit may be linked with more than one network (nonsiblings sharing a household). If the population of selection units is stratified, a network may also intersect more than one stratum.

Because of the unequal selection or inclusion probabilities, the sample mean does not form an unbiased estimator of the population mean with such a design. Unbiased estimators for such designs were given by Birnbaum and Sirken (1965). In one of the estimators termed the *multiplicity estimator* each observation is divided by its multiplicity. Since multiplicity is proportional to the draw-by-draw selection probability, such an estimator is akin to the Hansen–Hurwitz estimator. The Horvitz–Thompson estimator for network sampling, in which each person's inclusion probability is determined by the multiplicities, was also given by Birnbaum and Sirken. Subsequent papers on network sampling (Nathan 1976; Sirken 1970, 1972a,b) have concentrated on the multiplicity estimator. Levy (1977) and Sirken and Levy (1974) examined ratios of multiplicity estimators, which could be used, for example, to estimate the proportion of an ethnic group with a rare disease. Czaja et al. (1986) evaluated the effects of reporting errors through the linkages in network sampling—cases in which, for example, the patient's household may be more reliable at reporting the disease than is a relative's household. References to many innovative applications of network sampling are found in Kalton and Anderson (1986) and Sudman et al. (1988). Faulkenberry and Garoui (1991) discuss network sampling estimators in the context of area sampling methods used in agricultural surveys.

The network sampling design was first used not to increase efficiency but because it unavoidably arose in the sampling situation (a patient having records at more than one medical center). Later papers on the subject recognized its potential for giving lower variance estimates than conventional procedures and for increasing the "yield" of the survey; that is, the total number of people in the sample with the disease or other characteristic. In Section 15.4 we take a look at the wider field of link-tracing designs, in which social or other links between units are followed in obtaining the sample.

## 15.1   ESTIMATION OF THE POPULATION TOTAL OR MEAN

Let the value of the variable of interest for the $i$th observational unit in a population be denoted $y_i$. In a survey to estimate the prevalence of a disease or other characteristic, $y_i$ is an indicator variable, equal to 1 if the unit has the characteristic and zero otherwise. The variable of interest $y_i$ need not be an indicator variable; it could, for example, be the cost of medical treatment for the disease for the $i$th person. Let $N$ denote the number of observational units in the population. The population total is $\tau = \sum_{i=1}^{N} y_i$. Let $m_i$ be the multiplicity of the $i$th observational unit, that is, the number of selection units to which that observational unit is linked. The number of selection units in the population will be denoted $M$. The population mean per selection unit is $\mu = \tau/M$. Next we consider a sampling design in which a simple random sample (without replacement) of $n$ selection units is obtained and every observational unit linked to any selection unit is included in the sample.

### Multiplicity Estimator

The draw-by-draw selection probability $p_i$ for the $i$th observational unit is the probability that any one of the $m_i$ selection units to which it is linked is selected, that is,

$$p_i = \frac{m_i}{M} \tag{1}$$

An unbiased estimator of the population total $\tau$ may be formed by dividing each observed y-value by the associated selection probability. The *multiplicity estimator* thus obtained is

$$\hat{\tau}_m = \frac{M}{n} \sum_{i \in s} \frac{y_i}{m_i} \tag{2}$$

in which s is the sequence of observational units in the sample, including repeat selections. An observational unit may be selected more than once, even though selection units are sampled without replacement, because the observational unit may be linked to more than one selection unit. The expected number of times the $i$th observational unit is selected is $np_i$.

The notation for the multiplicity estimator may be simplified in a way that renders the statistical properties of the multiplicity estimator transparent. For the $j$th selection unit in the population, define the variable $w_j$ to be the sum of the $y_i/m_i$ for all observational units linked with selection unit $j$, that is,

$$w_j = \sum_{i \in A_j} \frac{y_i}{m_i} \tag{3}$$

where $A_j$ is the set of observational units that are linked to selection unit $j$.

With this notation, the multiplicity estimator may be written

$$\hat{\tau}_m = \frac{M}{n} \sum_{j=1}^{n} w_j \tag{4}$$

Thinking of $w_j$ as a new variable of interest associated with the $j$th selection unit, the multiplicity estimator is just $M\bar{w}$, where $\bar{w}$ is the sample mean of a simple random sample of size $n$. Thus, from the basic results on simple random sampling,

$$\mathrm{var}(\hat{\tau}_m) = \frac{M(M - n)}{n} \sigma_w^2 \tag{5}$$

where

$$\sigma_w^2 = \frac{1}{M - 1} \sum_{j=1}^{M} (w_j - \mu)^2 \tag{6}$$

in which $\mu = \tau/M$ is the population mean per selection unit.

An unbiased estimator of this variance is

$$\widehat{\mathrm{var}}(\hat{\tau}_m) = \frac{M(M - n)}{n} s_w^2 \tag{7}$$

where

$$s_w^2 = \frac{1}{n-1} \sum_{j=1}^{n} (w_j - \bar{w})^2 \tag{8}$$

For estimating the population mean per selection unit, $\hat{\mu}_m = \hat{\tau}_m/M$, $\text{var}(\hat{\mu}_m) = \text{var}(\hat{\tau}_m)/M^2$, and $\widehat{\text{var}}(\hat{\mu}_m) = \widehat{\text{var}}(\hat{\tau}_m)/M^2$.

**Horvitz–Thompson Estimator**

The probability that the $i$th observational unit is included in the sample is the probability that one or more of the $m_i$ selection units to which it is linked is selected. Since the inclusion probabilities are identical for all observational units in a network, the problem can be simplified by changing notation to be in terms of networks rather than individual observational units.

Recall that a network is composed of all observational units having the same linkage configuration. The population can be partitioned into $K$ networks, which will be labeled $1, \ldots, K$. Let $y_k^*$ now denote the total of the $y$-values over all the observational units in the $k$th network, and let $m_k^*$ denote the common multiplicity for any observational unit within this network.

The inclusion probability for the $k$th network, which is in fact the inclusion probability for any of the observational units within this network, is

$$\pi_k = 1 - \binom{M - m_k^*}{n} \bigg/ \binom{M}{n} \tag{9}$$

that is, 1 minus the probability that the entire simple random sample of $n$ selection units is selected from the $M - m_k$ selection units that are not linked with network $k$.

Let $\kappa$ denote the number of distinct networks of observational units included in the sample. The Horvitz–Thompson estimator of the population total is

$$\hat{\tau}_\pi = \sum_{k=1}^{\kappa} \frac{y_k^*}{\pi_k} \tag{10}$$

The Horvitz–Thompson estimator is an unbiased estimator which, unlike the multiplicity estimator, does not depend on the number of times any unit is selected.

The $m_{kl}^*$ denote the number of selection units that are linked to *both* networks $k$ and $l$. The probability that both networks $k$ and $l$ are included in the sample is

$$\pi_{kl} = \pi_k + \pi_l - 1 + \binom{M - m_k^* - m_l^* + m_{kl}^*}{n} \bigg/ \binom{M}{n} \tag{11}$$

The usual variance formulas for the Horvitz–Thompson estimator then apply, giving

$$\text{var}(\hat{\tau}_\pi) = \sum_{k=1}^{K}\left(\frac{1-\pi_k}{\pi_k}\right)y_k^{*2} + \sum_{k=1}^{K}\sum_{l\neq k}\left(\frac{\pi_{kl}-\pi_k\pi_l}{\pi_k\pi_l}\right)y_i^{*}y_j^{*} \tag{12}$$

An unbiased estimator of this variance is

$$
\begin{aligned}
\widehat{\text{var}}(\hat{\tau}_\pi) &= \sum_{k=1}^{K}\left(\frac{1-\pi_k}{\pi_k^2}\right)y_k^{*2} + \sum_{k=1}^{K}\sum_{l\neq k}\left(\frac{\pi_{kl}-\pi_k\pi_l}{\pi_k\pi_l}\right)\frac{y_k^{*}y_l^{*}}{\pi_{kl}} \\
&= \sum_{k=1}^{K}\left(\frac{1}{\pi_k^2}-\frac{1}{\pi_k}\right)y_k^{*2} + \sum_{k=1}^{K}\sum_{l\neq k}\left(\frac{1}{\pi_k\pi_l}-\frac{1}{\pi_{kl}}\right)y_k^{*}y_l^{*} \tag{13}
\end{aligned}
$$

For estimating the population mean per selection unit, $\hat{\mu}_\pi = \hat{\tau}_\pi/M$, $\text{var}(\hat{\mu}_\pi) = \text{var}(\hat{\tau}_\pi)/M^2$, and $\widehat{\text{var}}(\hat{\mu}_\pi) = \widehat{\text{var}}(\hat{\tau}_\pi)/M^2$.

**Example 1.**  The computations of the network sampling estimators are demonstrated in the following simple example. In a survey to estimate the prevalence of a disease in a city, a simple random sample of $n = 100$ households is selected by simple random sampling, and adult residents of selected households report on their siblings living in the city as well as on themselves. Households are selection units, adult people are the observational units, and the variable of interest $y_i$ equals 1 if the person has the disease and zero otherwise. There are $M = 5000$ households in the city.

The 100 households in the sample are arbitrarily ordered to put interesting cases (those involving nonzero $y$-values) first. In sample household 1 live two adults, a man and a woman. The man has one sibling living in the city in another household. The man does not have the disease ($y_1 = 0$), but his sibling does ($y_2 = 1$). Together these two siblings form a network (network 1) with multiplicity $m_1 = 2$. The woman in household 1, who has the disease ($y_3 = 1$), has two siblings in separate households, one with the disease ($y_4 = 1$) and the other without ($y_5 = 0$). These three siblings form a network (network 2) with multiplicity $m_2 = 3$. The household of this last sibling (person 5) was also selected in the random sample, so that each of the three siblings in that network was selected or reported on twice. In this household (household 2) lives also the sibling's spouse, who does not have the disease ($y_6 = 0$) and who has no siblings in the city. That spouse (person 6) forms individually a network of multiplicity $m_3 = 1$. In household 3 lives one adult alone, who has the disease ($y_7 = 1$) and has no siblings in the city, so that person forms the fourth network, with multiplicity $m_4 = 1$. In the other 97 households selected, no one has the disease nor do any of their siblings. The fact that the $y$-values are zero for all of these cases will simplify our calculations.

For the multiplicity estimator, we will run through the household selections computing the value of $w_j$ for each [using Equation (3)]. For the first household, $w_1 = 1/2 + 2/3 = 7/6$. For the second household, $w_2 = 2/3 + 0/1 = 2/3$. For the third, $w_3 = 1/1 = 1$. For each of the other 97 households in the sample, $w_j = 0$.

The multiplicity estimator of the total number $\tau$ of people with the disease in the city [using Equation (4)] is

$$\hat{\tau}_m = \frac{5000}{100}\left(\frac{7}{6} + \frac{2}{3} + 1 + 0 + \cdots + 0\right) = 5000(0.02833) = 141.7$$

The sample mean of the $w$-variables is 0.02833 and their sample variance is $s_w^2 = 0.02753$. The estimated variance of the estimator [from Equation (7)] is

$$\widehat{\text{var}}(\hat{\tau}_m) = \frac{(5000)(5000 - 100)}{100}(0.02753) = 6745$$

and the estimated standard error is about 82.

For the Horvitz–Thompson estimator, the first four of the distinct networks in the sample have total $y$-values $y_1^* = 1$, $y_2^* = 2$, $y_3^* = 0$, and $y_4^* = 1$. From Equation (9) the inclusion probability for the first network is

$$\pi_1 = 1 - \binom{5000 - 2}{100}\bigg/\binom{5000}{100} = 1 - \frac{4998!}{100!\,4898!}\frac{100!\,4900!}{5000!}$$
$$= 1 - \frac{4899(4900)}{4999(5000)} = 1 - 0.9603961 = 0.0396039$$

For the second network,

$$\pi_2 = 1 - \binom{5000 - 3}{100}\bigg/\binom{5000}{100} = 1 - 0.9411805 = 0.0588195$$

For the third and fourth networks, each with multiplicity 1,

$$\pi_3 = \pi_4 = \frac{100}{5000} = 0.02$$

From Equation (10) the Horvitz–Thompson estimate of the number with the disease is

$$\hat{\tau}_\pi = \frac{1}{0.0396039} + \frac{2}{0.0588195} + \frac{0}{0.02} + \frac{1}{0.02} + 0 + \cdots + 0 = 109.3$$

In estimating the variance of the Horvitz–Thompson estimator, only the networks with positive totals contribute nonzero terms. From Equation (11) the relevant joint inclusion probabilities are

$$\pi_{12} = 0.0396039 + 0.0588195 - 1 + \left( \frac{5000 - 2 - 3 + 1}{100} \right) \Big/ \left( \frac{5000}{100} \right)$$

$$= 0.020769$$

$$\pi_{14} = 0.0396039 + 0.02 - 1 + \left( \frac{5000 - 2 - 1}{100} \right) \Big/ \left( \frac{5000}{100} \right) = 0.0007844$$

$$\pi_{24} = 0.0588195 + 0.02 - 1 + \left( \frac{5000 - 3 - 1}{100} \right) \Big/ \left( \frac{5000}{100} \right) = 0.0011651$$

The estimated variance [from Equation (13)], is

$$\widehat{\mathrm{var}}(\hat{\tau}_\pi) = \left( \frac{1}{0.0396039^2} - \frac{1}{0.0396039} \right) + \left( \frac{1}{0.0588195^2} - \frac{1}{0.0588195} \right) 2^2$$

$$+ \left( \frac{1}{0.02^2} - \frac{1}{0.02} \right)$$

$$+ 2\Bigg[ \left( \frac{1}{0.0396039(0.0588195)} - \frac{1}{0.020769} \right) 2$$

$$+ \left( \frac{1}{0.0396039(0.02)} - \frac{1}{0.0007844} \right)$$

$$+ \left( \frac{1}{0.0588195(0.02)} - \frac{1}{0.0011651} \right) 2 \Bigg]$$

$$= 5617$$

giving an estimated standard error of 75.    □

## 15.2   DERIVATIONS AND COMMENTS

To show that the multiplicity estimator is unbiased for the population mean, let $z_i$ be the number of times the $i$th observational unit is selected and write the multiplicity estimator as

$$\hat{\tau}_m = \sum_{i=1}^{N} \frac{y_i}{np_i} z_i \tag{14}$$

where $p_i = m_i/M$. The random variable $z_i$ has a hypergeometric distribution with expected value $np_i$. Thus, the expected value of $\hat{\tau}_m$ is

$$\mathrm{E}(\hat{\tau}_m) = \sum_{i=1}^{N} \frac{y_i}{np_i} \mathrm{E}(z_i) = \sum_{i=1}^{N} y_i = \tau \tag{15}$$

Given the unbiasedness of $\hat{\tau}_m$, the variance results follow from basic results of simple random sampling in terms of the new variables of interest $w_j$ defined for the selection units.

The multiplicity estimator has the form of a Hansen–Hurwitz estimator, with the $y$-value for each unit multiplied by the number of times that unit is selected and divided by the selection probability for that unit. Perhaps more to the point, each $y_i$ is divided by $np_i$, the expected number of times the unit is selected. The unbiasedness holds even though the number of observational units in the sample is a random variable in network sampling. The Hansen–Hurwitz estimator was originally introduced for sampling with unequal but known probabilities with replacement. Because of the without-replacement sampling of selection units in network sampling, a finite population correction factor enters the variance expressions in network sampling. In network sampling, the selection and inclusion probabilities are not in general known except for units in the sample.

The inclusion probabilities are obtained using the result from probability theory that if $A$ is any event and $A^c$ it complement, $p(A) = 1 - p(A^c)$. The joint inclusion probabilities are based on the result that if $A$ and $B$ are any two events, the probability that both $A$ and $B$ occur is the sum of the two individual probabilities minus the probability of either $A$ or $B$, that is, $p(A \cap B) = p(A) + p(B) - p(A \cup B)$.

## 15.3   STRATIFICATION IN NETWORK SAMPLING

When the selection units of the population are stratified, a complication arises in that a given observational unit may be linked to selection units in more than one stratum. Then observations in different strata are not independent as in conventional stratified sampling.

Suppose that the $M$ selection units in the population are partitioned into $L$ strata, with $M_h$ selection units in stratum $h$, and suppose that a stratified random sample is selected with sample size $n_h$ in stratum $h$, for $h = 1, \ldots, L$. For each selection unit in the sample, all observational units linked to it—regardless of which strata they are in—are included in the sample. Let $A_{hj}$ be the set of observational units linked to the $j$th selection unit in stratum $h$. For the $i$th observational unit, let $m_i$ be the number of selection units—which may be from more than one stratum—to which it is linked. For the $j$th selection unit in stratum $h$, a new variable of interest $w_{hj}$ is defined by $w_{hj} = \sum_{i \in A_{hj}} y_i / m_i$. Define the sample mean of the $w$-variables in stratum $h$ to be $\bar{w}_h = (1/n_h) \sum_{j=1}^{n_h} w_{hj}$.

The stratified multiplicity estimator (Birnbaum and Sirken 1965) has the form

$$\hat{\tau}_m = \sum_{h=1}^{L} M_h \bar{w}_h \tag{16}$$

It is an unbiased estimator of the population total, with variance

$$\text{var}(\hat{\tau}_m) = \sum_{h=1}^{L} \frac{M_h(M_h - n_h)}{n_h} \sigma_{wh}^2 \tag{17}$$

in which $\sigma_{wh}^2$ is the finite population variance of the $w$-values within stratum $h$. An unbiased estimate of this variance is obtained by replacing $\sigma_{wh}^2$ with $s_{wh}^2$, the sample variance of the $w$-values within stratum $h$.

Note that while $\hat{\tau}_m$ is unbiased for the overall population total $\tau$, an individual term $M_h \bar{w}_h$ is not in general unbiased for a relevant population total within stratum $h$. That is because $\bar{w}_h$ may be based partly on $y$-values of observational units associated with strata other than $h$. For example, if selection units are households, which are stratified by geographic region, and observational units are people linked by sibling relationships, selection of a household in one stratum may result in reporting on siblings in one or more other strata. The $y$-values for each of these siblings are combined in the value $w_{hj}$ for that household.

For an observational unit $i$ linked to a sample selection unit in stratum $h$, the weight given to the value $y_i$ in the estimator above [Equation (16)] is $M_h/n_h m_i$. Thus, for an observational unit linked to selection units in more than one stratum, the weight given to its $y$-value may vary depending on the stratum in which the selection is made to which it is linked.

To avoid this seemingly arbitrary dependence on the stratum through which a given observation is reported, an alternative estimator may be considered. In the Hansen–Hurwitz estimator, each $y$-value is divided by the expected number of times it is selected under the design. Let $m_{hi}$ denote the number of selection units in stratum $h$ to which the $i$th observational unit is linked. Let $z_{hi}$ be the number of selection units in the sample linked to observational unit $i$. The random variable $z_{hi}$ has a hypergeometric distribution with expected value $n_h m_{hi}/M_h$. The total number of times observational unit $i$ is selected is the sum, over the $L$ strata, of the $z_{hi}$. The expected number of times observational unit $i$ is selected is thus $\sum_{h=1}^{L} n_h m_{hi}/M_h$.

An unbiased estimator perhaps closer to the spirit of the Hansen–Hurwitz estimator is obtained by dividing each $y$-value by this expectation. To do this, define the new variable $w'_{hj}$ for the $j$th selection unit of stratum $h$ by

$$w'_{hj} = \frac{n_h}{M_h} \sum_{i \in A_j} \frac{y_i}{\sum_{h=1}^{L} n_h m_{hi}/M_h}$$

and let $\bar{w}'_h$ denote the sample mean of the $w'$-values within stratum $h$. An unbiased estimator of $\tau$ is

$$\hat{\tau}_p = \sum_{h=1}^{L} M_h \bar{w}'_h$$

The variance formulas [Equation (17) and its estimator] for the stratified multiplicity estimator hold for this alternative estimator when $w'$-values are substituted for $w$.

For the Horvitz–Thompson estimator with stratified network sampling, let $m_{hk}^*$ denote the number of selection units in stratum $h$ linked to the $k$th network in the population, and let $m_{hkl}^*$ denote the number linked to both networks $k$ and $l$. The inclusion probabilities are

$$\pi_k = 1 - \prod_{h=1}^{L} \binom{M_h - m_{hk}^*}{n_h} \bigg/ \binom{M_h}{n_h}$$

$$\pi_{kl} = \pi_k + \pi_l - 1 \prod_{h=1}^{L} \binom{M_h - m_{hk}^* - m_{hl}^* + m_{hkl}^*}{n_h} \bigg/ \binom{M_h}{n_h}$$

With these inclusion probabilities, the usual Horvitz–Thompson formulas (Chapter 6) hold.

## 15.4 OTHER LINK-TRACING DESIGNS

Any design in which links or connections between units are used in obtaining the sample can be referred to as a *link-tracing design*. In addition to the network sampling designs described in previous sections, the literature on link-tracing designs includes procedures variously termed *snowball sampling, chain-referral sampling, random walks, web crawls*, and *adaptive sampling*. A wide range of innovative link-tracing procedures have been used in actual studies to obtain samples from hidden, hard-to-access, and elusive populations.

For example, in studies of injecting drug use and other risk behaviors in relation to the transmission of HIV and hepatitis C infections, initial respondents may be asked to identify drug-injecting or sexual partners, who are then added to the sample. In a study to examine the relation of network structure and risk behaviors, such as needle sharing among drug injectors in the Bushwick section of Brooklyn, New York, initial respondents were used as "auxiliary recruiters" to bring members of their networks into the study (Friedman et al. 1997; Neaigus et al. 1995, 1996). In a long-term study on the heterosexual transmission of HIV infection (Rothenberg et al. 1995), the target population of interest consisted of commercial sex workers, their paying and nonpaying partners, persons who use injectable drugs, and the sexual partners of drug users in the Colorado Springs area. Persons in the purposively selected initial sample were interviewed and, in addition to their individual characteristics, identities of their sexual partners were obtained. Persons named by two or more respondents were also located and interviewed.

Link-tracing designs are also involved in sampling and searching the Internet (Henzinger et al. 2000; Lawrence and Giles 1998). Even the largest search engine indexes list only a fraction of the Web sites in existence at any given time, so that obtaining a sample of Web sites for studies of World Wide Web characteristics, such as the proportion of sites that are devoted to a given topic, involve "crawling" or otherwise searching through the Web by following links from one site to another. The probability that a site is included in the sample depends on such factors as the number of other sites that link to it, how many other sites those sites link to, and the probability that the other sites are included in the sample. Estimates of Internet characteristics based naively on sample means of sample characteristics, without taking the design or search procedure into account, therefore tend to be biased.

The network sampling designs described in this chapter are particularly nice because the links between units are symmetric and it is possible to trace the links to include in the sample every unit in a given network. Thus, the inclusion probabilities can be computed for every unit in the sample and design-unbiased estimators formed. In many real situations, however, the links followed may be asymmetric— for example, one person will lead investigators to a second but the second, if asked, will not lead investigators to the first—and for practical reasons it may not be possible to follow links to encompass an entire connected network. In response to the inherent difficulties of the real situations, a rather wide range of design and model-based inference methods for link-tracing designs have been developed.

Mathematically, the situation can be conceptualized as sampling in a *graph*. In graph theory, a graph consists of a set of nodes, representing people or other units, and a set of edges or arcs representing the links or social connections between the nodes. Visually, the graph can be depicted as a set of small circles representing the nodes and lines or arrows representing the links. The whole graph represents the population of interest, such as a population of people with its social structure of relationships. The problem is that we can observe only a sample from the graph—a sample of nodes and a sample of links.

The term *snowball sampling* has been applied to two types of procedures related to network sampling. In one type (see Kalton and Anderson 1986), a few identified members of a rare population are asked to identify other members of the population, those so identified are asked to identify others, and so on, for the purpose of obtaining a nonprobability sample or for constructing a frame from which to sample. In the other type (Goodman 1961), individuals in the sample are asked to identify a fixed number of other individuals, who in turn are asked to identify other individuals, for a fixed number of stages, for the purpose of estimating the number of "mutual relationships" or "social circles" in the population.

Some properties of sample statistics from graphs were investigated by Bloemena (1969). Snowball designs were developed in the graph setting with a variety of initial probability sampling designs and any numbers of links and waves by Frank (1971, 1977a,b, 1978a,b, 1979a, 1997), who obtained a variety of design- and model- based methods for estimating graph quantities from the sample data. Snijders (1992) used the same term, *snowball sampling*, to include designs in which only a subsample of links from each node is traced. The case in which only one of the links from a node is selected at random and followed to another node, and then one of its links selected, and so on, was called a *random walk* by Klovdahl (1989). Link-tracing sampling methods in which there is only one link from each node have been termed *chains* (Erickson 1979). Frank and Snijders (1994) describe model- and design-based estimation of a hidden population size, that is, the number of nodes in the graph, based on snowball samples. Additional practical and statistical issues in sampling from social networks with various types of snowball, chain referral, and other link-tracing designs are discussed in Frank (1997b, 1981, 1988), Frank and Harary (1982), Granovetter (1976), Jansson (1997), Karlberg (1997), Morgan and Rytina (1977), Robins (1998), Spreen (1992, 1998), Spreen and Zwaagstra (1994), van Meter (1990), Wasserman and Faust (1994), and Watters and Biernacki (1989). Maximum-likelihood estimation for a wide variety of link-tracing designs is described in Thompson and Frank (2000).

Another link-tracing procedure for which design-based estimators are available is adaptive cluster sampling (Thompson 1990, 1997a,b, Thompson and Seber 1996; see also Chapter 24), which has been formulated in the graph setting as well as the spatial setting. Following selection of an initial sample of nodes by any of a number of initial designs, the decision on whether to follow links from a node or not depends on the value of a variable of interest observed for the node. For example, in an epidemiological study of a sexually transmitted disease, sexual or social links may be followed only from respondents who have been infected. Design-unbiased estimation methods have been worked out for a wide variety of adaptive cluster sampling strategies. Recent reviews of the field of link-tracing sampling are found in Thompson (1997c), and Thompson and Frank (2000).

## EXERCISES

1. Using the network sample of Example 1, estimate the total amount spent on medical care for the disease investigated. The dollar amounts for the seven people linked to the three households listed in the example are $y_1 = 0$, $y_2 = 3400$, $y_3 = 5000$, $y_4 = 6600$, $y_5 = 0$, $y_6 = 0$, and $y_7 = 12,000$. None of people linked to the other 97 households in the sample required any medical treatment for the disease. Use the multiplicity estimator and give an estimate of its variance.

2. With the data of Exercise 1, estimate the total amount spent on medical care for the disease using the Horvitz–Thompson estimator, and estimate its variance.

CHAPTER 16

# Detectability and Sampling

In the basic sampling framework, it is assumed that the variable of interest is recorded without error for each unit in the sample. In many actual situations, however, this is hardly the case. In surveys of most bird species, it is unlikely that every bird in a selected plot will be detected. In aerial surveys of large mammals, some animals in the area observed may remain unsighted. In a trawl survey for fish or other marine species, not every individual in the path of the net is caught. Similarly, when soil or ore samples are assessed for discrete mineral objects such as diamonds, some of the objects in the sample may be missed. In archaeological surveys in which sample plots or trenches are searched for artifacts, some artifacts in the sample plots may remain undiscovered. In surveys of human populations also, some individuals in sampled units may remain undetected.

The probability that an object in a selected unit is observed—whether seen, heard, caught, or detected by some other means—is termed its *detectability*. In this chapter, to avoid endless references to nameless "objects" and because much of the study of the detectability problem has been associated with ecological surveys, the individual objects in the population will be referred to as "animals."

## 16.1 CONSTANT DETECTABILITY OVER A REGION

Imagine that the detectability for a given species is some constant probability $p$ throughout a region of area $A$. Such a situation may be reasonable to assume in the case of a survey from a high-altitude aircraft evenly covering a whole study region or in the case of a survey of seabird cliff nesting sights from an offshore vessel.

Let $y$ represent the number of animals observed in the region, while the number actually there is $\tau$, the population total. Wherever an animal is in the region, its probability of detection is $p$. If, in addition, detections are independent—detection of one animal not affecting detection of another—the number observed $y$ has a binomial distribution.

The expected number observed is

$$E(y) - \tau p \tag{1}$$

and its variance is

$$\text{var}(y) = \tau p(1 - p) \tag{2}$$

If the probability of detection $p$ is known, an unbiased estimator of the population total $\tau$ is

$$\hat{\tau} = \frac{y}{p} \tag{3}$$

The variance of this estimator is

$$\text{var}(\hat{\tau}) = \tau\left(\frac{1-p}{p}\right) \tag{4}$$

An unbiased estimator of this variance is

$$\widehat{\text{var}}(\hat{\tau}) = \frac{y(1-p)}{p^2} \tag{5}$$

The *density* of the population is defined to be $D = \tau/A$, the number of animals per unit area in the study region. An unbiased estimate of density is

$$\hat{D} = \frac{y}{pA} \tag{6}$$

having variance

$$\text{var}(\hat{D}) = \frac{\tau}{A^2}\left(\frac{1-p}{p}\right)$$

and estimated variance

$$\widehat{\text{var}}(\hat{D}) = \frac{y}{A^2}\left(\frac{1-p}{p^2}\right)$$

*Example 1.* The locations of brant nests in a study region of 400 by 1600 meters in the Yukon–Kuskokwim Delta of Alaska (data from Anthony 1990) are shown in Figure 16.1. The nests have been censused with a combination of aerial photography and ground studies, and the 76 nests in the figure will be taken to be the actual population total $\tau$ in the study region.

Suppose that a survey methodology is proposed, for example, aerial observation alone, for which the detectability is known to be $p = 0.9$. From such a survey, the total number of nests in the study region would be estimated by dividing the

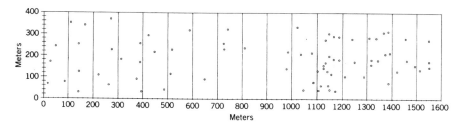

**Figure 16.1.** Locations of brant nests in a study region of 400 by 1600 meters in the Yukon–Kuskokwim Delta of Alaska. (From Anthony 1990.)

observed number $y$ by 0.9, and the variance of the estimator [from Equation (4)] would be

$$\text{var}(\hat{\tau}) = 76\left(\frac{1 - 0.9}{0.9}\right) = 8.4$$

The standard error is $\sqrt{8.4} = 2.9$.                                                                ☐

## 16.2  ESTIMATING DETECTABILITY

In most situations, the probability of detection $p$ would not be known but would be estimated by one method or another. For example, for an aerial survey, $p$ could be estimated by comparing the number of animals seen from the air to the number found by ground crews, either in the same study area or in similar areas. Other methods of estimating detectability include double sampling; capture–recapture methods and methods using radio-collared animals; distance-based methods such as those associated with line transects; net mesh size experiments; and regression methods that relate the number of animals detected to factors such as aircraft speed.

Based on the method by which $p$ was estimated, an estimate of var$(\hat{p})$ is usually available. For example, suppose that $p$ had been estimated as a ratio $\hat{p} = \bar{x}/\bar{y}$ in an independent study, with a random sample of $n$ plots on which $x_i$ is the number of animals detected on the $i$th plot by the standard survey method and $y_i$ is the actual number present based on intensive searching of the plot. In estimating population total $\tau$ with $y/\hat{p}$ or density $D$ with $y/\hat{p}A$, one may think of dividing the number of animals observed $y$ by $\hat{p}$ or of multiplying $y$ by the reciprocal $r = 1/\hat{p}$. The variance of $r$ may be estimated as $\widehat{\text{var}}(r) = s_d^2/n\bar{x}^2$, where $s_d^2$ is the usual residual sample variance used in ratio estimation, $s_d^2 = \sum_{i=1}^{n}(y_i - \hat{p}x_i)^2/(n-1)$. The variance of $\hat{p}$ may be estimated by reversing the roles of $x$ and $y$ in the ratio estimation.

Double-sampling ratio estimation methods apply when an accurate count is made for a subsample of the units surveyed by the usual survey method. For every unit in a sample of size $n'$, the number $x_i$ of animals detected is recorded. For a subsample of $n$ of the sample units, an accurate count $y_i$ of animals present is determined. The ratio $r = \bar{y}/\bar{x}$ from the subsample estimates the reciprocal $1/p$ of detectability. Estimation and variance formulas are given in Chapter 14.

Capture–recapture methods may be used to estimate detectability for a given species. For example, to estimate catchability $p$ in a fish trawl survey, a known number of fish are captured, marked, and released back into the population. Then, using the usual trawl survey methods and taking the area swept by the net into account, the ratio of marked fish caught to the total number of marked fish in the population is used as an estimate of catchability in further surveys. The variance associated with this estimate is given in Chapter 18. For populations in which radio-collared animals are present, detectability may be estimated from the proportion of radio-collared animals detected by the usual survey methods. Capture–recapture estimation methods can also be used in surveys in which two independent observers record or map detections; animals detected by one observer serve as marked animals to the other.

Multiple regression methods [Caughley (1974) and Caughley et al. (1976), summarized in Pollock and Kendall (1987) and Seber (1982, p. 457)] have been suggested for correcting for imperfect detectability in aerial surveys. Since observed density of animals in an aerial survey decreased with such factors as aircraft speed and altitude, the method involves regressing observed density on such variables and extrapolating to estimate actual density at the point where these variables are zero. Ramsey et al. (1987) combined a general regression model for such variables with the distance-based methods of line transects and variable circular plots.

Methods for estimating detectability in aerial surveys are reviewed in Pollock and Kendall (1987). Additional discussion of double-sampling methods for populations with detectability problems is found in Becker and Reed (1990), Eberhardt and Simmons (1987), and Rivest and Crépeau (1990).

## 16.3 EFFECT OF ESTIMATED DETECTABILITY

For now, let us assume that an estimate $\hat{p}$ of the detection probability has been obtained, that $\hat{p}$ is approximately unbiased for $p$, and that the estimate $\hat{p}$ is uncorrelated with the number $y$ of animals observed in the present survey. Then, using Taylor series approximations, an estimator of the population total $\tau$ is

$$\hat{\tau} - \frac{y}{\hat{p}}$$

and an approximate formula for the variance of this estimator is

$$\text{var}(\hat{\tau}) \approx \frac{1}{p^2}\left[\text{var}(y) + \tau^2\text{var}(\hat{p})\right] = \tau\left(\frac{1-p}{p}\right) + \frac{\tau^2}{p^2}\text{var}(\hat{p}) \tag{7}$$

Thus, the estimation of the detectability adds a term to the variance. To estimate this variance, estimators for $\tau$ and $\text{var}(\hat{p})$ would be substituted in the formula.

*Example 2: Estimated Detectability.*   Suppose that the detectability in the proposed survey methodology for the brant nests of Example 1 had been independently estimated, by comparing aerial observations with more exhaustive air and ground searches, to be $\hat{p} = 0.9$ with an estimation variance of $\text{var}(\hat{p}) = 0.000625$, so that the standard error of $\hat{p}$ is $\sqrt{0.000625} = 0.025$ and a 95% confidence interval for the true detectability would be approximately from 0.85 to 0.95. Then the variance of the estimator $\hat{\tau}$ would be [by Equation (7)]

$$\text{var}(\hat{\tau}) \approx 2.9 + \frac{76^2}{0.9^2}(0.000625) = 8.4 + 4.5 = 12.9 \qquad \square$$

## 16.4   DETECTABILITY WITH SIMPLE RANDOM SAMPLING

Now suppose that a simple random sample without replacement of $n$ units is selected from a population of $N$ units, and that animals within a selected plot are detected with probability $p$, detections being independent. Examples in which such a situation could apply include aerial transect surveys of large mammals in which visibility along the transect flown is roughly constant, surveys from offshore of selected sections of cliffs for nesting sites, and trawl surveys of fish or shellfish in which constant catchability applies to the area swept by the net. In any such situation, the properties of observations, and hence of the estimators, will depend both on the sampling design and the detectability.

Let $Y_i$ denote the number of animals actually in unit $i$ and $y_i$ be the number detected by the observer. The population total is $\tau = \sum_{i=1}^{N} Y_i$ and the population mean is $\mu = \tau/N$. The values $Y_i, Y_2, \ldots, Y_N$ are considered fixed. Given that the unit $i$ is in the sample, $y_i$ is a binomial random variable with expected value $E(y_i) = Y_i p$ and variance $\text{var}(y_i) = Y_i p(1-p)$. An estimate of the number of animals actually in unit $i$ is $Y_i = y_i/p$. Conditional on unit $i$ being in the sample, $E(\hat{Y}_i) = Y_i$ and $\text{var}(\hat{Y}_i) = Y_i(1-p)/p$.

With a simple random sample of $n$ units, an estimator of the population total $\tau$ is

$$\hat{\tau} = N\frac{\bar{y}}{p}$$

where $\bar{y} = (1/n)\sum_{i=1}^{n} y_i$, the sample mean of the observed values.

The estimator $\hat{\tau}$ is an unbiased estimator of the population total $\tau$ and has variance

$$\text{var}(\hat{\tau}) = N^2\left[\left(\frac{N-n}{N}\right)\frac{\sigma^2}{n} + \left(\frac{1-p}{p}\right)\frac{\mu}{n}\right] \qquad (8)$$

where the population variance $\sigma^2$ is

$$\sigma^2 = \frac{1}{N-1}\sum_{i=1}^{N}(Y_i - \mu)^2$$

The first term in the variance [Equation (8)] is due to sampling only $n$ units out of the $N$ in the population, and the second term is due to the imperfect detectability. Notice the lower detectability leads to higher variance in the estimator: The term $(1-p)/p$ decreases as the detection probability $p$ increases.

An unbiased estimator of the variance of $\hat{\tau}$ is

$$\widehat{\mathrm{var}}(\hat{\tau}) = N^2\left[\left(\frac{N-n}{N}\right)\frac{s^2}{p^2 n} + \left(\frac{1-p}{p^2 N}\right)\bar{y}\right]$$

where $s^2$ is the sample variance of the observed $y$-values, that is,

$$s^2 = \frac{1}{n-1}\sum_{i=1}^{n}(y_i - \bar{y})^2$$

For estimating the mean $\mu$, use $\hat{\mu} = \bar{y}/p$ and divide the leading $N^2$ out of the variance formulas.

***Example 3: Simple Random Sample with Detectability.*** Suppose that the 1600-meter-long brant nest study region is divided into $N = 16$ plots, so that each plot in Figure 16.1 extends 100 meters horizontally and 400 meters vertically. The population $y$-values, giving the number of nests in each of the rectangular plots, are (from left to right in Figure 16.1) 4, 5, 4, 5, 3, 3, 1, 3, 1, 2, 5, 18, 4, 10, 4, and 4. The population mean of these 16 values is $\mu = 4.75$ and the finite-population variance is $\sigma^2 = 16.73$.

With perfect detectability, an estimate of the total number of nests in the study region based on a simple random sample of $n = 8$ of the plots would be $\hat{\tau} = 16\bar{y}$ with variance

$$\mathrm{var}(\hat{\tau}) = 16(16 - 8)\frac{16.73}{8} = 267.7$$

or a standard error of 16.4.

With detectability $p = 0.9$ in each plot selected, the estimate $\hat{\tau} = 16\bar{y}/0.9$ would have variance [Equation (8)]

$$\mathrm{var}(\hat{\tau}) = 267.7 + (16)^2\frac{(1 - 0.9)(4.75)}{0.9(8)} = 267.7 + 16.9 = 284.6$$

or a standard error of 16.9.                                                                            ☐

## 16.5 ESTIMATED DETECTABILITY AND SIMPLE RANDOM SAMPLING

Suppose that the probability $p$ of detection is not known, but an estimate $\hat{p}$ is obtained that is approximately unbiased for $p$ and is uncorrelated with $\bar{y}$. Assume also that an estimate $\widehat{\text{var}}(\hat{p})$ of the variance of $\hat{p}$ is available. For an estimator of the population total, consider

$$\hat{\tau} = \frac{N\bar{y}}{\hat{p}}$$

With the estimated detectability in the denominator, $\hat{\tau}$ is no longer unbiased for $\tau$ (although it may be approximately so). By Taylor's theorem, an approximate formula for the variance of $\hat{\tau}$ is

$$\text{var}(\hat{\tau}) \approx \frac{N^2}{p^2}[\text{var}(\bar{y}) + \mu^2 \, \text{var}(\hat{p})]$$

$$= N^2 \left[ \left( \frac{N-n}{N} \right) \frac{\sigma^2}{n} + \left( \frac{1-p}{p} \right) \frac{\mu}{n} + \frac{\mu^2}{p^2} \text{var}(\hat{p}) \right] \qquad (9)$$

The third term in the variance is due to estimating the detectability.

An estimate of variance would be obtained by using

$$\widehat{\text{var}}(\hat{\tau}) = \frac{N^2}{\hat{p}^2} \left[ \left( \frac{N-n}{N} \right) \frac{s^2}{n} + \left( \frac{1-\hat{p}}{N} \right) \bar{y} + \frac{\bar{y}^2}{\hat{p}^2} \widehat{\text{var}}(\hat{p}) \right]$$

In some applications, it is not detectability $p$ directly but its reciprocal $\beta = 1/p$ that is estimated. Let $\hat{\beta}$ be an approximately unbiased estimate of $\beta$. If detectability is known, $\hat{\tau}$ may be written $\hat{\tau} = N\bar{y}\beta$. If detectability is estimated, $\hat{\tau} = N\bar{y}\hat{\beta}$. By Taylor's theorem, an approximate variance formula is

$$\text{var}(\hat{\tau}) \approx N^2 \left[ \left( \frac{N-n}{N} \right) \frac{\sigma^2}{n} + \frac{(\beta-1)\mu}{n} + \frac{\mu^2}{\beta^2} \text{var}(\hat{\beta}) \right]$$

An estimate of this variance is

$$\widehat{\text{var}}(\hat{\tau}) = N^2 \left[ \frac{\hat{\beta}^2(N-n)}{N} \left( \frac{s^2}{n} \right) + \frac{\hat{\beta}(\hat{\beta}-1)\bar{y}}{N} + \bar{y}^2\widehat{\text{var}}(\hat{\beta}) \right]$$

***Example 4: Random Sampling and Estimated Detectability.*** With the detectability of $p = 0.9$ of the brant nest example independently estimated with variance $\text{var}(\hat{p}) = 0.000625$ and the simple random sample of 8 of the 16 plots in the study region, the variance of the estimator $\hat{\tau} = 16\bar{y}/\hat{p}$ is approximately [by Equation (9)]

$$\text{var}(\hat{\tau}) \approx 267.7 + 16.9 + 16^2 \left(\frac{4.75^2}{0.9^2}\right)(0.000625)$$

$$= 267.7 + 16.9 + 4.5$$

$$= 289.1 \qquad \qquad \square$$

## 16.6  SAMPLING WITH REPLACEMENT

If simple random sampling with replacement is used, the estimator $\hat{\tau} = N\bar{y}/p$, with known $p$, is unbiased for $\tau$ with variance

$$\text{var}(\hat{\tau}) = N^2 \left[\frac{\sigma^2}{n} + \frac{(1-p)\mu}{pn}\right]$$

Let $\hat{\tau}_i = Ny_i/p$. Then $\hat{\tau}$ is a sample mean of $n$ independent, identically distributed random variables $\hat{\tau}_1, \ldots, \hat{\tau}_n$, so an unbiased estimate of the variance of $\hat{\tau}$ is

$$\widehat{\text{var}}(\hat{\tau}) = \frac{1}{n(n-1)} \sum_{i=1}^{n} (\hat{\tau}_i - \hat{\tau})^2$$

Notice that as in so many similar situations, with-replacement sampling leads to slightly higher variance, by giving up a finite-population correction factor, but leads to simpler variance estimation.

When $p$ is estimated, the term $(n^2\mu^2/p^2)\text{var}(\hat{p})$ is added to the variance and $(N^2\bar{y}^2/p^4)\widehat{\text{var}}(\hat{p})$ to the estimated variance, as before.

## 16.7  DERIVATIONS

**Unbiasedness of $\hat{\tau}$.** Given the sample $s$ of $n$ units,

$$E(\hat{\tau} \mid s) = E\left(\frac{N}{np} \sum_{i=1}^{n} y_i \mid s\right) = \frac{N}{np} \sum_{i=1}^{n} pY_i = \frac{N}{n} \sum_{i=1}^{n} Y_i$$

Unconditionally, $E(N/n)\sum_{i=1}^{n} Y_i = \tau$ by results of simple random sampling. Thus, $E(\hat{\tau}) = E[E(\hat{\tau} \mid s)] = \tau$, so that $\hat{\tau}$ is unbiased for $\tau$. $\qquad \square$

**Variance of $\hat{\tau}$.** Conditional on the sample $s$, the variance of $\hat{\tau}$ is

$$\text{var}(\hat{\tau} \mid s) = \frac{N^2}{p^2n^2} \sum_{i=1}^{n} \text{var}(y_i) = \frac{N^2}{p^2n^2} \sum Y_i p(1-p) = \frac{N^2}{n^2}\left(\frac{1-p}{p}\right) \sum_{i=1}^{n} Y_i$$

since, conditional on $s$, the $y_i$'s are independent. Unconditionally,

$$E[\text{var}(\hat{\tau} \mid s)] = \frac{N^2(1-p)}{pn} E\left(\frac{1}{n}\sum_{i=1}^{n} Y_i\right) = \frac{N^2(1-p)}{p}\left(\frac{\mu}{n}\right)$$

by simple random sampling results.
   Also by simple random sampling,

$$\text{var}[E(\hat{\tau} \mid s)] = \text{var}\left(\frac{N}{n}\sum_{i=1}^{n} Y_i\right) = N^2\left(\frac{N-n}{N}\right)\frac{\sigma^2}{n}$$

Thus,

$$\text{var}(\hat{\tau}) = \text{var}[E(\hat{\tau} \mid s)] + E[\text{var}(\hat{\tau} \mid s)] = N^2\left[\left(\frac{N-n}{N}\right)\frac{\sigma^2}{n} + \left(\frac{1-p}{p}\right)\frac{\mu}{n}\right]$$

which completes the derivation.                                                                        □

   ***Estimated Variance of*** $\hat{\tau}$. The proof will use conditioning on $\mathbf{y} = y_1, \ldots, y_N$, a realization of the observed values as if all $N$ units of the population were sampled. Conditional on $\mathbf{y}$, simple random sampling of $n$ units gives

$$E(\bar{y} \mid \mathbf{y}) = \frac{1}{N}\sum_{i=1}^{N} y_i$$

and

$$\text{var}(\bar{y} \mid \mathbf{y}) = \frac{N-n}{Nn(N-1)}\sum_{i=1}^{N}\left(y_i - \frac{1}{N}\sum_{i=1}^{N} y_i\right)^2$$

Also by simple random sampling,

$$E\left[\left(\frac{N-n}{Nn}\right)s^2 \mid \mathbf{y}\right] = \text{var}(\bar{y} \mid \mathbf{y})$$

Unconditionally,

$$E(\bar{y}) = \frac{1}{N}\sum_{i=1}^{N} E(y_i) = \frac{1}{N}\sum_{i=1}^{N} pY_i = p\mu$$

and

$$\text{var}[\text{E}(\bar{y} \mid \mathbf{y})] = \text{var}\left(\frac{1}{N}\sum_{i=1}^{N} y_i\right) = \frac{p(1-p)}{N^2}\sum_{i=1}^{N} Y_i = \frac{p(1-p)\mu}{N}$$

because of the independent binomial distributions of the $y_i$, so that

$$\text{E}\left[\frac{(1-p)\bar{y}}{N}\right] = \text{var}[\text{E}(\bar{y} \mid \mathbf{y})]$$

Also unconditionally,

$$\text{E}\left[\left(\frac{N-n}{Nn}\right)s^2\right] = \text{E}[\text{var}(\bar{y} \mid \mathbf{y})]$$

Since $\text{var}(\bar{y}) = \text{var}[\text{E}(\bar{y} \mid \mathbf{y})] + \text{E}[\text{var}(\bar{y} \mid \mathbf{y})]$, an unbiased estimate of $\text{var}(\bar{y})$ is thus

$$\widehat{\text{var}}(\bar{y}) = \left(\frac{N-n}{Nn}\right)s^2 + \left(\frac{1-p}{N}\right)\bar{y}$$

Since $\hat{\tau} = (N/np)\bar{y}$, an unbiased estimate of $\text{var}(\hat{\tau})$ is $(N^2/p^2)\widehat{\text{var}}(\bar{y})$, and the derivation is complete.  □

**With Estimated Detectability.** Let $U$ be a random variable with mean $\mu_u$ and variance $\sigma_u^2$, and let $V$ be another random variable with mean $\mu_v$ and variance $\sigma_v^2$. The first-order Taylor's series approximations for the variance of the product and of the ratio of the two random variables are

$$\text{var}(UV) = \mu_v^2\sigma_u^2 + \mu_u^2\sigma_v^2 + 2\mu_u\mu_v\,\text{cov}(U, V)$$

$$\text{var}\left(\frac{U}{V}\right) = \left(\frac{1}{\mu_v^2}\right)\sigma_u^2 + \left(\frac{\mu_u^2}{\mu_v^4}\right)\sigma_v^2 - 2\left(\frac{\mu_u}{\mu_v^3}\right)\text{cov}(U, V)$$

The results with estimated detectability use these approximations with $\bar{y}$ as $U$ and $\hat{p}$ or $\hat{\beta}$ as $V$, with the covariance term zero.  □

## 16.8   UNEQUAL PROBABILITY SAMPLING OF GROUPS WITH UNEQUAL DETECTION PROBABILITIES

Up to now we have considered simple random sampling of units, observing objects such as individual animals, each with equal detection probability. A generalization of this situation was considered in Steinhorst and Samuel (1989), in which the units may be selected by any sampling design with known inclusion probabilities, the objects are groups of animals for which the number in the group is recorded, and the detection probability may differ for different objects, depending, for example, on group size. In fact, the variable of interest $y_{ij}$ of the $j$th object in the $i$th unit may be any type of variable—continuous, discrete, or indicator.

Let $\pi_i$ be the probability that unit (plot) $i$ is included in the sample, and let $\pi_{ii'}$ be the probability that both units $i$ and $i'$ are included. With the $j$th object (group) in the $i$th unit (plot) is associated a variable $y_{ij}$, which may, for example, be the number of animals in group $ij$. The probability of detection for the $j$th object of the $i$th unit is $g_{ij}$. Let $M_i$ denote the number of objects in the $i$th unit, and let $m_i$ be the number of these that are detected. The number of distinct units in the sample is $\nu$. Let $\tau_i = \sum_{j=1}^{M_i} y_{ij}$ be the total of the $y$-values (number of animals) in unit $i$.

The objective is to estimate the population total $\tau = \sum_{i=1}^{N} \sum_{j=1}^{M_i} y_{ij}$, for example the total number of animals in the population. An unbiased estimator of $\tau$, based on the Horvitz-Thompson method, is

$$\hat{\tau} = \sum_{i=1}^{\nu} \frac{1}{\pi_i} \sum_{j=1}^{m_i} \frac{y_{ij}}{g_{ij}}$$

The variance of the estimator is

$$\mathrm{var}(\hat{\tau}) = \sum_{i=1}^{N} \left( \frac{1 - \pi_i}{\pi_i} \right) \tau_i^2 + \sum_{i=1}^{N} \sum_{i' \neq 1}^{N} \left( \frac{\pi_{ii'} - \pi_i \pi_{i'}}{\pi_i \pi_{i'}} \right) \tau_i \tau_{i'}$$
$$+ \sum_{i=1}^{N} \frac{1}{\pi_i} \sum_{j=1}^{M_i} \left( \frac{1 - g_{ij}}{g_{ij}} \right) y_{ij}^2$$

An unbiased estimator of this variance is

$$\widehat{\mathrm{var}}(\hat{\tau}) = \sum_{i=1}^{\nu} \left( \frac{1 - \pi_i}{\pi_i^2} \right) \hat{\tau}_i^2 + \sum_{i=1}^{\nu} \sum_{i' \neq 1} \left( \frac{\pi_{ii'} - \pi_i \pi_{i'}}{\pi_{ii'} \pi_i \pi_{i'}} \right) \hat{\tau}_i \hat{\tau}_{i'}$$
$$+ \sum_{i=1}^{\nu} \frac{1}{\pi_i} \sum_{j=1}^{m_i} \left( \frac{1 - g_{ij}}{g_{ij}^2} \right) y_{ij}^2$$

where

$$\hat{\tau}_i = \sum_{j=1}^{m_i} \frac{y_{ij}}{g_{ij}}$$

## 16.9 DERIVATIONS

The properties of $\hat{\tau}$ are most easily derived by writing the estimator in the form

$$\hat{\tau} = \sum_{i=1}^{N} \frac{I_i}{\pi_i} \sum_{j=1}^{M_i} \frac{y_{ij} z_{ij}}{g_{ij}}$$

where the indicator variable $I_i = 1$ if unit $i$ is included in the sample and $I_i = 0$ otherwise and $z_{ij} = 1$ indicates detection of object $ij$ and $z_{ij} = 0$ otherwise. Formally, $I_i = 1$ with probability $\pi_i$, $z_{ij} = 1$ with probability $g_{ij}$, and the $z$'s are assumed independent of the $I$'s. The $ij$th object is observed only if both $I_i = 1$ and $z_{ij} = 1$.

Unbiasedness of the estimator follows, since $E(I_i) = \pi_i$, $E(z_{ij}) = g_{ij}$, and $E(I_i z_{ij}) = \pi_i g_{ij}$. The formula for the variance is obtained by conditioning on the sample $s$ selected, that is by conditioning on the outcome of the vector $\mathbf{I} = \{I_1, \ldots, I_N\}$. The derivation is given in Steinhorst and Samuel (1989).

For the estimator of the variance of $\hat{\tau}$, it is easiest to condition on the set $\mathbf{z}$ of all the $z_{ij}$ in the population and use the decomposition

$$\mathrm{var}(\tau) = E[\mathrm{var}(\hat{\tau} \mid \mathbf{z})] + \mathrm{var}[\hat{\tau} \mid \mathbf{z}]$$

Here, the formality of $z_{ij}$ being defined even for objects in units not in the sample is useful, even though the concept of detections in units not in the sample is purely hypothetical.

Conditional on $\mathbf{z}$, the conditional expectation of $\hat{\tau}$ is

$$E(\hat{\tau} \mid \mathbf{z}) = \sum_{i=1}^{N} \sum_{j=1}^{M_i} \frac{y_{ij} z_{ij}}{g_{ij}}$$

Conditional on $\mathbf{z}$, $\hat{\tau}$ is a Horvitz–Thompson estimator of $\sum_{i=1}^{N} \sum_{j=1}^{M_i} y_{ij} z_{ij} / g_{ij}$ with conditionally unbiased estimator of variance

$$v = \sum_{i=1}^{\nu} \left( \frac{1 - \pi_i}{\pi_i^2} \right) \hat{\tau}_i^2 + \sum_{i=1}^{\nu} \sum_{i' \neq 1} \left( \frac{\pi_{ii'} - \pi_i \pi_{i'}}{\pi_{ij} \pi_i \pi_{i'}} \right) \hat{\tau}_i \hat{\tau}_{i'}$$

so that $E(v \mid \mathbf{z}) = \mathrm{var}(\hat{\tau} \mid \mathbf{z})$. Unconditionally, $E(v) = E[\mathrm{var}(\hat{\tau} \mid \mathbf{z})]$.

The variance of the conditional expectation is

$$\mathrm{var}[E(\hat{\tau} \mid \mathbf{z})] = \mathrm{var}\left( \sum_{i=1}^{N} \sum_{j=1}^{M_i} \frac{y_{ij} z_{ij}}{g_{ij}} \right)$$

$$= \sum_{i=1}^{N} \sum_{j=1}^{M_i} \frac{y_{ij}^2}{g_{ij}^2} \mathrm{var}(z_{ij})$$

$$= \sum_{i=1}^{N} \sum_{j=1}^{M_i} \frac{y_{ij}^2}{g_{ij}^2} (1 - g_{ij})$$

since the $z_{ij}$ are independent Bernoulli random variables, each with a different mean $g_{ij}$ and variance $g_{ij}(1 - g_{ij})$.

The variance $\mathrm{var}[(\mathrm{E}(\hat{\tau}\mid \mathbf{z}))]$ is thus a population total of the variables

$$w_{ij} = \frac{y_{ij}^2}{g_{ij}}(1 - g_{ij})$$

An unbiased estimator of this total, by the Horvitz–Thompson method, is

$$\sum_{i=1}^{N}\frac{I_i}{\pi_i}\sum_{j=1}^{M_i}\left(\frac{w_{ij}z_{ij}}{g_{ij}}\right) = \sum_{i=1}^{\nu}\frac{i}{\pi_i}\sum_{j=1}^{m_i}\left(\frac{1-g_i}{g_i^2}\right)y_{ij}^2$$

Adding this part to $v$ gives the unbiased estimator $\widehat{\mathrm{var}}(\hat{\tau})$. Steinhorst and Samuel (1989) give a slightly different formula for an estimator of variance, with no claim of unbiasedness. Results on detectability in sampling more general than those in this chapter are given in Thompson and Seber (1994).

## EXERCISES

1. In an aerial survey of a region in interior Alaska, 82 moose were detected. Intensive supplemental studies determined the probability of detection to be 0.89. Estimate the total number of moose in the study region and estimate the variance of that estimate. [The data numbers for this exercise and Exercise 3 are from Reed (1990); additional structure of the survey has been ignored for this exercise.]

2. Suppose that detectability is $p = 0.25$ throughout a region of $A = 100$ square kilometers and that $y = 60$ animals are seen during the survey. Estimate the number and density of animals in the region and estimate the variance of each estimator.

3. A simple random sample of $n = 20$ plots are selected for an aerial survey and the following numbers of moose are detected: 0, 0, 0, 0, 1, 2, 6, 11, 5, 1, 9, 3, 1, 0, 10, 4, 7, 22, 0, 0. Assuming detectability of 0.89, estimate the mean number of moose and give an estimate of variance. The study region contains $N = 100$ plots.

4. Suppose that a simple random sample of $n = 5$ plots is selected from a study area of $N = 100$ plots and that the numbers of animals detected in the four plots are 10, 7, 0, 0, and 5, but that the probability of detection for any animal in a selected plot is $p = 0.80$. Estimate the total number of animals in the study region and estimate the variance of the estimator.

## CHAPTER 17

# Line and Point Transects

In a line transect survey of an animal or plant species, an observer moves along a selected line and notes the location relative to the line of every individual of the species detected. It typically occurs in such surveys that more individuals are detected close to the line than far away from it, not because abundance is higher near the line but because the probability of detection is higher near the line than far from it. To estimate the abundance or density of the species in the study area from one or more such transects, this nonconstant detectability must be taken into account.

Line transect methods have been used for many types of populations, including bird, mammal, and plant species as well as other objects for which detectability depends on location relative to the observer. For convenience, the individuals in the population will be referred to generically as "animals." For surveys of some species, the observer walks along the transect. Line transect methods have also been applied to aerial surveys, surveys from research vessels, and sightings of animals from automobiles.

A line transect is characterized by a *detectability function* giving the probability that an animal (or plant) at a given location is detected. In most situations, the probability of detection can be expected to decrease as distance from the transect line increases. In many cases, detectability on the line itself can be assumed perfect. In other cases, avoidance by the animals of the observer can result in detectability reaching a maximum at some distance from the line.

Reviews and references on transect methods include Buckland et al. (1992), Burnham et al. (1980), Eberhardt (1978a), Gates (1979), Ramsey et al. (1988), and Seber (1982, 1986, 1992).

In this chapter we look first at some of the density estimation methods that can be used with line transect data, initially without regard to the sampling procedure for selecting transects. Then different designs for selecting a sample of transects are examined. For estimating the variance of the population density estimators, estimators relying on sampling procedures rather than model assumptions about the population are emphasized. Some underlying ideas about sampling and estimation with line transects are examined rather closely. Most of the line transect density or abundance estimators are based on average detectability, effective area observed, or density of detections along the line. Estimators based on individual detection probabilities are also covered.

Detectability functions are useful for evaluating many survey methods in addition to line transects. One may think of detectability units, characterizing the methods and the locations selected for making observations of an elusive population, as a generalization of the units of classical survey sampling. Some general results on design and estimation in terms of detectability functions are given in Sections 17.11 through 17.14. Section 17.15 is devoted to variable circular plots or point transects in which the observer is stationary for a specified amount of time at each selected site and locations or distances associated with detections are recorded.

## 17.1 DENSITY ESTIMATION METHODS FOR LINE TRANSECTS

Figure 17.1 depicts observations of animals or other objects from a segment of a line transect. The perpendicular distances from the objects to the transect are indicated with dashed lines. Given the set of distances of observed animals from one or

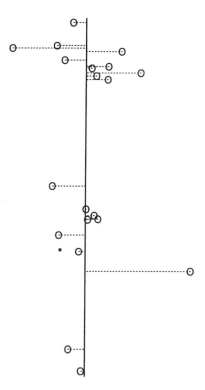

**Figure 17.1.** Observations of animals or other objects from a transect line.

more transect lines, for which detectability is virtually perfect near the line but decreases with distance from the line, it may not be immediately apparent how to estimate the abundance or density of the animals in the population. We will start with a couple of simple, commonsense methods for estimating population density or abundance from such data, progressing to more advanced methods, noting the similarity of the ideas underlying both the simple and the elaborate methods.

In the methods that follow, the object is to estimate the density of animals or other objects in a study region of area $A$. For the $i$th transect in the sample, the variable of interest $y_i$ is the number of animals observed. The sample size $n$ refers to the number of transects selected (not to the variable of interest). The total number of animals in the study region is denoted $\tau$, and the density of animals is $D = \tau/A$. Burnham et al. (1980, p. 33) suggest that the data should include at least 40 detections to provide reliable estimates. For illustrative purposes, a smaller number is used in the examples accompanying the following methods.

## 17.2   NARROW-STRIP METHOD

Although the detectability of animals far away from the transect line may be imperfect, there may be some narrow strip along the line in which detectability is virtually perfect. Then by using only those observations within the strip and ignoring the more distant observations, one may consider the strip a conventional unit or plot and estimate the population total or density in the usual way.

Let $L$ denote the length of the transect and let $w_0$ be the maximum distance from the line to which detectability is assumed perfect. Then the width of the strip is $2w_0$ and its area is $2w_0L$. Let $y_0$ be the number of animals detected within the narrow strip. To estimate the density $D$, that is, the number of animals per unit area, one may use the number of animals in the narrow strip divided by its area:

$$\hat{D} = \frac{y_0}{2w_0L} \tag{1}$$

If the study region has area $A$, the total number of animals in the study region is estimated as

$$\hat{\tau} = A\hat{D} = \frac{Ay_0}{2w_0L} \tag{2}$$

The distance $w_0$ is generally smaller than the maximum distance to which animals have been detected, and hence the number of animals $y_0$ used to estimate density is generally fewer than the total number $y$ detected. Various methods have been proposed for choosing the distance $w_0$ to which detection is assumed perfect. One way is to examine a histogram of the distance data and look for a distance at which the relative frequency of observations drops off sharply.

***Example 1: Narrow-Strip Method.*** On a line transect of length $L = 100$ meters, a total of $y = 18$ birds were detected at the following distances (in meters) from the transect line: 0, 0, 1, 3, 7, 11, 11, 12, 15, 15, 18, 19, 21, 23, 28, 33, 34, 44. It is desired to estimate the density of birds in the study region.

Plotting the numbers of birds detected in each 10-meter distance interval (Figure 17.2), we find that 5 were seen within 10 meters of the line, 7 were seen between 10 and 20 meters, 3 between 20 and 30 meters, 2 between 30 and 40 meters, and 1 between 40 and 50 meters. Choosing $w_0 = 20$ as the distance beyond which sightings drop off markedly, the narrow strip has width $2w_0 = 40$ meters. The number of birds detected within this strip was $y_0 = 12$.

The estimate of population density [from Equation (1)] is

$$\hat{D} = \frac{12}{2(20)(100)} = 0.003$$

so that the estimate is 0.003 bird per square meter or 30 birds per hectare.

Although the narrow-strip method is very simple, it is not entirely satisfying, first because not all observations obtained are used, second because the determination of the width of the narrow strip seems somewhat arbitrary, and third because detectability may in fact decrease smoothly with distance so that the narrow strip with perfect detectability really has width zero. ☐

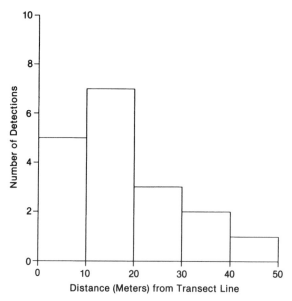

**Figure 17.2.** Example histogram of number of birds detected in each 10-meter distance interval from the transect.

## 17.3   SMOOTH-BY-EYE METHOD

In making a histogram to approximate a probability or probability density function $f$, one first chooses an interval width and then sets the height $\hat{f}$ of the histogram for a given distance $x$ by the following formula:

$$\hat{f}(x) = \frac{\text{number of observations in the interval containing } x}{(\text{total number of observations})(\text{interval width})} \qquad (3)$$

Note that, in keeping with its probability interpretation, the area under the histogram adds to one. Using the data of Example 1 with an interval width of 20 meters (the width used in the narrow-strip method), the height of the histogram for the first interval, and specifically for distance $x = 0$, using Equation (3), would be

$$\hat{f}(0) = \frac{12}{18(20)} = 0.0333$$

since 12 out of the 18 birds were seen in the first 20-meter interval.

The narrow strip used only the data of this first interval. For the interval of the narrow strip, the histogram height [Equation (3)] is $\hat{f}(0) = y_0/(yw_0)$, so that the narrow-strip estimate of $D$ may be written in terms of $\hat{f}(0)$ as

$$\hat{D} = \frac{\hat{f}(0)y}{2L} = \frac{0.0333(18)}{2(100)} = 0.003$$

The histogram for distance $x$ from the transect line may be viewed as approximating a smooth probability density function $f(x)$ that would describe the distribution of detection distances that would be obtained if one ran an infinite number of randomly selected transect lines for the species in question.

Looking at the histogram with 10-meter intervals (Figure 17.3), it is easy to imagine that a better estimate might be obtained of $f(0)$, the value of the true, smooth density of detections at zero distance from the transect line. The height of the histogram for the first interval, in which 5 birds were seen, is $5/[18(10)] = 0.028$. For the second interval, in which 7 were detected, the height is $7/[18(10)] = 0.039$. Similarly, for the remaining three intervals, the heights are 0.017, 0.011, and 0.006.

The observed distribution of detections, as depicted in the histogram (Figure 17.3), actually increases a bit with distance before decreasing. Suppose, however, that the true density of detections decreases smoothly with distance, reflecting decreasing detectability, and that the irregularities in the histogram are due to random chance and the small number of observations. Then a better estimate of $f(0)$, the theoretically true density of detections at distance zero, might be obtained by fitting a smooth, decreasing curve to the histogram.

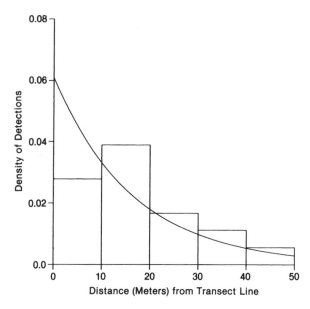

**Figure 17.3.** Density of detections with fitted exponential curve.

*Example 2: Smooth-by-Eye-Method.* Fitting such a curve by eye to the histogram (Figure 17.3) and seeing where the fitted curve intersected the vertical axis, the author obtained the estimate $\hat{f}(0) = 0.036$, which is higher than the histogram at the first interval but lower than the second.

The estimate of bird population density based on this smoothed-by-eye estimate of density at zero is

$$\hat{D} = \frac{\hat{f}(0)y}{2L} = \frac{0.036(18)}{2(100)} = 0.0032$$

or 32 birds per hectare.

The author's smoothed-by-eye curve is not illustrated; the reader is encouraged to make his or her own smooth-by-eye estimate from the density histogram of Figure 17.3. Since the choice of interval width and the smoothing are subjective, one person's estimate may differ from another's. The following methods reduce this subjectivity to some extent, but are based on very much the same idea.          □

## 17.4 PARAMETRIC METHODS

When animals are observed in a strip plot—that is, every animal within the strip of half-width $w$ and length $L$ is observed—the estimate of density is $\hat{D} = y/2Lw$, the number of animals observed divided by the area of the plot. When animals are observed from a line transect with a detectability function $g(x)$ having perfect detectability on the line and decreasing with distance $x$ away from the line, the distances to observed animals from randomly placed transects will tend to have a probability density $f(x)$ of the same shape as the detectability function but scaled so that the area under the probability density function $f$ equals 1. With perfect detectability on the line, the density estimate is

$$\hat{D} = \frac{y\hat{f}(0)}{2L} \qquad (4)$$

and the crux of the problem is estimating $f(0)$, the density at zero distance from the line.

One can imagine an equivalent strip plot, with perfect detectability out to some distance $w$, in which the same number of animals would be seen, on average, as are seen from the transect with decreasing detectability. The relationship between the line transect and the effectively equivalent strip plot is

$$f(0) = \frac{1}{w} \qquad (5)$$

and $w$ is called the *effective half-width* of the transect. In terms of effective half-width, the density estimate based on an estimate $\hat{w}$ of $w$ is

$$\hat{D} = \frac{y}{2L\hat{w}} \qquad (6)$$

Thus, one may equivalently proceed either to estimate $f(0)$ or to estimate $w$.

When a specific parametric form—that is, a functional form depending on unknown parameters—is assumed for the detectability function $g(x)$, statistical methods such as maximum likelihood may be used to estimate the unknown parameters and thereby obtain an estimate of $f(0)$ or of $w$. Some classes of parametric models are examined in Buckland (1985), Burnham et al. (1980), Pollock (1978), Quinn and Gallucci (1980), and Ramsey (1979). Two of the simplest will be used as examples here.

The advantage of assuming a simple form for the detectability curve is that it leads to simple estimators of population density—estimators that are best in some sense if the assumption is true. The disadvantage is that the assumed class of curves may not have the flexibility to represent the true detectability realistically enough. Two examples of parametric detectability functions, the exponential and the half-normal, are described here mainly because they lead to simple estimators of density.

The exponential class of detectability functions is $g(x) = \exp(-x/w)$. The larger the value of the parameters $w$, the higher the detectability of animals far from the transect line. The maximum likelihood estimator for $w$ (Ramsey 1979) is $\hat{w} = \bar{x}$, that is, the average distance of detection.

*Example 3: Exponential Detectability.* With the data of the bird example (Example 1), the average detection distance is $\bar{x} = 16.39$ meters. The estimate [Equation (6)] of population density is

$$\hat{D} = \frac{18}{2(16.39)(100)} = 0.055$$

or 55 birds per hectare. The fitted exponential curve $\hat{g}(x)/\hat{w}$ is shown in Figure 17.3.  □

Although the exponential model leads to an extremely simple estimator, it is not considered realistic for most real populations and does not in practice lead to good estimation results. Several authors (Buckland 1985; Burnham et al. 1980; Eberhardt 1978a) have argued that the detectability function should have a "shoulder," that is, be level or have zero derivative in the immediate vicinity of the transect line. The simplest model with such a shoulder is the half-normal. The half-normal detectability function is

$$g(x) = \exp\left(\frac{-\pi x^2}{4w^2}\right) \tag{7}$$

The maximum likelihood estimate of $w$ is

$$\hat{w} = \sqrt{\frac{\pi}{2y}\sum_{i=1}^{y} x_i^2} \tag{8}$$

*Example 4: Half-Normal Detectability.* With the bird data from Example 1, the average squared detection distance is $(1/n)\sum x_i^2 = (1/18)(0^2 + \cdots + 44^2) = 417.5$. The estimate of $w$ from Equation (8) is

$$\hat{w} = \sqrt{\left(\frac{3.1417}{2}\right)(417.5)} = 25.61$$

The estimate of density from Equation (6) is

$$\hat{D} = \frac{18}{2(25.61)(100)} = 0.0035$$

or 35 birds per hectare. The fitted half-normal curve $\hat{g}(x)/\hat{w}$ is shown in Figure 17.4.  □

More complicated models, with greater flexibility to fit real data, have been examined, but estimation with such models is somewhat complicated. The most

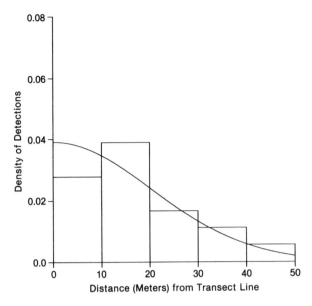

**Figure 17.4.** Density of detections with fitted half-normal curve.

adaptable models of all are the *nonparameteric models*, which essentially are smoothing techniques used to estimate $f(0)$.

## 17.5   NONPARAMETRIC METHODS

To avoid assumptions about the shape of the unknown detectability functions, nonparametric density estimation methods can be used. Using observations of random variables from a probability density function $f$, the methods use smoothing techniques to estimate the value $f(x)$ of the density function at any given value of $x$. With line transect sampling, the probability density function of interest is the density of observed detection distances. Two examples of nonparametric methods are given here; others are discussed in Buckland (1985), Burnham et al. (1980), and Johnson and Routledge (1985). Detectability on the transect line is assumed perfect, so that the estimate has the form $\hat{D} = y\hat{f}(0)/2L$.

### Estimating $f(0)$ by the Kernel Method

In the extensive statistical literature on the estimation of probability density functions (PDFs), the dominant trend is kernel estimation, a nonparametric smoothing approach (see, e.g., Silverman 1986). Application of the methodology to line transect estimation was suggested by Seber (1986) and has been utilized by Quang (1992) for the closely related problem of estimation in variable circular plot surveys.

The method employs a *kernel function* $K(x)$, which integrates to 1; that is,

$$\int_{-\infty}^{\infty} K(x)\,dx = 1$$

The kernel estimator of the PDF $f$ at $x$ is

$$\hat{f}(x) = \frac{2}{yh}\sum_{j=1}^{y} K\left(\frac{x - x_j}{h}\right) \tag{9}$$

where $h$ is called the *window width* and $x_j$ is the value of the $i$th observation (i.e., the distance from the transect line to the $j$th animal) and $y$ is the number of observations (i.e., the number of animals detected). The coefficient 2 arises when the density of unsigned distance, without regard to which side of the line, is used.

To estimate $f(0)$ with a symmetric kernel, the estimator becomes

$$\hat{f}(0) = \frac{2}{yh}\sum_{j=1}^{y} K\left(\frac{x_j}{h}\right) \tag{10}$$

With the normal kernel, for example,

$$K\left(\frac{x_j}{h}\right) = \frac{1}{\sqrt{2\pi}} e^{(1/2)(x_j/h)^2} \tag{11}$$

Silverman (1986, p. 48) gives a simple rule for choosing the window width $h$:

$$h = 0.9ay^{-1/5} \tag{12}$$

where $a = \min(s, Q/1.34)$, in which $s$ is the sample standard deviation of the $x$'s observed and $Q$ is their interquartile range. But when dealing with positive distances only, one should use the median distance in place of the interquartile range.

The estimator of the population density of animals is then

$$\hat{D} = \frac{y\hat{f}(0)}{2L} \tag{13}$$

***Example 5: Normal Kernal.***   In the bird data from Example 1, the median absolute distance is 15, and $15/1.34 = 11.19$. Since 11.19 is less than the sample standard deviation $s = 12.56$, Silverman's rule [Equation (12)] for choosing window width $h$ gives

$$h = 0.9(11.19)(18)^{-1/5} = 5.65$$

The normal kernel estimate of $f(0)$, from Equations (10) and (11), is

$$\hat{f}(0) = \frac{2}{18(5.65)\sqrt{2\pi}}[e^{-0^2/2(5.65)^2} + \cdots + e^{-44^2/2(5.65)^2}] = 0.0376$$

The estimate of bird density [(Equation (13)] is

$$\hat{D} = \frac{y\hat{f}(0)}{2L} = \frac{18(0.0376)}{2(100)} = 0.0034$$

bird per square meter, or 34 birds per hectare.                                    □

## Fourier Series Method

The Fourier series method of estimating $f(0)$ is

$$\hat{f}(0) = \frac{1}{w^*} + \sum_{k=1}^{M} \hat{A}_k \tag{14}$$

where $w^*$ is the maximum distance at which animals can be observed and the coefficients $\hat{A}_k$ are given by

$$\hat{A}_k = \frac{2}{yw^*}\left[\sum_{i=1}^{y} \cos\left(\frac{k\pi x_i}{w^*}\right)\right] \tag{15}$$

The number $m$ of terms to use in the approximation is somewhat arbitrary, but the following rule of thumb has been recommended (Burnham et al. 1980): Starting with $m = 1$, choose the first whole number $m$ such that

$$\frac{1}{w^*}\sqrt{\frac{2}{y+1}} \geq |\hat{A}_{m+1}| \tag{16}$$

In determining the maximum detectability distance $w^*$, Burnham et al. (1980) and Crain et al. (1979) recommend using some distance less than the greatest distance at which an animal was detected, throwing out the largest 1 to 3% of the observed distances as outliers (see also Burnham et al. 1981; Quang 1990).

*Example 6: Fourier Series Method.*   In applying the Fourier series method to the bird data, the largest observation (from the array in Example 1), the detection at 44 meters, is thrown out as an outlier and the next largest, 34 meters, is used as $w^*$. The inequality of the rule of thumb [Equation (16)] is satisfied for the value $m = 1$, so only one term is needed. Thus only one coefficient, $\hat{A}_1$, needs to be computed, but it involves 17 terms (the number of observations after discarding the largest). The coefficient $a_1$ is calculated from Equation (15):

$$\hat{A}_1 = \frac{2}{17(34)} \left[ \cos \frac{1(3.1417)(0)}{34} + \cdots + \cos \frac{1(3.1417)(34)}{34} \right]$$
$$= 0.0091$$

The estimate of $f(0)$ is calculated from Equation (14):

$$\hat{f}(0) = \frac{1}{34} + 0.0091 = 0.0385$$

The estimate of population density from Equation (13) is

$$\hat{D} = \frac{17(0.0385)}{2(100)} = 0.0033$$

or 33 birds per hectare. ☐

## 17.6 DESIGNS FOR SELECTING TRANSECTS

The sampling design in a line transect study is the procedure by which the transect locations are selected. Desired properties such as unbiasedness or approximate unbiasedness of density estimators and estimates of variance will be based as much as possible on the design rather than on assumptions about the population.

However many animals may be seen from a transect, a single transect is still a sample of size 1. A more precise estimate of abundance or density in a study area would be expected from a probability sample of $n$ transects, particularly if the animals are very unevenly distributed over the study region.

Variance estimates based on a sample of several transects are to be preferred to "analytical" estimates based on observations within a single transect, a recommendation emphasized by a number of authors (see Burnham and Anderson 1976, p. 329; Eberhardt 1978b; Overton 1969; and Seber 1982, p. 467). Procedures for estimating the variance of estimators from data within a single transect invariably rely on model assumptions about the distribution of animals, the typical assumption for such procedures being that animals are uniformly and independently located in the study region. So far, we are assiduously avoiding any such assumptions! Barry and Welsh (2001) examine the interplay between design, model, and estimating the detectability function with line transects and note in particular the problem of implicitly assuming independence when evaluating the effectiveness of methods.

Other aspects of the observational method, such as speed at which a line transect is traversed, affect the shape of the detectability functions and hence the properties of the observations and estimators. The effects of such choices are examined in Chapter 22.

## 17.7 RANDOM SAMPLE OF TRANSECTS

A random sample of $n$ transects in the study area will be selected as follows. A straight baseline of length $B$ is drawn across (or below) the study region on a map. The study area need not be regular in shape. The length of the baseline is the width of the perpendicular projection onto the line of every point in the study area—that is, the width of the shadow cast by the study area onto the line. A random sample of $n$ transect locations $v_1, v_2, \ldots, v_n$ is selected from the uniform distribution on the interval $[0, B]$. Transect lines perpendicular to the baseline are then located through each selected point $v_1$. The transects either run completely across the study area (Figure 17.5), or if a maximum transect length $L$ is desired that is less than the distance across the study region, parallel baselines may be drawn distance $L$ apart and starting points selected from the total length of baseline (Figure 17.6). Note that sampling is with replacement, although since transect locations are selected from a continuous distribution, there is zero probability of selecting precisely the same transect twice.

In selecting a random sample of transects, some biases may be introduced due to boundary problems—that is, slightly lower average detection probability for an animal near the boundary of the study area. Such biases would tend to be small in any

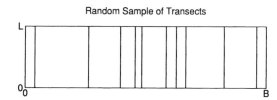

**Figure 17.5.** Random sample of 10 line transects in a study region.

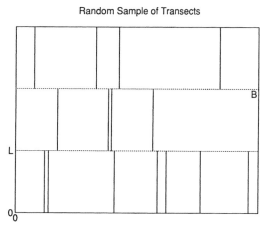

**Figure 17.6.** Random sample of 15 line transects of length $L$ in a study region of width wider than $L$.

case when the study area is large relative to the effective width of transects. In Section 17.11 it is shown how such biases can be eliminated completely if desired. In the present section these boundary-induced biases are ignored.

## Unbiased Estimator

Because of irregularities in the shape of the study region, the length $L_i$ of the $i$th transect is a random variable, with expected value $E(L_i) = A/B$, where $A$ is the area of the study region and $B$ is the length of the baseline. Let $y_i$ denote the number of animals seen from the $i$th transect.

If the effective width $w$ or the density $f(0)$ are known, an unbiased estimator of density, based on the $i$th transect, is

$$\hat{D}_i = \frac{B}{A}\left(\frac{y_i}{2w}\right) = \frac{B}{A}\left(\frac{y_i f(0)}{2}\right) = \frac{y_i f(0)}{2E(L)}$$

Hence an unbiased estimator based on the $n$ transects is

$$\hat{D} = \frac{1}{n}\sum_{i=1}^{n}\hat{D}_i = \frac{B}{A}\left(\frac{\bar{y}}{2w}\right) = \frac{B}{A}\left(\frac{\bar{y}f(0)}{2}\right)$$

where $\bar{y} = (1/n)\sum_{i=1}^{n} y_i$ is the sample mean of the numbers observed.

When $w$ or $f(0)$ in the expression for $\hat{D}$ are estimated, for example, by one of the methods just given, the estimated value $\hat{w}$ or $\hat{f}(0)$ is substituted in the expression for $\hat{D}$ and the unbiasedness holds only approximately.

If individual estimates $\hat{w}_1, \ldots, \hat{w}_n$ or $\hat{f}_1(0), \ldots, \hat{f}_n(0)$ are made independently for each transect and $\hat{D}_i = By_i/2A\hat{w}$ or $\hat{D}_i = By_i\hat{f}_i(0)/2A$, then $\hat{D}_i, \ldots, \hat{D}_n$ are independent and an unbiased estimator of variance is

$$\widehat{\text{var}}(\hat{D}) = \frac{1}{n(n-1)}\sum_{i=1}^{n}(\hat{D}_i - \hat{D})^2$$

However, one often gets a better estimate of $w$ or $f(0)$ by pooling the distance data from all transects in the survey. With $\hat{D}_i = By_i/2A\hat{w}$ or $\hat{D}_i = By_i\hat{f}(0)/2A$ using the pooled estimates, the $\hat{D}_i$ are not independent and $\widehat{\text{var}}(\hat{D})$ tends to underestimate the true variance of the estimator.

With the pooled estimates, a better estimate of variance could be obtained through a resampling method such as the bootstrap or jackknife. For the bootstrap method (Efron 1982; Efron and Gong 1983), the sample of $n$ transects is treated as a population in itself. A bootstrap sample is obtained by selecting $n$ of these transects from the sample at random with replacement, and bootstrap estimate $\hat{D}^{*1}$ is computed from the bootstrap sample by the same method as used for the original sample. Note that the bootstrap sample may differ from the original sample because of the with-replacement sampling. Repeating the procedure to obtain $M$ independent bootstrap values $\hat{D}^{*1},\ldots,\hat{D}^{*M}$, the bootstrap estimate of variance is

$$\widehat{\text{var}}_b(\hat{D}) = \frac{1}{M-1}\sum_{m=1}^{M}(\hat{D}^{*m} - \hat{D}_b)^2$$

where $\hat{D}_b = (1/M)\sum_{m=1}^{M}\hat{D}^{*M}$.

The jackknife estimate is obtained by systematically deleting one transect at a time from the sample. Let $\hat{D}_{(i)}$ be the estimate obtained from the $n-1$ remaining transects in the sample after deleting the $i$th transect, and let $\hat{D}_{(\cdot)} = (1/n)\sum_{i=1}^{n}\hat{D}_{(i)}$. Note that for each of the $n$ jackknife samples, each consisting of $n-1$ transects, the entire process of making a pooled estimate of $w$ or $f(0)$ and then estimating density is repeated. The jackknife estimate of variance is

$$\widehat{\text{var}}_j(\hat{D}) = \frac{n-1}{n}\sum_{i=1}^{n}[\hat{D}_{(i)} - \hat{D}_{(\cdot)}]^2$$

Note that for each method, the resampling is in terms of transects, which are independent because of the initial random selection, not in terms of observed distances, which cannot be assumed independent without assumptions about the population itself.

Approximate $100(1-\alpha)\%$ confidence intervals based on the jackknife method are usually of the form $\hat{D} \pm t\sqrt{\widehat{\text{var}}_j(\hat{D})}$, where $t$ is the upper $\alpha/2$ point of the $t$-distribution with $n-1$ degrees of freedom. With the bootstrap method, confidence intervals are typically based on percentiles of the distribution of the bootstrap estimates $\hat{D}^{*m}$. Buckland (1982) used bootstrap methods based on distance observations to obtain confidence intervals for line transect estimates.

From any estimate $\hat{D}$ of population density with estimated variance $\widehat{\text{var}}(\hat{D})$, an estimate of the total abundance in the study region is $\hat{\tau} = A\hat{D}$ with variance estimate $A^2\widehat{\text{var}}(\hat{D})$.

### Ratio Estimator

When the study area is irregular in shape, the transects will be of differing lengths. When this is the case, a ratio estimate based on the lengths may be preferred. The ratio estimator is

$$\hat{D}_r = \frac{\sum_{i=1}^{n} L_i\hat{D}_i}{\sum_{i=1}^{n} L_i} = \frac{\sum_{i=1}^{n} y_i}{2\hat{w}\sum_{i=1}^{n} L_i} = \frac{\sum_{i=1}^{n} y_i}{2\sum_{i=1}^{n} L_i}\hat{f}(0)$$

where $\hat{D}_i = y_i/2L_i\hat{w} = y_i\hat{f}(0)/2L_i$.

As a ratio estimator, $\hat{D}_r$ is not unbiased in the design sense, even when $w$ or $f(0)$ is known. It would be model-unbiased, however, under assumptions of a linear relationship between expected number of animals seen and transect length.

A variance estimator of adjusted ratio type is

$$\widehat{\mathrm{var}}_1(\hat{D}_r) = \frac{1}{L^2 n(n-1)} \sum_{i=1}^{n} \left(\frac{y_i}{2\hat{w}} - \hat{D}_r L_i\right)^2$$

$$= \frac{1}{L^2 n(n-1)} \sum_{i=1}^{n} \left(\frac{y_i \hat{f}(0)}{2} - \hat{D}_r L_i\right)^2$$

where $L = (1/n) \sum_{i=1}^{n} L_i$, the average length of the $n$ transects in the sample.

A model-based ratio variance estimator has also been suggested (Buckland 1982; Burnham and Anderson 1976; Burnham et al. 1980; Seber 1979, 1982, p. 463):

$$\widehat{\mathrm{var}}_2(\hat{D}_r) = \frac{1}{Ln(n-1)} \sum_{i=1}^{n} L_i(\hat{D}_i - \hat{D}_r)^2 = \frac{1}{Ln(n-1)} \sum_{i=1}^{n} \frac{[y_i \hat{f}_i(0)/2 - L_i \hat{D}_r]^2}{L_i}$$

For the jackknife estimate of variance when $w$ or $f(0)$ is estimated, Buckland (1982) suggests the following slight modification of the length-weighted estimator given in Burnham et al. (1980). Let $\hat{D}_{(i)}$ be the ratio estimate of density with the $i$th transect deleted, and define the pseudovalue $\hat{D}^{(i)} = [nL\hat{D}_r - (nL - L_i)\hat{D}_{(i)}]/L_i$. The jackknife estimate of variance is

$$\widehat{\mathrm{var}}_j(\hat{D}_r) = \frac{1}{Ln(n-1)} \sum_{i=1}^{n} L_i[\hat{D}^{(i)} - \hat{D}_r]^2$$

## 17.8   SYSTEMATIC SELECTION OF TRANSECTS

Many researchers will prefer a systematic selection of transects to avoid the uneven coverage of the study region obtained with random sampling. Figure 17.7 shows a systematic sample of 10 transect locations, evenly spaced from a single randomly selected location along the initial tenth of the baseline. With such a sample, the following results on unbiasedness or approximate unbiasedness of density estimators

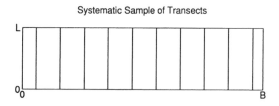

Systematic Sample of Transects

**Figure 17.7.** Systematic sample of line transects.

will hold, but not the results on unbiasedness of or approximate unbiasedness of estimators of variance.

An unbiased estimator of variance is available for a systematic sample with more than one randomly selected starting point, but the resulting coverage is not as "even" as with the single starting point. For many natural populations, variance estimators designed to be used with simple random sampling will tend to be conservative—tending to overestimate the actual variance—when used with systematic sampling with a single starting point.

## 17.9  SELECTION WITH PROBABILITY PROPORTIONAL TO LENGTH

Transect lines may be selected with probability proportional to length by selecting $n$ points independently from a uniform distribution over the whole study area. This may be accomplished by enclosing the study area in a rectangle and picking random coordinate pairs until $n$ locations within the study area are obtained. For each point selected, a transect is selected through the point and perpendicular to the baseline. A transect in a wide section of the study area—a long transect—has a greater probability of selection because more points in the study area lead to its selection. The probability density function for transect location $v$ along the baseline is $L(v)/A$.

For a single transect selected by the design above, consider the estimator

$$\hat{D} = \frac{y}{2wL(v)}$$

where $w$ is the effective half-width of the transect, $v$ the intersection point of that transect with the baseline, and $L(v)$ the width of the study area—the length of the transect—at that point.

Let $z_j$ be an indicator variable equal to 1 if the $j$th animal of the population is detected and zero otherwise. Conditional on the selected starting point $v$, the expected value of the estimator is

$$\mathrm{E}\left[\frac{y}{2wL(v)}\right] = \mathrm{E}\left[\frac{\sum_{j=1}^{\tau} z_j}{2wL(v)} \,\middle|\, v\right] = \frac{\sum_{j=1}^{\tau} g(v - x_j)}{2wL(v)}$$

Unconditionally under the design, the expected value is

$$\mathrm{E}(\hat{D}) = \frac{1}{2w} \sum_{j=1}^{\tau} \int_{-\infty}^{\infty} \frac{g(v - x_j)}{L(v)} \left[\frac{L(v)}{A}\right] dv = \frac{\tau}{A} = D$$

Thus, $\hat{D}$ is unbiased for $D$, assuming that $w$ or $f(0)$ is known. When an estimate of $w$ or $f(0) = 1/w$ is substituted, the unbiasedness is approximate.

Denote by $\hat{D}_i$ the estimator above for the $i$th transect in the sample. Each of the $n$ estimators is unbiased for $D$, so that their average

$$\hat{D}_p = \frac{1}{n}\sum_{i=1}^{n}\hat{D}_i$$

is unbiased for $D$.

Since the $n$ starting locations were selected independently and detections are independent, the $\hat{D}_i$ are independent and identically distributed random variables. An unbiased estimator of the variance of their sample mean $\hat{D}_p$ is therefore

$$\widehat{\operatorname{var}}(\hat{D}_p) = \frac{1}{n(n-1)}\sum_{i=1}^{n}(y_i - \hat{D})^2$$

When $w$ or $f(0) = 1/w$ are estimated, the unbiasedness of the estimator is only approximate and the estimator of variance will be unbiased only with individual, independent estimates $\hat{w}_i$ or $\hat{f}_i(0)$ for each transect. The methodology for estimating $w$ or $f(0)$ and obtaining variance estimates with pooled estimators under sampling with probability proportional to length is not yet well developed.

## 17.10   NOTE ON ESTIMATION OF VARIANCE FOR THE KERNEL METHOD

If the observed distances were independent, as well as identically distributed, one could obtain an estimate of the variance of $\hat{D}$ from the kernel estimator, since the terms $K(x_j/h)$ would be independent and identically distributed. It would then be straightforward to estimate the variance of $\hat{D}$ or $\hat{f}(0)$ using the sample variance of the $K(x_j/h)$. However, the observed distances, although identically distributed due to random location of the transect, are not independent without additional assumptions about the population itself. Independence of the observed distance requires the assumption that the spatial distribution of the animals in the population is random. Any tendency of the animals to aggregate, to defend territories, or to be affected by a patchy environment will lead to correlations between distances to animals observed from the same transect. With extremely patchy populations, one may, for example, detect animals mostly at short distances from one transect and at long distances from another. These correlations in observed distances will occur even though detections may be conditionally independent—that is, given the locations of two animals, detection of one from a given transect does not affect detection of the other.

If $n$ transects are selected by random sampling, then if $f(0)$ is known, an unbiased estimator of the population density $D$ is

$$\hat{D} = \frac{1}{n}\sum_{i=1}^{n}\hat{D}_i = \frac{\bar{y}f(0)}{2E(L)}$$

where $\hat{D}_i = y_i f(0)/2E(L)$, $y_i$ is the number of animals detected from transect $i$ and $E(L)$ is the expected value of transect length (see Section 17.11).

An unbiased estimator of the variance of $\hat{D}$, assuming independence, is

$$\widehat{\text{var}}(\hat{D}) = \frac{1}{n(n-1)} \sum_{i=1}^{n} (\hat{D}_i - \hat{D})^2$$

If it is unrealistic to assume that the animals in the population are independently located, however, we must look to the sampling design for help in estimating variance.

If $f(0)$ is to be estimated by the kernel method, one can either determine a window width $h_i$ separately for each transect or determine a single window width $h$ from all of the distance data. In the first case, let

$$\hat{D}_i = \frac{1}{h_i E(L)} \sum_{j=1}^{n} K\left(\frac{x_j}{h_i}\right)$$

and let

$$\hat{D} = \frac{1}{n} \sum_{i=1}^{n} \hat{D}_i$$

Since density estimators in general are not unbiased, $\hat{D}$ is not unbiased for $D$, although it is approximately unbiased. However, the variance estimator

$$\widehat{\text{var}}(\hat{D}) = \frac{1}{n(n-1)} \sum_{i=1}^{n} (\hat{D}_i - \hat{D})^2$$

is unbiased for the variance of $\hat{D}$, because the $\hat{D}_i$ are independent and identically distributed due to the random selection on the $n$ transect locations.

If the $n$ transects have similar detectability conditions, a better estimate of $f(0)$ may be obtained by combining all the distance data from the survey. Let $h$ be the window width used. The estimator is

$$\hat{D} = \frac{\bar{y}\hat{f}(0)}{2E(L)} = \frac{1}{nhE(L)} \sum_{i=1}^{n} \sum_{j=1}^{y_i} K\left(\frac{x_{ij}}{h}\right)$$

where $x_{ij}$ is the distance to the $i$th transect line of the $j$th animal detected from that transect. Define

$$\hat{D}_i = \frac{1}{hE(L)} \sum_{j=1}^{y_i} K\left(\frac{x_{ij}}{h}\right)$$

Then $\hat{D} = (1/n)\sum_{i=1}^{n}\hat{D}_i$ and the variance estimator

$$\widehat{\mathrm{var}}(\hat{D}) = \frac{1}{n(n-1)}\sum_{i=1}^{n}(\hat{D}_i - \hat{D})^2$$

is unbiased for the variance of $\hat{D}$ if the window width $h$ is fixed—not determined from the data. Some bias is introduced into the variance estimator when $h$ is determined from the data. This bias would tend to be small if $h$ is determined from a great many observations and hence has a small variance. This bias could be reduced by using a bootstrap or jackknife estimate of variance.

## 17.11  SOME UNDERLYING IDEAS ABOUT LINE TRANSECTS

### Line Transects and Detectability Functions

Letting $x$ denote a location in the study area, the detectability function $g(x)$ gives the probability that an animal at location $x$ is detected by the observer. Figure 17.8 shows a line transect detectability function with exponential profile. Figure 17.9 shows a line transect detectability function with half-normal profile. In each case, the detectability function has a ridge, usually assumed equal to 1 in height, directly over the transect line and decreases with distance away from the line. When the detectability curve is symmetric about the line and constant along the line, as is usually assumed with line transects, it is sufficient to consider only the profile, as in the curves of Figures 17.3 and 17.4. However, many of the basic results on observations made with detectability functions hold no matter what the shape of the detectability curve (see, e.g., Sections 17.14 and 17.15 and Chapter 22).

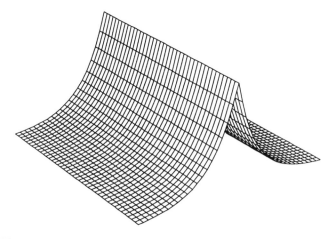

**Figure 17.8.** Detectability function of a line transect with exponential profile.

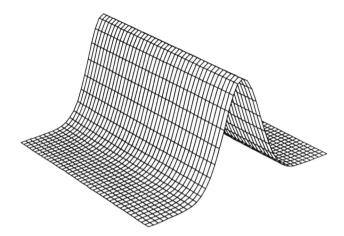

**Figure 17.9.** Detectability function of a line transect with half-normal profile.

With transects selected perpendicular to a baseline, it will be sufficient that $x$ denote the projection of the animal's location onto the baseline. Note that $g(x)$ is a conditional probability of detection, given that an animal is at location $x$, and says nothing about the probability that an animal is at that location in the first place. In practice, it is common to let $x$ represent the perpendicular distance from the transect line rather than a coordinate-point location in the study area. The distance measured from a transect can be signed as positive or negative, depending on which side of the transect the animal is on. However, because the detectability function is usually assumed to be symmetric, it will subsequently be convenient to work with the absolute or unsigned distances.

The population total $\tau$ is the number of animals in the study region. The locations of the $\tau$ individuals in the study region will be denoted $x_1, x_2, \ldots, x_\tau$. To avoid any assumptions about a process or model giving rise to the population, the $\tau$ locations will be viewed as given and fixed. This viewpoint would be equivalent to the fixed-population approach of classical sampling if the observations were made through sample plots, in which every animal or plant could be counted (perfect detectability), rather than transects with associated detectability functions. Consideration of models giving rise to the spatial patterns of the population is postponed until later.

Detections are assumed to be conditionally independent—detection of one animal not affecting the probability of detection for any other.

### Single Transect

Consider a single line transect laid out in a region $A$, which is small enough so that the probability of detection $g(x)$ from the transect is positive everywhere in $A$. The position of the transect is taken as given, so that no sample selection probabilities are involved. Let $g(x)$ be the detectability function associated with the transect. In concept, $g(x)$ may be viewed as a surface over the entire two-dimensional region, of height everywhere between 0 and 1. In many specific line transect applications, however, $g(x)$ is viewed as a function of distance only, so that $x$ represents perpendicular distance from the line rather than location. Typically, $g(x) = 1$ directly over the transect line—corresponding to perfect detectability for any animal on the line—with $g(x)$ decreasing symmetrically toward zero to either side of the line. It will be useful to proceed as if the detectability function were known before examining the methods of estimating detectability. The actual number of animals in the region $A$ is $\tau$, while the number detected from the transect is denoted $y$.

## Average Detectability

If the transect is selected by a probability sampling design, an animal at location $x$ has an average detection probability under the design. Let $\rho(x)$ denote the average detectability under the design at location $x$. The sample is determined by selecting a location $v$ along the baseline and running the transect perpendicular to the baseline at $v$. The detectability at $x$ given the transect is located at $v$ is denoted $g_v(x)$, so that $\rho(x) = E[g_v(x)]$. Given the $\tau$ animals in the population at locations $x_1, \ldots, x_\tau$ and given the transect location $v$, the conditional expectation of the number $y$ of animals detected is

$$E(y \mid s) = E\left( \sum_{j=1}^{\tau} z_j \mid v \right) = \sum_{j=1}^{\tau} g_v(x_j)$$

The unconditional expectation is

$$E(y) = \sum_{j=1}^{\tau} \rho(x_j)$$

If the transect could be selected in such a way that average detectability $\rho(x)$ was constant everywhere in the study area, then $E(y) = \tau \rho$ and an unbiased estimator of $\tau$ would be

$$\hat{\tau} = \frac{y}{\rho}$$

where $\rho$ denotes the constant average detectability.

## Random Transect

Many line transect methods in the literature are justified by reference to a transect "selected at random," so an expression for $\hat{\tau}$ of the form above will be unbiased. The interest in random selection is motivated by the desire to obtain unbiasedness without having to appeal to assumptions about the distribution of the population itself—without having to assume, for example, that the animals are uniformly distributed in the study region or that their locations are a realization of a stationary stochastic process. Such assumptions are deemed unrealistic for many natural populations.

The essence of a *randomly selected transect* is the attainment of constant average detectability over the study region. This goal is not easy to obtain exactly for line transect sampling, because of problems with transects located near the boundary of the study region. Suppose that the study region is a rectangle of width $W$ and length $L$, and that the starting location $v$ for a transect of length $L$ is selected from a uniform distribution on the interval $[0, W]$. The transect line is then positioned perpendicular to the base of the rectangle. For the $j$th animal in the population, let $x_j$ denote its projected position along the base of the rectangle, so that $x_j$ is a value between 0 and $W$. The distance from the animal to the transect is $|v - x_j|$ and the detectability is given by $g(v - x_j)$, where $g$ is a symmetric function about zero.

With such a design, since $v$ has uniform probability density on the interval $[0, W]$, the average detectability for an animal at location $x_j$ is

$$\rho(x_j) = \frac{1}{W} \int_0^W g(v - x_j) \, dv$$

But the integral above depends on the location $x_j$ of the animal. Suppose, for example, that there is some maximum distance $w_{max}$ beyond which no detection can occur. Then $g(x) = 0$ for all $x > w_{max}$, so that no animals farther than distance $w_{max}$ from the transect are detected. Then for all locations at least distance $w_{max}$ from either side boundary of the study region, $\rho(x)$ is a constant—call it $\rho$. For an animal on the side boundary, $\rho(x_j) = (1/2)\rho$. For animals within distance $w_{max}$ of the side boundary, the average detectability will be something intermediate between $(1/2)\rho$ and $\rho$.

Thus, the expected number of animals seen is

$$E(y) = \sum_{j=1}^{\tau} \rho(x_j) \leq \tau\rho$$

so that, with $\hat{\tau} = y/\rho$,

$$E(\hat{\tau}) \leq \tau$$

and the estimator would tend to underestimate the true number in the population.

Fortunately, in real situations, the boundary region may be very small relative to the whole study region, so that the induced bias would be small. One could in principle eliminate the bias by "wrapping around" the study rectangle, connecting the two sides. To apply that in practice, however, would require that whenever a transect was selected at distance $v < w$ from one boundary, the observers would run an additional transect at distance $v$ from the other boundary—outside the study region! The plan is not likely to become popular in practice.

## Average Detectability and Effective Area

Because of the random selection of transects, each transect location has a uniform distribution on the interval $[0, B]$, and the average detectability $\rho$ at any location $x$ in the study region is

$$\rho = \frac{1}{B} \int_{-w_{max}}^{w_{max}} g(v) \, dv = \frac{2w}{B}$$

where $w_{max}$ is a distance beyond which detectability is zero (and with any minor differences for points near the boundary avoided by the wraparound concept), and

$$w = \int_0^{w_{max}} g(v) \, dv$$

represents the area under the detectability curve.

Since detectability $g(v)$ is zero beyond distance $w_{max}$ from the line, the integral $w$ can be written

$$w = \int_0^{\infty} g(v) \, dv$$

The integral $w$ is referred to as the *effective half-width* of the transect, because of the interpretation that the same number of animals would be expected to be seen in a strip plot of half-width $w$—that is, with perfect detectability to distance $w$ to either side of the line and zero detectability beyond that. The *effective area* covered by the transect is thus $2Lw$.

The estimator

$$\hat{\tau}_i = \frac{y_i}{\rho} = \frac{y_i B}{2w}$$

is unbiased for the population total $\tau$ by the result in the section above on a single random transect. Hence an unbiased estimator based on all $n$ transects is

$$\hat{\tau} = \frac{1}{n} \sum_{i=1}^{n} \hat{\tau}_i = \frac{\bar{y}}{\rho} = \frac{\bar{y} B}{2w}$$

where $\bar{y} = (1/n) \sum_{i=1}^{n} y_1$ is the sample mean of the numbers observed. Also, the estimator $\hat{D} = \hat{\tau}/A$ is unbiased for population density $D = \tau/A$.

Because of the independent selection of the $n$ starting locations and the independence of detections, the $n$ estimators $\hat{\tau}_1, \ldots, \hat{\tau}_n$ are independent and identically distributed. Hence, an unbiased estimator of the variance of their sample mean $\hat{\tau}$ is

$$\widehat{\mathrm{var}}(\hat{\tau}) = \frac{1}{n(n-1)} \sum_{i=1}^{n} (\hat{\tau}_i - \hat{\tau})^2$$

### Effect of Estimating Detectability

The results above hold when the effective half-width $w$ is known. In real situations, a major endeavor in line transect methodology is the estimation of $w$ or of $1/w$ or some other aspect of detectability. When $w$ or its reciprocal is estimated, the unbiasedness of the results above holds only approximately, and an additional term— based on Taylor's approximation—enters the variance expression.

Suppose that either $w$ is replaced by an asymptotically unbiased estimator $\hat{w}$, or $1/w$ is replaced by an unbiased estimator $\widehat{1/w}$ and that $\bar{y}$ is uncorrelated with $\hat{w}$ or $\widehat{1/w}$. Then $\hat{D}_1 = (B/2A)(\bar{y}/\bar{w})$ contains the ratio of two random variables, while $\hat{D}_2 = (B/2A)[\bar{y}(\widehat{1/w})]$ contains the product of two random variables.

Using Taylor's theorem, $\hat{D}_1$ and $\hat{D}_2$ are approximately unbiased for $D$ and approximate variance formulas are

$$\mathrm{var}(\hat{D}_1) \approx \frac{B^2}{(2Aw)^2} \mathrm{var}(\bar{y}) + \frac{D^2}{w^2} \mathrm{var}(\bar{w})$$

$$\mathrm{var}(\hat{D}_2) \approx \frac{B^2}{(2Aw)^2} \mathrm{var}(\bar{y}) + w^2 D^2 \mathrm{var}(\widehat{1/w})$$

Equivalent expressions are given in Burnham et al. (1980), Seber (1982), and elsewhere. For each of the variances, the first term is $\mathrm{var}(B\bar{y}/2Aw)$, the variance that would apply if $w$ were known. Thus, estimation of the effective half-width or equivalent aspect of detectability results in an additional component of variance.

### Probability Density Function of an Observed Distance

When the data from a line transect are examined, typically there are more observations of animals a short distance from the line than far from it. When the transect lines are located at random, one can assume that this distribution of sighting distances is due to the detectability function decreasing as a function of distance from the line, rather than to any peculiarity in the distribution of animals in the study area. In this section, the probability density function of the distance from the line of an observed animal will be derived based on the random selection of transect lines, without any assumptions about the distribution of animals in the study area. The locations of animals in the study area will be taken as given and fixed during the time of the survey.

The main results of this section have been obtained in Burnham and Anderson (1976), Ramsey (1979), and Seber (1973), under the assumption that the animals are randomly distributed in the study area—that is, the locations of animals are independent and the expected number of animals per unit area is constant throughout the study region. Burnham and Anderson and Ramsey assume, in addition, random selection of a transect. Since the distributions of many animal populations are anything but random, we will determine in this section which of the results hold based on the design alone.

Because of the random location of the transect perpendicular to a baseline of length $B$, the probability density function of the distance of the $j$th animal from the transect is the uniform density $1/B$. The conditional probability of detecting the $j$th animal, given that it is distance $d$ from the line, is $g(d)$. The probability that the distance of the $j$th animal in the study region from the line is in the interval $(d, d + \Delta d)$, where $\Delta d$ is a small positive number, is $\Delta d/B$. The unconditional probability that the $j$th animal in the study area is detected is the average detection probability

$$\rho = \frac{1}{B} \int g(x)\, dx = \frac{2w}{B}$$

where $w$ is the effective half-width of the transect.

The probability that the distance of the $j$th animal from the line is in the interval $(d, d + \Delta d)$ *and* the animal is detected is $g(d)\Delta d/B$. Thus, the conditional probability that the distance is in the interval $(d, d + \Delta d)$ given that the animal is detected is $g(d)\Delta d/B\rho = g(d)\Delta d/2w$. The probability density function for the signed distance $x$ from the $j$th animal to the line, given that the animal is detected, is $g(x)/2w$, so that the probability density function for the absolute distance is

$$f(x) = \frac{g(x)}{w}$$

This distribution is the same for each of the $\tau$ animals in the population, due to the random selection of the transect. However, the distance from animal $j$ to the transect is not independent of the distance from animal $k$ to the transect, the joint probabilities depending on the relative locations of the two animals.

Thus, the absolute distances $x_1, \ldots, x_y$ of the $y$ animals from the transect are identically distributed with PDF $f$, but are *not* independent without additional assumptions about a model giving rise to the distribution of animals in the population.

The expected number of animals seen from the randomly selected transect line is

$$E(y) = E\left[E\left(\sum_{j=1}^{\tau} z_j \mid v\right)\right] = \sum_{j=1}^{\tau} E[g(v - x_j)] = \tau\rho = \tau\frac{2w}{B}$$

where $z_j$ is 1 if the $j$th animal is detected and zero otherwise, and $x_j$ is the projected location of animal $j$ along the baseline.

When the assumption $g(0) = 1$ of perfect detectability along the transect line is made, then

$$f(0) = \frac{1}{w}$$

Under this assumption,

$$E(y) = \frac{2\tau}{Bf(0)}$$

Thus, if the density $f$ of the observed distances is known for zero distance, an unbiased estimator of the number $\tau$ of animals in the study area is

$$\hat{\tau} = \frac{yf(0)B}{2}$$

An unbiased estimator of population density $D$ is

$$\hat{D} = yf(0)\frac{B}{2A}$$

where $A$ is in the area of the study region.

Suppose that the length of a transect located perpendicular to point $v$ along the baseline is $L(v)$. If the transect traverses the entire study region, its length will vary if the region is irregularly shaped. If the study region is wider than the length of a transect, the baseline can be continued in parallel segments so that every point in the study region is on some potentially selected transect. The area of the study region may be written

$$A = \int_0^B L(v)\,dv$$

The ratio $A/B$ is the expected length $E(L)$ of a transect:

$$E(L) = \int_0^B L(v)\frac{1}{B}\,dv = \frac{A}{B}$$

Thus, an unbiased estimator of density $D$, if $f(0)$ is known, is

$$\hat{D} = \frac{yf(0)}{2E(L)}$$

If transect length is a constant $L$, the estimator is

$$\hat{D} = \frac{yf(0)}{2L}$$

## 17.12   DETECTABILITY IMPERFECT ON THE LINE OR DEPENDENT ON SIZE

Estimators of the form $\hat{D} = yf(0)/2L$ are appropriate when objects on the line are sure to be detected, that is, when $g(0) = 1$. When $g(0) < 1$, the estimator should have the form $\hat{D} = yf(0)/2g(0)L$, with estimates inserted for $f(0)$ and $g(0)$ or their ratio. More generally, if $g(x)$ is known for some distance $x$, not necessarily zero, an estimate is given by $\hat{D} = yf(x)/2g(x)L$. Line transects in which $g(0)$ is not 1 are considered in Pollock and Kendall (1987), Quang and Lanctot (1991), Schweder (1989), Schweder et al. (1991), and Zahl (1989).

   For animals that occur in groups, the probability of detecting the group may be dependent on group size. If any individual of such a group is detected, the number of individuals in the group may be counted. For such "size-based" situations, the groups may be considered the objects detected. Methods of estimation for size-based line transect studies are given in Drummer and McDonald (1987), Otto and Pollock (1990), Quang (1991), and Quinn (1981).

## 17.13   ESTIMATION USING INDIVIDUAL DETECTABILITIES

The following result gives a way to obtain an unbiased estimate of the total number $\tau$ of animals in $A$: Let the estimator $\hat{\tau}$ be given by

$$\hat{\tau} = \sum_{j=1}^{y} \frac{1}{g(x_j)}$$

where $x_j$ is the location of the $j$th animal observed and $g(x_j)$ is the probability of detection at that location. If $g(x) > 0$ for all locations $x$ in $A$, then $\hat{\tau}$ is an unbiased estimator of $\tau$, with variance

$$\text{var}(\hat{\tau}) = \sum_{j=1}^{\tau} \frac{1 - g(x_j)}{g(x_j)}$$

An unbiased estimator of this variance is

$$\widehat{\text{var}}(\hat{\tau}) = \sum_{j=1}^{\tau} \frac{1 - g(x_j)}{g^2(x_j)}$$

*Derivation of Result Above.*    For the $j$th individual in the population, with location $x_j$, define the random variable $z_j$ to be 1 if that individual is detected and 0 otherwise. The $z_j$ are independent—but not identically distributed—Bernoulli random variables, with $E(z_j) = g(x_j)$ and $var(z_j) = g(x_j)[1 - g(x_j)]$. The estimator $\hat{\tau}$ can alternatively be written

$$\hat{\tau} = \sum_{j=1}^{\tau} \frac{z_j}{g(x_j)}$$

Hence the expected value is

$$E(\hat{\tau}) = \sum_{j=1}^{\tau} \frac{E(z_j)}{g(x_j)} = \sum_{j=1}^{\tau} \frac{g(x_j)}{g(x_j)} = \tau$$

so $\hat{\tau}$ is unbiased for $\tau$.

Because of the independence of the Bernoulli trials, the variance of $\hat{\tau}$ is

$$var(\hat{\tau}) = var\left[\sum_{j=1}^{\tau} \frac{z_j}{g(x_j)}\right] = \sum_{j=1}^{\tau} \frac{g(x_j)[1 - g(x_j)]}{g^2(x_j)} = \sum_{j=1}^{\tau} \frac{1 - g(x_j)}{g(x_j)}$$

The variance estimator can be written alternatively as

$$\widehat{var}(\hat{\tau}) = \sum_{j=1}^{\tau} \frac{1 - g(x_j)}{g^2(x_j)} z_j$$

and its unbiasedness for $var(\hat{\tau})$ follows since $E(z_j) = g(x_j)$, so that the derivation is complete.                                                                         □

### Estimation of Individual Detectabilities

To use the estimator above, one would have to know or to estimate the detection probability $g(x_j)$ for each individual detected. Further, it is unlikely that the study region is so small that all detection probabilities are greater than zero for a given transect. However, a probability selection of the transect location can ensure that all points in the study region have positive detection probability.

Hayne (1949) proposed an estimator of the above form for use in line transect surveys in which each animal has a "flushing radius." If the observer comes within the flushing distance $r_j$ of the $j$th animal in the population, the animal will flush and hence be detected. Note that $r_j$ is the radial distance from the observer to the animal when detected. Suppose that a transect is selected at random (i.e., a point is selected with uniform probability along the width $W$ of a rectangular study area), and the transect is located perpendicular to the baseline and through the point. Assume that the transect runs the whole length $L$ of the study area. If the transect intersects a circle of radius $r_j$ centered on the location $x_j$ of the $j$th animal, the $j$th animal will flush and be detected as the observer walks the transect. The probability that the transect intersects the circle is $g(x_j) = 2r_j/W$, provided that the animal's location $x_j$ is not too close to the side boundary of the study region. The *Hayne estimator* of the total is

$$\hat{\tau}_H = \frac{W}{2} \sum_{j=1}^{y} \frac{1}{r_j}$$

Since the study area of the study region is $LW$, the corresponding estimator of density is

$$\hat{D}_H = \frac{1}{2L} \sum_{j=1}^{y} \frac{1}{r_j}$$

Studies have shown the Hayne estimator to be sensitive to errors in measuring the distances $r_j$ and to departures from the assumption of circular flushing region and to have larger variance than other estimators (see Gates 1979, p. 111); Seber 1982, pp. 39–40). Burnham (1979) generalized the Hayne method to allow for elliptic flushing regions. The methods based on flushing distances are more closely related to line intercept methods than to most other line transect methods. Most estimators used in line transect methods do not, in fact, endeavor to divide by individual detection probabilities but use, instead, average detection probabilities or the equivalent quantities effective areas observed or detection density on the line.

## 17.14 DETECTABILITY FUNCTIONS OTHER THAN LINE TRANSECTS

Detectability functions can characterize many observational methods used in surveys of populations of elusive objects. As such, detectability functions are a generalization of the units of classical survey sampling. For a survey in which the study region is divided into plots and every object is observed within a selected plot, a plot is characterized by a detectability function that equals 1 over the plot and zero elsewhere. A detectability function that equals some constant $p$ (less than 1) over a plot and zero elsewhere characterizes an observational unit in an aerial survey in which detectability can be assumed constant over a selected plot or a trawl fishery survey in which fish in the path of the net are caught with probability $p$. In line transect surveys, the detectability function is usually assumed laterally symmetric and decreasing with distance from the transect line. With variable circular plots, the detectability function is usually assumed radially symmetric and decreasing with distance from the center.

Associated with a selected site or observational unit is a detectability function $g$ giving the probability of detection for any object (animal) in the study region as a function of the location of the object relative to the site. Let $\mathbf{x} = \{x_1, x_2\}$ denote a rectangular coordinate location relative to the site. The effective area observed from the site is

$$a = \int \int g(x_1, x_2) \, dx_1 \, dx_2$$

For a plot, the effective area is the actual area of the plot. For a line transect, the effective area is the volume under the detectability function, as in Figure 17.8 or 17.9.

A site may be randomly selected by selecting a location from a uniform distribution over the study region of area $A$. For simplicity, consider a rectangular study region. To avoid the issue of unequal average detectability for objects near the boundary of the study region, we assume that for a site near the boundary the detectability function can be continued at the opposite side of the study region. Conceptually, the study area is folded into a torus (doughnut) by joining the right to the left side and then the top to the bottom. To accomplish this amazing feat in reality would require observations from sites outside the opposite edges of the study region, from which only detections within the study region are recorded. In practice, if the study area is large relative to the maximum detection distance, the boundary issues will be unimportant. Random selection of a site gives the same average detectability $\rho = a/A$ to every point in the study region.

With a randomly selected site, the expected value of the number $y$ of animals detected is $E(y) = \tau \rho = aD$. Thus, an unbiased estimator of the total number $\tau$ of animals in the study region is

$$\hat{\tau} = \frac{y}{\rho}$$

and an unbiased estimator of animal density is

$$\hat{D} = \frac{y}{a}$$

With a random sample of $n$ sites, with a detectability function of the same shape centered at each, the estimator is $\hat{D} = \bar{y}/a$, where $\bar{y}$ is the sample mean of the numbers of animals observed.

An estimator of the form $\hat{D} = \bar{y}/a$ is appropriate no matter what the shape of the detectability function, provided that a sampling design is used which gives every location in the study region equal average detectability. The estimator is design-unbiased under such a design; no assumptions are made about the spatial distribution of the population. When the effective area $a$ cannot be determined with certainty, it or its reciprocal must be estimated. With $a$ or $1/a$ replaced by a consistent estimator, the design unbiasedness of $\hat{D}$ is approximate only and an additional component of variance, associated with the estimation of $a$ or $1/a$, is introduced.

In Chapter 22 the unified framework of detectability functions as observational units will be used to compare mean square errors of different observational methods—each with the same effective area and including plots of different shapes, line transects of different outlines, variable circular plots, and plots of constant detectability—under an assumed model.

## 17.15   VARIABLE CIRCULAR PLOTS OR POINT TRANSECTS

In many surveys of birds and other hard-to-detect animals, a sample of sites is selected in the study region, and at each site selected, the observer spends a specified amount of time and records every individual of the species detected. If distance or location relative to the observer is recorded for each detection, methods similar to those for line transects can be used to estimate population density. Because the probability of detection typically depends on distance from the point at which the observer is stationed, the observational method is referred to as *variable circular plots* or *point transects*. Discussions of statistical aspects of the method are found in Buckland (1987), Burnham et al. (1980, p. 195), Quang (1992), Ramsey and Scott (1979), and Ramsey et al. (1987, 1988).

In variable circular plot surveys, it is usually assumed that the detectability function is radially symmetric, so that the probability of detection depends only on the distance from the observer and not on the direction. An example of such a function, with a half-normal profile, is shown in Figure 17.10.

With a radially symmetric detectability function, it is convenient to change to polar coordinates. Let $r$ be the radial distance from the observer and $\theta$ the angle to the object detected. The rectangular coordinates $w_1$ and $w_2$ may be obtained

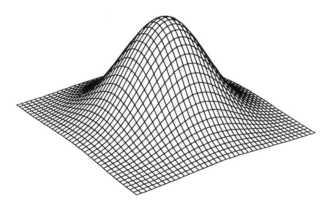

**Figure 17.10.** Detectability function of a variable circular plot with half-normal shape.

as $w_1 = r\cos(\theta)$ and $w_2 = r\sin(\theta)$. The effective area is

$$a = \int\int g(x_1, x_2)\, dx_1\, dx_2$$
$$= \int\int rg[r\cos(\theta), r\sin(\theta)]\, dr\, d\theta$$

Since $g$ does not depend on $\theta$ and $\int d\theta = 2\pi$, the effective area may be written

$$a = 2\pi \int_0^\infty rg_r(r)\, dr$$

where $g_r(r)$ is the probability of detection for an object at distance $r$ from the observer.

With random sampling, the probability density function of observed distance given detection of an individual is

$$f(r) = \frac{2\pi r g_r(r)}{a}$$

so that

$$\frac{f(r)}{r} = \frac{2\pi g_r(r)}{a}$$

With the assumption $g_r(0) = 1$,

$$\lim_{r \to 0} \frac{f(r)}{r} = \frac{2\pi}{a}$$

Since $f(0) = 0$, the limit of $f(r)/r$ is the derivative at zero $f'(0)$. Thus, estimation of the slope at zero of the density of observed distances is equivalent to estimating the reciprocal of effective area, and an estimate of density is

$$\hat{D} = \frac{\bar{y}\hat{f}'(0)}{2\pi}$$

Quang (1992) uses kernel methods to estimate $f'(0)$. Burnham et al. (1980), Ramsey and Scott (1979), and Ramsey et al. (1987) point out that when each observed radial distance $r$ is transformed to the area variable $z - \pi r^2$, the estimator is $\hat{D} = \bar{y}\hat{f}_z(0)$, where $f_z(z)$ is the density function of the observed areas. The problem is then identical to that of line transects, in which the object is to estimate density at zero. However, Buckland (1987) warns that some of the usual line transect methods do not work well with the area data; as one alternative, Buckland introduces binomial methods in which every detection distance is classified into one of two distance categories.

## EXERCISE

1. On a line transect survey, the observer recorded the following distances (in meters) from detected birds to the transect line: 0, 1, 1, 2, 3, 3, 4, 6, 7, 7, 9, 12, 17. The transect line is 50 meters long. Estimate the density of the bird population by three of the following methods. List key assumptions that go with each method you use.

   (a) Narrow strip
   (b) Smooth-by-eye
   (c) Exponential detectability
   (d) Half-normal detectability
   (e) Normal kernel
   (f) Fourier series

# CHAPTER 18

# Capture–Recapture Sampling

In capture–recapture sampling to estimate the total number of individuals in a population, an initial sample is obtained and the individuals in that sample are marked or otherwise identified. A second sample is obtained independently, and it is noted how many of the individuals in that sample are marked. If the second sample is representative of the population as a whole, the sample proportion of marked individuals should be about the same as the population proportion of marked individuals. From this relationship, the total number of individuals in the population can be estimated.

Capture–recapture methods have been used to estimate the abundance of animal populations, including bird, mammal, fish, reptile, insect, and other species, to estimate the detectability of animals for other survey methods, and to estimate survival and other population parameters. Capture–recapture methods have also been used to estimate the abundance of elusive human populations such as the homeless, to adjust for census undercounts of minority groups, and to estimate the number of vital events such as accidents in a population.

The animals or other individuals need not literally be captured or marked or recaptured. If it is possible to identify individual animals by natural markings, then two independent sighting surveys may be carried out, and the number of individuals sighted in both surveys is the number of "recaptures." Similarly, if a number of animals in a population have been fitted with radio transmitters and hence have known locations, then in a survey in which observers detect animals by some means independently of the transmitters, the number of transmitter-fitted animals detected is the number of recaptures. For other species, however, it may be necessary to capture the animals by such means as traps or nets, and to mark them with bands, tags, coded wire implants, paint, or streamers.

For human populations, the two samples often consist of two lists. For instance, the first list may be from the census data and the second list may be data from a follow-up survey. Or the first list may be health department records of accidents, and the second list may be insurance company records.

In more complex capture–recapture animal studies, animals may be captured and released on several different occasions, with the capture history of any animal in the sample identifiable from the previous marks. Complicating factors include capture probabilities that vary from animal to animal or from sample to sample, mortality caused by tagging, mortality between sample times, births, immigration and emigration from the study area, and animals becoming "trap happy" or "trap shy" through the handling procedure.

Reviews and basic references on capture–recapture methods include Cormack (1979), Otis et al. (1978), Pollock (1981, 1991), Pollock et al. (1990), and Seber (1973, 1982, 1986, 1992). Seber (1973) classified the methods of mark–recapture according to whether the population was closed--no change in the population during the period of the study--or open--allowing for births, deaths, immigration, and emigration--and according to whether there is a single mark–release or several mark–releases. Recent articles on the use of capture–recapture methods for elusive human populations include Cowan and Malec (1986), Freedman (1991), Sudman et al. (1988), and Wolter (1986, 1991).

In the following summary of simple capture–recapture methods, a notation is used to facilitate consideration of sampling design aspects, while keeping subscripts to a minimum. The total number of individuals in the population is denoted $\tau$, while the total number of marked individuals in the population, which is also the number of individuals in the initial sample, is $X$. The number of individuals in the second sample is $y$, of which $x$ are marked. A sample of size $n$ refers to a selection of $n$ units--whether plots, strips, trawl paths, trap attraction regions, line transects, variable circular plots, or other detectability units--by which individuals in the population are observed.

## 18.1  SINGLE RECAPTURE

In a simple capture–recapture survey of the animal population, an initial sample of $X$ animals is captured, marked, and released back into the population. A second sample, of $y$ animals, is then captured independently, of which some number $x$ are observed to be marked. If the second sample is representative of the population as a whole, the proportion of marked animals in the sample will be about the same as the proportion of the whole population in the sample. The total number $\tau$ of animals in the population may then be estimated by assuming that the proportion of marked animals in the second sample is representative of the proportion of marked animals in the population, that is, by setting

$$\frac{x}{y} = \frac{X}{\tau} \tag{1}$$

and solving for the unknown population size $\tau$. Equivalently, the proportion of the marked animals in the population that is captured in the second sample should approximately equal the proportion of the population as a whole captured in the second sample, that is,

$$\frac{x}{X} = \frac{y}{\tau} \tag{2}$$

Solving either equation for the unknown population total $\tau$ gives the Petersen estimator

$$\hat{\tau} = \frac{y}{x}X \tag{3}$$

An estimator of the variance of $\hat{\tau}$ (Sekar and Deming 1949) is

$$\widehat{\text{var}}(\hat{\tau}) = \frac{Xy(X-x)(y-x)}{x^3} \tag{4}$$

A simple, approximate $100(1-\alpha)\%$ confidence interval is the standard

$$\hat{\tau} \pm z\sqrt{\widehat{\text{var}}(\hat{\tau})}$$

where $z$ is the upper $\alpha/2$ point of the normal distribution.

Because the number $x$ of marked animals in the second sample may be zero, the estimator $\hat{\tau}$ does not have a finite variance. Therefore, the following modified estimator $\tilde{\tau}$ was proposed by Chapman (cf. Seber 1982, p. 60):

$$\tilde{\tau} = \frac{(X+1)(y+1)}{x+1} - 1 \tag{5}$$

An approximately unbiased estimator of the variance of this modified estimator (Seber 1970, 1982, p. 60) is

$$\widehat{\text{var}}(\tilde{\tau}) = \frac{(X+1)(y+1)(X-x)(y-x)}{(x+1)^2(x+2)} \tag{6}$$

A simple, approximate $100(1-\alpha)$ confidence interval for the Chapman estimator is

$$\tilde{\tau} \pm z\sqrt{\widehat{\text{var}}(\tilde{\tau})} \tag{7}$$

where $z$ is the upper $\alpha/2$ point of the normal distribution.

Alternative confidence interval procedures proposed for these estimators include exact methods under an assumed distribution, Monte Carlo or bootstrap intervals (Buckland 1980, 1984), intervals based on the approximate distribution of the reciprocal or other transformation of the estimator (Jensen 1989; Manly 1984; Seber 1973), and likelihood-based methods (McDonald and Palanacki 1989; see also Otis et al. 1978, and reviews in Seber 1982, 1986, 1992). Large sample properties of estimators in capture–recapture are reviewed in Sen (1988).

*Example 1: Single Recapture.* In a field study $X = 300$ mice are caught in traps, tagged, and released. A few days later the researchers return to the study area and independently capture $y = 200$ mice, of which they find that $x = 50$ have tags.

The Petersen estimate of the number of mice in the population [from Equation (3)] is

$$\hat{\tau} = \frac{200}{50}(300) = 1200$$

The estimated variance for the Petersen [from Equation (4)] is

$$\widehat{\mathrm{var}}(\hat{\tau}) = \frac{300(200)(300-50)(200-50)}{50^3} = 18,000 \qquad (8)$$

giving a standard error of $\sqrt{18,000} = 134.2$. An approximate 95% confidence interval is $1200 \pm 1.96(134.2) = 1200 \pm 263 = (937, 1463)$.

The Chapman estimate [from Equation (5)] is

$$\tilde{\tau} = \frac{301(201)}{51} - 1 = 1185.3 \approx 1185$$

The estimated variance [from Equation (6)] is

$$\widehat{\mathrm{var}}(\tilde{\tau}) = \frac{301(201)(250)(150)}{51^2(52)} = 16,774.5$$

giving a standard error of $\sqrt{16,774.5} \approx 129.5$. ☐

## 18.2 MODELS FOR SIMPLE CAPTURE–RECAPTURE

In the general multinomial model for a single capture and recapture of a closed population, the capture or detection history of any animal in the population can be categorized into exactly one of four categories: detected on both first and second occasions, detected on first but not on second, detected on second but not on first, detected on neither occasion. Thus, a multinomial model, with the four probabilities for the four cells adding to 1, applies to the capture history of each animal. If, in addition, on each sampling occasion the detection outcomes for different individuals are independent, the model for the numbers of individuals with each capture history will be a product of multinomials. In the general model, the probability of detection may be different for different individuals and for different sampling occasions. The general model contains too many parameters in relation to the number of observations, so that further restrictions are needed for effective estimation of $\tau$ or detection probabilities.

One such restricted model assumes that detection probability is the same for each individual in the population during a sampling occasion but may differ for the two samples. Independence between the two sampling occasions is also assumed. With this model, the maximum likelihood estimator of population total is the integer part of the Petersen estimator $\hat{\tau} = Xy/x$. The maximum likelihood estimator (MLE) of the probability $p_1$ of capture in the first sample is $\hat{p}_1 = x/y$. The MLE of capture probability for the second sample is $\hat{p}_2 = x/X$. Even if capture probabilities are different for different individuals at the first sample but equal at the second sample, the estimator is still the Petersen estimator (Ahlo 1990).

If a single capture probability $p$ applies to both samples and to all individuals, with independence between samples, the maximum likelihood estimators of $p$ and $\tau$ are $\hat{p} = 2x/(X + y)$ and $\hat{\tau} = (X + y)/2\hat{p}$.

Another of the models allows for behavioral response to capture, so that an individual's probability of capture at the second sampling event depends on whether the individual was captured previously. Of the models allowing for different capture probabilities for each individual, one model assumes that for each individual the probability is the same on each sampling occasion. Estimation with this model is practical only with more than two occasions or when capture probabilities are restricted to depend on auxiliary variables.

If the numbers $X$ and $y$ of individuals in the two samples are fixed and the second sample is a simple random sample (without replacement) of the individuals in the population, the number $x$ of marked animals in the second sample has a hypergeometric distribution. With equal capture probabilities among individuals, this is the conditional distribution under the multinomial model of $x$ given $X$ and $y$. Under this model, the maximum likelihood estimator of $\tau$ is again the integer part of the Petersen estimator. *Inverse sampling* refers to designs in which $x$ is fixed in advance, so that $y$ is a random variable—sampling continues until the second sample contains the prescribed number $x$ of marked individuals.

Under the Poisson model, the numbers of individuals in the capture-history categories are assumed to be independent random variables. It follows that under this model, the total number $\tau$ of individuals in the population is also random.

Reviews and descriptions of models in capture–recapture surveys are found in Cormack (1979), Otis et al. (1978), Pollock et al. (1990), Seber (1982, 1986, 1992), and Wolter (1986). General statistical treatments of categorical count data include Agresti (1984, 1990) and Bishop et al. (1975).

## 18.3   SAMPLING DESIGN IN CAPTURE–RECAPTURE: RATIO VARIANCE ESTIMATOR

In the models above, the capture or detection of one individual is assumed independent of the capture of other individuals, or, in the case of the hypergeometric model, the second sample is a simple random sample without replacement of individuals. When the observations come from a sample of units such as groups of names, plots, trap locations, paths of nets, line transects, or variable circular plots, the sample does not consist of a simple random sample of individuals nor are individuals, detected independently. For such cases, models of the foregoing types are useful for assessing the nonsampling variability. The sampling variability, which depends on the way the sample is obtained and is influenced by such factors as spatial heterogeneity and between-group differences, may, however, account for the largest part of the variance of $\hat{\tau}$.

Ideally, one would like the animals in the population surveyed by capture–recapture to behave like beans in a bowl. To estimate the number of beans in the bowl, take out a set number, mark them with a pen, return them to the bowl, and stir thoroughly. That way, the marked beans will be evenly mixed in with the unmarked beans, and the second sample can be considered a random sample of beans from the population, each bean having the same probability of being selected, and each combination of that number of beans having the same probability of being selected. For an animal population in which it is possible to select a truly simple random sample, the hypergeometric model for the number of marked animals in the second sample may apply almost as well as to the number of marked beans. The multinomial model may apply reasonably well to a wildlife survey in which the samples are selected by canvassing a study region from a helicopter, landing to mark every animal detected on the first sampling occasion, and noting how many of those observed on the second occasion are marked, so that neither sample size is fixed. With such a methodology, it may be reasonable to assume that the inclusion of one animal in a sample is independent of the inclusion of any other animals, so that the product-of-multinomials model applies.

Now imagine a simple capture–recapture survey of an animal population in which the original sample is obtained with traps at set locations in a study region. Associated with each trap may be an attraction region of unknown shape and size, depending on such factors as the trap bait and wind direction and the movements of animals. The marked animals returned to the population may distribute themselves very unevenly in the study region, perhaps in relation to the locations in which they were trapped or in which they were released, with the result that the distribution of marked animals in the population may be atypical of the population as a whole. As a result, the proportion of animals with marks may be quite different in one part of the study area than in another.

All would still be according to standard models if for the second sample one could truly select a simple random sample of *individuals* from the population or have the detection of one individual independent of another. But in many situations that is not feasible. Instead, one may obtain the second sample of animals through a sample of area units, transects, trawls, traps, plots, or other type of detectability units. With such observational methods, individual animals are not independently selected, even if the units are. With each transect or other unit, a batch of individuals is obtained. With the marked animals distributed unevenly in the population, the observations from one transect or unit in the sample may contain a very high proportion of marked animals, while another, in a different part of the study region, may have a low proportion.

In the following, random selection of detectability units, with and without replacement, is applied to capture–recapture sampling. Random selection of observational units such as line transects and variable circular plots, with consideration of problems at the boundary of a study region, were discussed in Chapter 17. The somewhat similar problem of unknown overlap of trap attraction areas with a study region boundary is described in Otis et al. (1978, pp. 67–69).

## Random Sampling with Replacement of Detectability Units

Suppose that a random sample (with replacements, for simplicity) of $n$ transects, plots, or other type of units is selected. For the $i$th such unit in the sample, let $y_i$ denote the number of animals observed and $x_i$ the number of those that are marked. The total number of animals observed in the second sample is $\sum_{i=1}^{n} y_i$, of which $\sum_{i=1}^{n} x_i$ are marked. Let $\bar{y}$ and $\bar{x}$ denote the sample means as usual. The total $X$ of marked individuals in the population is known, assuming a closed population--it is the number marked and released at the first sampling occasion.

The Petersen estimate of the number of animals in the population is

$$\hat{\tau} = \frac{\sum_{i=1}^{n} y_i}{\sum_{i=1}^{n} x_i} X = \frac{\bar{y}}{\bar{x}} X \tag{9}$$

which is a classical ratio estimate.

The usual adjusted ratio estimate of variance would be

$$\widehat{\text{var}}(\hat{\tau}) = \left(\frac{X}{\bar{x}}\right)^2 \frac{1}{n(n-1)} \sum_{i=1}^{n} (y_i - rx_i)^2 \tag{10}$$

where $r = \bar{y}/\bar{x}$. A variance estimate based on the reciprocal of the sample ratio is given in Seber (1982, p. 114).

An approximate $100(1 - \alpha)\%$ confidence interval is given by

$$\hat{\tau} \pm t\sqrt{\widehat{\text{var}}(\hat{\tau})} \tag{11}$$

where $t$ is the upper $\alpha/2$ point of the $t$-distribution with $n - 1$ degrees of freedom. Coverage probability might be improved slightly by using the reciprocals of the endpoints of the normal confidence interval for $\hat{\tau}^{-1}$. [Under the hypergeometric model, slight improvement in coverage probabilities was achieved with reciprocal-based confidence intervals in capture–recapture simulations reported in Buckland (1984).] Confidence intervals could also be obtained with the jackknife or bootstrap methods, with the resampling done in terms of the $n$ independently selected transects or units, not in terms of the nonindependently occurring animals (see Sections 18.6 and 18.7).

The ratio estimator has low variance when the $y_i$ tend to be proportional to the $x_i$, which would tend to hold if the marked animals were very evenly distributed throughout the population. Uneven distribution of the marked animals would increase the variance. If the marked animals were so unevenly distributed in the study region that every one of the $x_i$-values of the $n$ units in the sample could simultaneously equal zero for some samples, the ratio estimator, like the usual Petersen estimator, would not have a finite variance. The more spread out the marked animals in the study region and the larger the units and the sample size, the less likely it is that this possibility would arise.

*Example 2: Random Sample with Replacement of Detectability Units.* In a bird population, 382 birds in the study area were trapped with mist nets and marked conspicuously with bands that could be spotted by observers with binoculars. The second sample was made by observers walking along $n = 8$ transects selected at random in the study area, who recorded the numbers of birds of the species they saw and noted whether they were banded. The $(y, x)$ data from the second sample, with $y$ the number of birds seen from the transect and $x$ the number of those that were marked, was (33, 20), (13, 2), (1, 0), (45, 15), (21, 5), (82, 39), (14, 4), and (0, 0).

Distances were not recorded, and nothing was known about the shape of the detectability function associated with a transect, except that with the random placement of the transects, average detectability was assumed to be equal for each individual in the population. The total number seen in the second sample was 209, of which 85 were marked. The sample ratio is $r = 0.209/85 = 2.459$. The ratio (Petersen) estimate of the population total [Equation (9)] was $\hat{\tau} = 2.459$ (382) $\approx$ 939. The estimated variance [Equation (10)] is

$$\widehat{\text{var}}(\hat{\tau}) = \left(\frac{382}{10.625}\right)^2 \frac{1}{8(7)} \{[33 - (2.459)(20)]^2 + \cdots + [0 - (2.459)(0)]^2\}$$

$$= 15{,}698$$

the square root of which is 125.

The usual multinomial model variance estimate [Equation (8)], on the other hand, would be $\widehat{\text{var}}(\hat{\tau}) = 382(209)(382 - 85)(209 - 85)/85^3 = 4789$, with a square root of 69. It appears that this estimator underestimates the variance by ignoring the spatial heterogeneity and the lack of independence among animals, both factors in the way the second sample was obtained.                                $\square$

## Random Sampling without Replacement

Without-replacement sampling is possible when individuals on a list are grouped into nonoverlapping units for selection or when a spatial study region is partitioned into area units, with constant probability of detection applying to all individuals in a selected unit. Capture–recapture designs in which the second sample consists of a random sample without replacement of $n$ units out of $N$ in the population and the probability of detection or capture is constant for all individuals in a sampled unit are described in Seber (1982, pp. 111–114) and Wolter (1986). The estimator of population total $\tau$ is the Petersen estimator. Two estimators of variance under this design were given by Wolter, one of which has the form

$$\widehat{var}(\hat{\tau}) = \left(\frac{X}{\bar{x}}\right)^2 \frac{N-n}{Nn(n-1)} \sum_{i=1}^{n} (y_i - rx_i)^2 + \frac{X\bar{y}(\mu_x - \bar{x})(\bar{y} - \bar{x})}{\bar{x}^3} \qquad (12)$$

where $\mu_x = X/N$. Wolter gives in addition a finite-population central limit theorem for the asymptotic normality of the Petersen estimator under simple random sampling without replacement of units.

## 18.4 ESTIMATING DETECTABILITY WITH CAPTURE–RECAPTURE METHODS

Capture–recapture methods for estimating population size also explicitly or implicitly come up with estimates or probabilities of detection or capture. For example, the Petersen estimator has the form, in common with many other estimators of population abundance or density in which detectability is imperfect,

$$\hat{\tau} = \frac{y}{\hat{p}} \qquad (13)$$

where $\hat{p} = x/X$ estimate the detection probability in the second sample. Capture–recapture methods can also be used to estimate detectability for other, independent surveys to be done with the sampling methods of the second sample.

With the hypergeometric distribution of $x$, the variance of the estimated detectability, given $X$ and $y$, is

$$var(\hat{p}) = \frac{y}{X^2} \left(\frac{X}{\tau}\right)\left(1 - \frac{X}{\tau}\right)\frac{\tau - y}{\tau - 1} \qquad (14)$$

With random selection (with replacement) of $n$ units––such as observation sites, trap sites, line transects, trawls, plots, or strips––and observation of $x_i$ marked animals at the $i$th unit, an estimate of the detectability per unit is $\hat{p} = \bar{x}/X$, where $\bar{x} = \sum_{i=1}^{n} x_i$. A design-based estimate of its variance is $\widehat{var}(\hat{p}) = s_x^2/X^2 n$, where $s_x^2 = \sum_{i=1}^{n}(x_i - \bar{x})^2/(n-1)$.

***Example 3: Estimating Detectability.*** In the mouse example (Example 1), the probability of capture in the second sample is estimated as $\hat{p} = 50/300 = 0.17$. An estimate of the variance of the estimator of detectability, using Equation (14) with $\hat{\tau}$ in place of $\tau$, is

$$\widehat{var}(\hat{p}) = \frac{200}{300^2}\left(\frac{300}{1200}\right)\left(1 - \frac{300}{1200}\right)\left(\frac{1200 - 200}{1200 - 1}\right) = 0.00035$$

giving a standard error of about 0.02.

A number of authors (Ahlo 1990; Overton 1969; Pollock and Otto 1983) have presented estimators in the form $\hat{\tau} = \sum_{i=1}^{T} z_i/\hat{\pi}_i$, where $z_i$ is an indicator variable equal to 1 if the $i$th individual in the population is included in one or more samples, and $\hat{\pi}_i$ is an estimate of the probability of inclusion (detection) of that individual in at least one of the samples. With known inclusion probabilities, the estimator would, as a Horvitz–Thompson estimator, be unbiased. With the probabilities estimated explicitly or implicitly by various methods, substantial biases can arise (Burnham and Overton 1978; Otis et al. 1978; Pollock and Otto 1983). Ahlo (1990) and Pollock et al. (1984) used logistic regression of capture probabilities with auxiliary variables in obtaining estimates of population size where individual probabilities may vary.                                                        □

## 18.5  MULTIPLE RELEASES

Capture–recapture methods extend to situations with more than two sampling occasions. Suppose that a series of $k$ independent samples containing set numbers $y_1, y_2, \ldots, y_k$ of animals are captured from a closed population (cf. Schnabel 1938). With the $i$th sample, the number of $x_i$ of marked animals is observed; then the animals are given new marks before release. (Note that $x_i = 0$.) Given the numbers of individuals in each sample, the model for the set of observed numbers marked is a product of hypergeometric distributions:

$$f(x_2, x_3, \ldots, x_k) = \prod_{i=2}^{k} \binom{X_i}{x_i} \binom{\tau - X_i}{y_i - x_i} \Big/ \binom{\tau}{y_i} \tag{15}$$

where $X_i$ is the number marked in the population when ( just before) the $i$th sample is taken.

The value of $\tau$ that maximizes the probability above is the maximum likelihood estimate of the population size. The solution must be obtained iteratively.

***Example 4: Multiple Recaptures.***  To estimate the number of fish in a small lake, $y_1 = 20$ fish of the species are caught with barbless hooks. Before release, each is marked with a small notch in one of the fins. The next day, $y_2 = 20$ fish are caught and marked with a different shape of small notch. Of these, $x_2 = 2$ are found to have the first mark already and so return to the water with both marks. On the third day, $y_3 = 20$ fish are caught and given a third mark, distinguishable from the other two. Of these, 15 had no marks, and $x_3 = 4$ had one or more marks as follows: 2 had only first-day marks, 1 had only second-day marks, and 1 had both first- and second-day marks.

The likelihood or probability function for the recapture numbers [using Equation (15)] is

$$f(x_2, x_3) = f(2, 4) = \frac{\binom{20}{2}\binom{\tau - 20}{18}}{\binom{\tau}{20}} \frac{\binom{38}{4}\binom{\tau - 38}{16}}{\binom{\tau}{20}}$$

Finding the value of $\tau$ that maximizes this likelihood is only a computational problem. One could proceed by trial and error or with a more efficient search procedure. (In fact, with three capture times a closed-form solution exists; see Seber 1982, p. 132.) The value of $\tau$ that maximizes the likelihood is $\hat{\tau} = 193$.

The data from the three sampling dates may also be summarized in terms of the counts of fish with each possible catch history. The catch history of any fish in the lake could be identified by its mark pattern. An unknown number of fish, with catch history denoted (000), have never been caught. The catch history of a fish caught on day 1 but not on day 2 or day 3 is denoted (100); the total number of such fish is 16--from the 20 caught on the first day are subtracted the 2 recaught on the second day and the 2 recaught for the first time on the third day. The catch history of a fish caught on the first and third days but not the second is denoted (101), and so on.

With three sample dates, the number of possible capture histories is $2^3 = 8$. Assuming a closed population, the number of fish with each catch history are as follows: Unknown (000), 16 (100), 17 (010), 1 (110), 16 (001), 2 (101), 1 (011), and 1 (111). The number of distinct fish handled during the three sampling dates is 54, the sum of the counts above. ☐

## 18.6 MORE ELABORATE MODELS

With a sequence of samples from a population with deaths, recruitment, immigration, and emigration, the underlying probability model involves many unknown parameters. A general model, allowing in addition for deaths induced by the handling and tagging procedure, was given by Jolly (1965) and Seber (1965) and extended in various ways by other authors (see Seber 1986). In the Jolly–Seber model, the probability of capture may vary from sample to sample, but is assumed the same for all animals at each capture period. Similarly, the probability of surviving from one sampling period varies with sampling period but not with individuals. Accidental death due to handling in capture can be incorporated, and the model can be generalized to allow for capture probabilities dependent on capture history. Maximum likelihood estimates of parameters such as survival rates may be obtained in addition to population size. The estimates are obtained iteratively and, because of the complexity of the model, demand rather large sample sizes. The underlying models used with the Jolly–Seber method are products of multihypergeometric or multinomial distributions. Improved methods for variance estimation and confidence intervals are given in Manly (1984) and Seber and Manly (1985). Nonparametric bootstrap methods for variance estimates and confidence intervals for mark–recapture are described in Buckland (1984, 1988); comprehensive reviews of the literature and methods are found in Seber (1982, 1986, 1992).

A relatively recent approach to analyzing mark–recapture experiments, which appears promising for open populations, uses log-linear models (Agresti 1990; Cormack 1980, 1981, 1985, 1989). With these models, the logarithm of the expected number of animals with a given capture history is a linear function of a set of biologically interpretable parameters.

**EXERCISE**

1. In a capture–recapture survey, 10,000 salmon were released after marking by inserting small wires. When 6000 fish from the population were subsequently caught, 400 of the wire tags were discovered by running the fish through a metal detection device.

   (a) Estimate the number of fish in the population.

   (b) Give an approximate 95% confidence interval for the estimate.

   (c) Suppose that the detection device missed some of the wire tags. Would overestimation or underestimation of the population tend to result?

   (d) If the wire tags caused significant numbers of fish to die soon after release, would overestimation or underestimation tend to result?

CHAPTER 19

# Line-Intercept Sampling

In line-intercept sampling, a sample of lines is selected in a study area, and whenever an object of the population is intersected by one or more of the sample lines, a variable of interest associated with that object is recorded. Consider, for example, an ecological habitat study in which the object is to estimate the total quantity of berries of a certain plant species in a study area. A random sample of $n$ lines, each of length $l$, is selected and drawn on a map of the study area. Field workers walk each of the lines and, whenever the line intersects a bush of the species, the berries of that bush are collected and their quantity $y_k$ measured.

With the design described above, a large bush has a higher probability of inclusion in the sample than does a small bush. Unbiased estimation of the population quantities depends on determining these probabilities.

## 19.1  RANDOM SAMPLE OF LINES: FIXED DIRECTION

In the simplest design, $n$ transect lines are selected at random by selecting $n$ positions along a baseline of length $b$ that traverses the width of the study region and running a transect across the study area perpendicular to the baseline at each of the selected positions.

Let $K$ denote the number of objects in the population. Associated with the $k$th object is a variable of interest $y_k$. The object is to estimate the population total $\tau = \sum_{k=1}^{K} y_i$ or the density per unit area $D = \tau/A$, where $A$ is the area of the study region.

On any given draw, the probability that the transect line selected intersects the $k$th object is proportional to the width $w_k$ along the baseline of the set of points for which the perpendicular intersects object $k$. Thus, $w_k$ is the width of the "shadow" cast by object $k$ on the baseline. The draw-by-draw selection probability is

$$p_k = \frac{w_k}{b} \tag{1}$$

Let $C_i$ be the set of objects of the population that are intersected by the $i$th transect line in the sample. For each of these intersected objects, divide the value of the variable of interest $y_k$ by the selection probability $p_k$ and define the new variable $v_i$ as the sum

$$v_i = \sum_{k \in C_i} \frac{y_k}{p_k} \tag{2}$$

The variable $v_i$ is itself an unbiased estimator of the population total $\tau$. The random sample of $n$ transects gives $v_1, v_2, \ldots, v_n$, which are independent and identically distributed. Their sample mean

$$\hat{\tau}_p = \frac{1}{n} \sum_{i=1}^{n} v_i \tag{3}$$

is thus an unbiased estimator of $\tau$ with variance $\mathrm{var}(\hat{\tau}_p) = (1/n)\,\mathrm{var}(v_i)$. The estimator $\hat{\tau}_p$ is akin to the Hansen–Hurwitz estimator in being based on draw-by-draw selection probabilities. In the present design, however, objects are not selected independently, since joint selections occur with lines that intersect more than one object.

Let $s_v^2$ denote the sample variance of the $v$'s, that is,

$$s_v^2 = \frac{1}{n-1} \sum_{i=1}^{n} (v_i - \hat{\tau}_p)^2 \tag{4}$$

An unbiased estimate of the variance of $\hat{\tau}$ is

$$\widehat{\mathrm{var}}(\hat{\tau}_p) = \frac{s_v^2}{n} \tag{5}$$

There is no finite population correction factor because selection of positions along the baseline is essentially with replacement. Even with distinct transect positions, a given object may be intersected by more than one transect line and hence be counted more than once in the estimator $\hat{\tau}$.

An estimate that depends only on the distinct objects intersected by the sample of transect lines may be obtained by the Horvitz–Thompson method. Let $\kappa$ be the number of distinct objects intersected. The probability that the $k$th object is included in the sample is

$$\pi_k = 1 - (1 - p_k)^n \tag{6}$$

The Horvitz–Thompson estimator is

$$\hat{\tau}_\pi = \sum_{k=1}^{\kappa} \frac{Y_k}{\pi_k} \tag{7}$$

The variance formulas for the Horvitz–Thompson estimator depend on the joint inclusion probabilities. Let $w_{kh}$ denote the width along the baseline of the set of positions from which the perpendicular line intersects both objects $k$ and $h$. The total width along the baseline from which either object $k$ or object $h$ or both are intersected is $w_k + w_h - w_{kh}$. The probability that both object $k$ and object $h$ are intersected at least once by the sample transects is

$$\pi_{kh} = \pi_k + \pi_h - 1 + \left(1 - \frac{w_k + w_h - w_{kh}}{b}\right)^n \tag{8}$$

The Horvitz–Thompson variance and estimated variance formulae [Equations (5) and (6)] may then be used for $\hat{\tau}_\pi$.

From each type of estimator of population total, an estimator of population density may be obtained as

$$\hat{D} = \frac{1}{A}\hat{\tau} \tag{9}$$

with

$$\text{var}(\hat{D}) = \frac{1}{A^2}\text{var}(\hat{\tau}) \tag{10}$$

If the study region is rectangular of width $b$ with each transect running constant length $l$ across it, the area $A = bl$. If the transect is of irregular shape, the length $l_i$ of the $i$th randomly selected transect is a random variable, with expected value $E(l_i) = A/b$ (Seber 1979). If the study region is farther across than the maximum length $l$ of a single transect, the baseline may be continued in parallel lines distance $l$ apart. The estimators given above are unbiased for $\tau$ or $D$ whether the lines are of equal length or not. Ratio and mean-of-ratios estimators based on transect length were examined by Seber (1979). In common with other ratio-type estimators, these estimators are slightly biased.

***Example 1.*** The data for this example are from Becker (1991) and Becker and Gardner (1990). To estimate the abundance of wolverines in a study region, selected transects are flown in appropriate weather conditions with observers in the aircraft looking for tracks in the snow. Once a set of tracks is encountered, it is followed in each direction and mapped. For the $k$th set of tracks, the variable of interest $y_k$ is the number of wolverines associated with that set.

The results from such a survey are shown in Figure 19.1, depicting a rectangular study region 36 miles by 20 miles in the Chugach Mountains of Alaska. The sampling designs consisted of $n = 4$ randomly selected positions for transects that were systematically arranged. The four random starting positions (A1, B1, C1, and D1 in the figure) were selected from the first 12 miles ($b = 12$) along

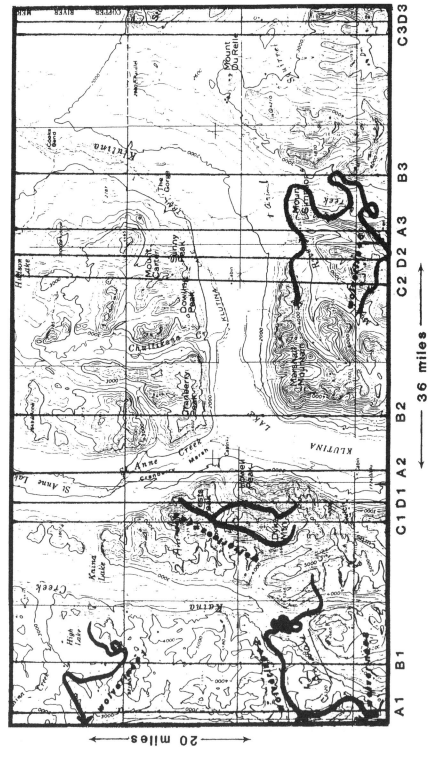

Figure 19.1. Wolverine tracks intersected by a systematic sample of transects with four randomly selected starting points. (From Becker 1990, 1991.)

248

the width of the study area. From each starting position selected, a transect was flown across the study region, with two more transect segments (e.g., A2 and A3 for starting position A1) added systematically at 12-mile intervals from the starting positions. Note that no complication is added by the fact that the design is a replicated systematic sample, since we have in effect a simple random sample of four transects selected within the first 12 miles, with each transect selected continued in three segments.

The selected transects intersected $\kappa = 4$ distinct sets of tracks, containing 6 wolverines. Numbering the sets of tracks from east to west and north to south, the numbers of wolverines were $y_1 = 1$, $y_2 = 2$, $y_3 = 2$, and $y_4 = 1$. The widths of the projections of the tracks onto the base of the study region are $w_1 = 5.25$ miles, $w_2 = 7.50$ miles, $w_3 = 2.40$ miles, and $w_4 = 7.05$ miles. Because of the random selection of starting positions in the first 12 miles, the selection probabilities [using Equation (1)] are $p_k = w_k/12$, giving $p_1 = 0.4375$, $p_2 = 0.625$, $p_3 = 0.2$, and $p_4 = 0.5875$.

The first transect intersects the first, second, and fourth set of tracks, so the variable $v_1$ [using Equation (2)] is $v_1 = (1/0.4375) + (2/0.625) + (1/0.5875) = 2.2857 + 3.2 + 1.7021 = 7.1878$. The second transect also intersects the first, second, and fourth set of tracks, so $v_2 = 7.1878$. The third transect intersects the third and fourth set of tracks, so $v_3 = (2/0.2) + (1/0.5875) = 10.0 + 1.7021 = 11.7021$. The fourth transect also intersects the third and fourth set of tracks, so $v_4 = 11.7021$.

The estimate [from Equation (3)] based on selection probabilities is

$$\hat{\tau}_p = \frac{1}{4}(7.1878 + 7.1878 + 11.7021 + 11.7021) = 9.44$$

or about 9 wolverines in the study region. The estimate variance [using Equation (5)] is

$$\widehat{\text{var}}(\hat{\tau}_p) = \frac{6.7930}{4} = 1.70$$

For the Horvitz–Thompson estimate, the inclusion probabilities derived from Equation (6) are $\pi_1 = 1 - (1 - 0.4375)^4 = 0.90$, $\pi_2 = 0.98$, $\pi_3 = 0.59$, and $\pi_4 = 0.98$. The Horvitz–Thompson estimate [using Equation (7)] is

$$\hat{\tau}_\pi = \frac{1}{0.90} + \frac{2}{0.98} + \frac{2}{0.59} + \frac{1}{0.97} = 7.57$$

or about 8 wolverines in the study region.

The width along the baseline from which both track sets 1 and 2 are intersected is $w_{12} = 5.25$. For the other combinations, $w_{13} = 0$, $w_{14} = 3.75$, $w_{23} = 0$, $w_{24} = 3.75$, and $w_{34} = 2.4$. The joint inclusion probabilities [using Equation (8)] are

$$\pi_{12} = 0.90 + 0.98 - 1 + \left(1 - \frac{5.25 + 7.5 - 5.25}{12}\right)^4 = 0.90$$

$$\pi_{13} = 0.90 + 0.59 - 1 + \left(1 - \frac{5.25 + 2.4 - 0}{12}\right)^4 = 0.51$$

$$\pi_{14} = 0.90 + 0.97 - 1 + \left(1 - \frac{5.25 + 7.05 - 3.75}{12}\right)^4 = 0.88$$

$$\pi_{23} = 0.98 + 0.59 - 1 + \left(1 - \frac{7.5 + 2.4 - 0}{12}\right)^4 = 0.57$$

$$\pi_{24} = 0.98 + 0.97 - 1 + \left(1 - \frac{7.5 + 7.05 - 3.75}{12}\right)^4 = 0.95$$

$$\pi_{34} = 0.59 + 0.97 - 1 + \left(1 - \frac{2.4 + 7.05 - 2.4}{12}\right)^4 = 0.59$$

The estimated variance for the Horvitz–Thompson estimator [Equation (6)] is

$$
\begin{aligned}
\widehat{\mathrm{var}}(\hat{\tau}_\pi) = &\left(\frac{1}{0.90^2} - \frac{1}{0.90}\right)1^2 + \left(\frac{1}{0.98^2} - \frac{1}{0.98}\right)2^2 \\
&+ \left(\frac{1}{0.59^2} - \frac{1}{0.59}\right)2^2 + \left(\frac{1}{0.97^2} - \frac{1}{0.97}\right)1^2 \\
&+ 2\left(\frac{1}{0.90(0.98)} - \frac{1}{0.90}\right)(1)(2) + 2\left(\frac{1}{0.90(0.59)} - \frac{1}{0.51}\right)(1)(2) \\
&+ 2\left(\frac{1}{0.90(0.97)} - \frac{1}{0.88}\right)(1)(1) + 2\left(\frac{1}{0.98(0.59)} - \frac{1}{0.57}\right)(2)(2) \\
&+ 2\left(\frac{1}{0.98(0.97)} - \frac{1}{0.95}\right)(2)(1) + 2\left(\frac{1}{0.59(0.97)} - \frac{1}{0.59}\right)(2)(1) \\
= &\ 5.27 \qquad\qquad\qquad\qquad\qquad\qquad\qquad\qquad\qquad\qquad \square
\end{aligned}
$$

## 19.2   LINES OF RANDOM POSITION AND DIRECTION

Now suppose that each sample line is selected completely at random in the study region. This may be accomplished by first selecting a location uniformly at random from the study region to be the midpoint of a transect of length $l$. Then, independently, an angle is chosen from a uniform distribution on $[0, \pi)$, giving the direction of the line. The problem of shorter lines near the boundary of the study region, which can lead to small biases in otherwise unbiased estimators, is usually dealt with, at least theoretically, by extending any cutoff portion of a selected line in another part of the study region (Kaiser 1983). In practice, the bias will be small if the study region is large in relation to the length of a transect line.

An object is completely intercepted by a sample line if extensions of the line in either direction do not intersect additional points of the object. The object is selected if it is completely intercepted. An object partially intercepted—as happens when an endpoint of the transect is within the object—is selected with probability 1/2 (the reason being to simplify computations of selection probabilities). The transect line is assumed to be longer than the maximum length, in the direction of the transect line, of any of the objects in the population.

Given the direction $\theta$ of the transect, the probability that the $k$th object is intersected is $p_k(\theta) = lw_k(\theta)/A$, where $w_k(\theta)$ is the width of object $k$ in the direction perpendicular to $\theta$, that is, the maximum distance between lines in the direction $\theta$ that intersect the object.

The unconditional probability of selection for object $k$ is $p_k = lc_k/A$, where $c_k = \mathrm{E}[w_k(\theta)]$, the expected value of $w_k(\theta)$ over the distribution of $\theta$.

Unbiased estimators of the population total $\tau$ may be obtained using either the conditional or the unconditional selection probabilities (Kaiser 1983). For the $i$th selected transect, define the new variables

$$v_i(\theta) = \sum_{k \in C_i} \frac{y_k}{p_k(\theta)} \tag{11}$$

$$v_i = \sum_{k \in C_i} \frac{y_k}{p_k} \tag{12}$$

Both $v_i(\theta)$ and $v_i$ are unbiased estimators of $\tau$.

With $n$ transects selected independently using the design above, the sample mean of the $v$'s is an unbiased estimator or $\tau$ with variance $(1/n)\mathrm{var}(v_i)$. Thus, two possible unbiased estimators are

$$\hat{\tau}_{p(\theta)} = \frac{1}{n} \sum_{i=1}^{n} v_i(\theta) \tag{13}$$

$$\hat{\tau}_p = \frac{1}{n} \sum_{i=1}^{n} v_i \tag{14}$$

An unbiased estimator of variance is $s_v^2/n$, where $s_v^2$ is the sample variance based on the corresponding $v$-values.

Further study is needed regarding the relative efficiency of $\hat{\tau}_{p(\theta)}$ and $\hat{\tau}_p$ (see Kimura and Lemberg 1981). The practical choice may depend on the relative ease of measuring $w_k(\theta)$ and $c_k$ for sampled objects. Kaiser (1983) gives the expected width of the $k$th object as $c_k = c_k^*/\pi$, where $c_k^*$ is the length of the perimeter of the smallest convex set containing object $k$ and suggests measuring $c_k^*$ by wrapping a rope tightly around object $k$ and measuring the length of rope (see also Kendall and Moran 1963, p. 58; Solomon 1978, p. 17).

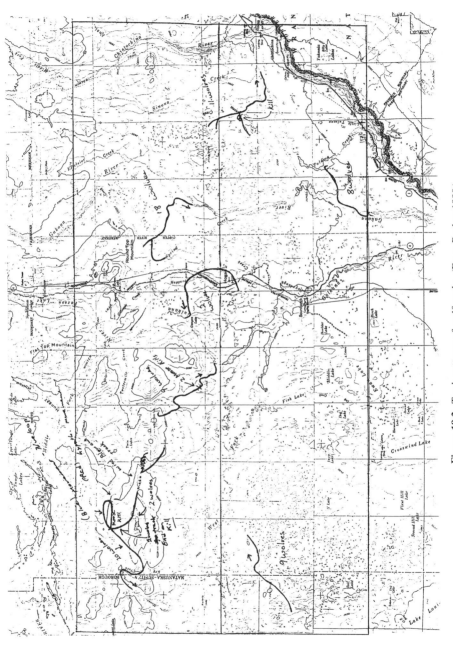

**Figure 19.2.** Track patterns of wolf packs. (From Becker 1990.)

252

For the Horvitz–Thompson estimator, the inclusion probability for the $k$th object is $\pi_k = 1 - (1 - p_k)^n$. The Horvitz–Thompson estimator is $\hat{\tau}_\pi = \sum_{k=1}^{\kappa} y_k / \pi_k$.

Line-intercept estimators for coverage and number of objects were discussed in Lucas and Seber (1977). Estimators for more general variables of interest were given in McDonald (1980). Jolly (1979) applied similar ideas to the problem of estimating the abundance of animals in large herds. Kaiser (1983) describes a design in which a location and direction are chosen at random and the line is extended across the study area, so that line length is a random variable. References to earlier work are found in the reviews of DeVries (1979) and Eberhardt (1978a). Line-intercept methods based on the length of interception and the number of interceptions have also been developed. Many of the ideas of geometric probability underlying line-intercept methods are described in Kendall and Moran (1963) and Solomon (1978).

## EXERCISES

1. Figure 19.2 shows track patters of wolf packs, mapped in a census of a study region in the Gulkana River area of Alaska (Becker 1990). The number $y_k$ of wolves in each pack is given near each set of tracks. Using the baseline (west–east), which starts along the base of the study region and continues through the middle of the region, select a simple random sample of three south–north transects. Estimate the total number of wolves in the region from your sample using $\hat{\tau}_p$. Estimate the associated variance.

2. Using the selected sample of Exercise 1, estimate the total number of wolves using $\hat{\tau}_\pi$. Estimate the associated variance.

# PART V

# Spatial Sampling

# CHAPTER 20

# Spatial Prediction or Kriging

In geological studies it is frequently desired to predict the amount of ore or fossil fuel that will be found at a site. The prediction may be based on values observed at other sites in the region, and these other sites may be spaced in the region irregularly . It may also be desired to predict or estimate the total amount of ore in the region. Similarly, in pollution monitoring, measurements of pollutant concentration at a few sites may be used to predict the concentration at a new site or to estimate the mean concentration over a larger area. In ecological studies also, observations of animals or plants at a sample of sites can be used to predict the abundance in the vicinity of a new site or in the entire study region.

In such situations it is useful to think of the value $y_t$ of ore, pollutant, or animal abundance at location $t$ as a random variable. From the observed values $y_1, \ldots, y_n$ at $n$ sites $t_1, \ldots, t_n$, we wish to estimate or predict the value of a new random variable $y_0$. The variable $y_0$ may be either the value of the variable at a new site or the total of the variable of interest over a larger geographic region. Since $y_0$ is viewed as a random variable rather than a parameter or fixed population characteristic, the inference problem is referred to as prediction rather than estimation, even though the prediction may be over space rather than time.

The spatial prediction problem and its solution—termed *kriging* in geostatistics—are essentially the same as the model-based prediction approach to survey sampling with auxiliary information, as described in Chapters 7 and 8. The prediction equations can be written equivalently either with covariances or with variances of differences (the variogram). The covariance approach has been traditional in statistics and time series, while the variogram has been traditional in geostatistics. In this chapter the covariance form is used first to emphasize the connection to the regression methods of survey sampling. The two approaches are exactly equivalent provided that the covariance function—or the variogram—is known. Slight differences can arise when the covariance function or variogram is estimated using visual methods.

General references on spatial prediction, kriging, and geostatistics include Cressie (1986, 1989, 1991), Hohn (1988), and Journel (1987, 1988, 1989).

References on spatial patterns and processes include Cox and Isham (1980), Cressie (1991), Diggle (1983), Matérn (1960, 1986), and Ripley (1981).

## 20.1 SPATIAL COVARIANCE FUNCTION

In both ecological and geological surveys, the values of the variable of interest at different sites are typically not independent of each other. Rather, the values at sites in close proximity tend to be related to each other. One summary of the relationship between the variables $y_1$ and $y_2$ associated with sites $t_1$ and $t_2$ is the covariance,

$$\text{cov}(y_1, y_2) = E[y_1 - E(y_1)][y_2 - E(y_1)] \tag{1}$$

When the covariance between two sites depends only on their relative positions, and not their exact locations within the study area, this relationship may be summarized with a *covariance function* $c(h)$, giving the covariance of the $y$-values for any two sites whose relative positions are separated by $h$, that is,

$$c(h) = \text{cov}(y_{t+h}, y_t) \tag{2}$$

For a two-dimensional study region, the location $t = (t_1, t_2)$ gives the coordinates in two dimensions of a site. The displacement $h$ between the sites is also vector-valued, so that the covariance function may depend on direction as well as distance between sites. If the expected value of the variable of interest is the same at every site and the covariance depends only on the displacement between sites, the process is called *second-order stationary*. If the covariance depends only on the distance $d$ between the sites and does not depend on direction, the process is said to be *isotropic*, and the covariance function will be written $c(d)$.

## 20.2 LINEAR PREDICTION (KRIGING)

Suppose that observations have been observed at a sample of $n$ sites and one wishes to predict what will be found at another site. For notational simplicity, write $y_1$ for the $y$-value observed at the $i$th site in the sample, for $i = 1, \ldots, n$, and write $c_{ij}$ for $\text{cov}(y_i, y_j)$, the covariance between the $y$-values at the $i$th and $j$th sites. Note that $c_{ii} = \text{var}(y_i)$. From observed $y$-values at sites $t_1, \ldots, t_n$, it is desired to predict the value of the random variable $y_0$ at the site $t_0$. For simplicity, the means $E(y_i)$ are assumed equal; a more general formulation is given in Section 20.5.

The object is to find a function $\hat{y}_0$ of the $n$ observed $y$-values that is unbiased for $y_0$, that is,

$$E(\hat{y}_0) = E(y_0) \tag{3}$$

and which minimizes the mean square prediction error

$$E(y_0 - \hat{y}_0)^2 \qquad (4)$$

In general, the best estimator is the conditional expectation of $y_0$ given the observed values $y_1, \ldots, y_n$. Determining $E(y_0 \mid y_1, \ldots, y_n)$ depends on the exact joint distribution of the $y$-values and may be difficult.

A generally more practical criterion is to find a *linear* function of the observed $y$-values that is unbiased and minimizes the mean square prediction error. Writing the linear estimator as

$$\hat{y}_0 = \sum_{i=1}^{n} a_i y_i \qquad (5)$$

the problem is to find the values $a_1, a_2, \ldots, a_n$ that minimize the mean square prediction error subject to unbiasedness.

The solution (obtained by the Lagrange multiplier method) is given in matrix form by

$$\mathbf{f} = \mathbf{G}^{-1}\mathbf{h} \qquad (6)$$

where the vectors $\mathbf{f}$ and $\mathbf{h}$ are

$$\mathbf{f} = \begin{pmatrix} a_1 \\ a_2 \\ \vdots \\ a_n \\ m \end{pmatrix}, \qquad \mathbf{h} = \begin{pmatrix} c_{10} \\ c_{20} \\ \vdots \\ c_{n0} \\ 1 \end{pmatrix}$$

and the matrix $\mathbf{G}$ is

$$\mathbf{G} = \begin{pmatrix} c_{11} & c_{12} & \cdots & c_{1n} & 1 \\ c_{21} & c_{22} & \cdots & c_{2n} & 1 \\ \vdots & \vdots & \ddots & \vdots & \vdots \\ c_{n1} & c_{n2} & \cdots & c_{nn} & 1 \\ 1 & 1 & \cdots & 1 & 0 \end{pmatrix}$$

The constant $m$, which is obtained along with the coefficients $a_i$, is the Lagrange multiplier and is used in calculating the mean square prediction error.

The best linear predictor $\hat{y}_0$ is called the *kriging predictor* in geology. The mean square prediction error of $\hat{y}_0$, also called the *kriging variance*, is

$$E(y_0 - \hat{y}_0)^2 = c_{00} - \sum_{i=1}^{n} a_i c_{i0} - m \qquad (7)$$

If the stochastic process giving rise to the population is Gaussian, that is, the joint distribution of any finite set of $y$-values is multivariate normal, then the best *linear* predictor $\hat{y}_0$ is the *best* predictor, having the lowest mean square prediction error of any function of the $n$ observed $y$-values.

The optimality of the prediction results depends on the covariance function being known. In practice, however, the covariance function is estimated from data—either data from the present survey or a larger store of data from past surveys. For a process that is stationary and isotropic, the covariance of sites distance $d$ apart can be estimated using a sample covariance based on the $n_d$ distinct pairs of sites in the data that are distance $d$ (or approximately distance $d$) apart. A simple covariance estimator is

$$\hat{c}(d) = \frac{1}{n_d}\sum (y_{t_i} - \bar{y})(y_{t_j} - \bar{y}) \tag{8}$$

where the summation is over the distinct pairs of observations that are approximately distance $d$ apart and $n_d$ is the number of such distinct pairs. A smooth curve may then be fitted to the estimated covariances, using a method such as nonlinear least squares, to obtain an estimated covariance function and hence an estimate of covariance for any distance.

**Example 1.** Data from Alaska Department of Fish and Game shrimp surveys in the vicinity of Kodiak Island, Alaska, were used to estimate the spatial covariance function, which in turn will be used to predict the amount of catch at a new location. The catch data, plotted by location on a chart of the study region, were originally recorded in pounds (lb), with distances measured in nautical miles (nmi.). A research vessel made tows of a trawl net approximately 1 nmi. apart in a grid pattern. Sample covariances were computed using pairs of data lumped into distance intervals. Then a curve of the form $a\exp(-bx)$ was fitted by nonlinear least squares to the covariance estimates. The fitted covariance function was

$$c(x) = 5.1e^{-0.49x} \tag{9}$$

Suppose that one tow has been made with a catch of $y_1 = 5.526$ (units are thousands of ponds) and a second tow 6 nmi. away produced $y_2 = 1.417$. What would be the predicted catch $y_0$ at a location 1 nmi. from the first tow and 5.4 nmi. from the second?

The variance is $c(0) = 5.1$. The covariance for the two tows 6 nmi. apart is $c_{12} = 5.1\{\exp[-0.49(6)]\} = 0.3$. The covariances with the new site are $c_{10} = 5.1\{\exp[-0.49(1)]\} = 3.1$ and $c_{20} = 0.4$. The prediction equation is

$$\begin{pmatrix} a_1 \\ a_2 \\ m \end{pmatrix} = \begin{pmatrix} 5.1 & 0.3 & 1 \\ 0.3 & 5.1 & 1 \\ 1 & 1 & 0 \end{pmatrix}^{-1} \begin{pmatrix} 3.1 \\ 0.4 \\ 1 \end{pmatrix}$$

$$= \begin{pmatrix} 0.104 & -0.104 & 0.5 \\ -0.104 & 0.104 & 0.5 \\ 0.5 & 0.5 & -2.7 \end{pmatrix} \begin{pmatrix} 3.1 \\ 0.4 \\ 1 \end{pmatrix} = \begin{pmatrix} 0.78 \\ 0.22 \\ -0.95 \end{pmatrix}$$

so $a_1 = 0.78$, $a_2 = 0.22$, and $m = -0.95$.

The predicted catch [using Equation (5)] is

$$\hat{y} = 0.78(5.526) + 0.22(1.1417) = 4.622$$

or 4622 lb at the new location.

The prediction mean square error [using Equation (7)] is

$$\hat{E}(y_0 - \hat{y}_0)^2 = 5.1 - 0.78(3.1) - 0.22(0.4) + 0.95 = 3.5$$

giving a root mean square error of 1.9 (i.e., 1900 lb).

The linear prediction method is scale-invariant, in that if the original data were transformed to kilograms and kilometers, the coefficients $a_i$ would be unaffected and the appropriate prediction in kilograms per kilometer would be obtained. Using the equivalents 1 nautical mile $\approx$ 1.15 mile, 1 kilometer $\approx$ 0.62 mile, and 1 kilogram $\approx$ 2.2046 pounds, the predicted catch equals 2097 kilograms (= 4.622/2.2046). □

## 20.3 VARIOGRAM

In the geological sciences, spatial variation has traditionally been summarized using the *variogram* in place of the covariance function. The *variogram* is defined as the variance of the difference of y-values at separate sites:

$$\text{var}(y_{t+h} - y_t) = 2\gamma(h) \tag{10}$$

The function $\gamma(h)$ is called the *semivariogram*. When the process is second-order stationary, the covariance function and the variogram contain equivalent information, since

$$\gamma(h) = c(0) - c(h) \tag{11}$$

Note that $c(0) = \text{var}(y_t)$, the variance of y at any site t. Cressie (1986, 1989, 1991) points out that the variogram exists even for some processes that are not second-order stationary, and hence is more general than the covariance function.

Assume that the process is also stationary in the mean, that is,

$$E(y_t) = E(y_s) \tag{12}$$

for any sites t and s in the study region. Then $\text{var}(y_{t+h} - y_t) = E(y_{t+h} - y_t)^2$, and a simple method for estimating the semivariogram is

$$2\hat{\gamma}(d) = \frac{1}{n_d} \sum (y_{t_i} - y_{t_j})^2 \tag{13}$$

where the summation is over all distinct pairs of sites in the sample that are distance $d$ apart and $n_d$ is the number of pairs that distance apart. For irregularly spaced data, pairs of sites that are approximately the same distance apart may be lumped together. Finally, a smooth curve is fitted to the variogram estimates to obtain values of the variogram for all distances. Estimation of the semivariance function $\gamma$ rather than of the covariance function $c$ is recommended by Cressie (1991), who points out that $\hat{\gamma}(d)$ is an unbiased estimator of $\gamma(d)$, while $\hat{c}(d)$ is not unbiased for $c(d)$ and that the semivariance estimates do better than the covariance estimates in the presence of an undetected linear trend in the variable of interest. Robust alternatives to $\hat{\gamma}(d)$ are also described in Cressie (1991).

The prediction equations may be written in terms of the semivariogram. Suppose again that observations have been made at a sample of $n$ sites, and one wishes to predict what will be found at another site. For notational simplicity, write $y_i$ for the $y$-value observed at the $i$th site in the sample, for $i = 1, \ldots, n$, and write $\gamma_{ij}$ for $\gamma(y_i - y_j)$, the semivariogram value for the difference between the $i$th and $j$th sites. From observed $y$-values at sites $t_1, \ldots, 4_n$, it is desired to predict the value of the random variable $y_0$ at the site $t_0$.

The object once again is to find a function $\hat{y}_0$ of the $n$ observed $y$-values that is unbiased for $y_0$, that is,

$$E(\hat{y}_0) = E(y_0) \tag{3}$$

and that minimizes the mean square prediction error

$$E(y_0 - \hat{y}_0)^2 \tag{4}$$

Writing the linear estimator as

$$\hat{y}_0 = \sum_{i=1}^{n} a_i y_i \tag{5}$$

the problem is to find values $a_1, a_2, \ldots, a_n$ that minimize the mean square prediction error subject to unbiasedness.

The solution is given in matrix form by

$$\mathbf{a} = \mathbf{\Gamma}^{-1} \gamma \tag{14}$$

where

$$\mathbf{a} = \begin{pmatrix} a_1 \\ a_2 \\ \vdots \\ a_n \\ m^* \end{pmatrix}, \quad \gamma = \begin{pmatrix} \gamma_{10} \\ \gamma_{20} \\ \vdots \\ \gamma_{n0} \\ 1 \end{pmatrix}$$

$$\mathbf{\Gamma} = \begin{pmatrix} \gamma_{11} & \gamma_{12} & \cdots & \gamma_{1n} & 1 \\ \gamma_{21} & \gamma_{22} & \cdots & \gamma_{2n} & 1 \\ \vdots & \vdots & \ddots & \vdots & \vdots \\ \gamma_{n1} & \gamma_{n2} & \cdots & \gamma_{nn} & 1 \\ 1 & 1 & \cdots & 1 & 0 \end{pmatrix}$$

The mean square prediction error of $\hat{y}_0$ is

$$E(y_0 - \hat{y}_0)^2 = \sum_{i=1}^{n} a_i \gamma_{i0} + m^* \tag{15}$$

Note that $m* = -m$, where $m$ is the constant obtained using the covariance-based prediction equation.

**Example 2: Reworked with Semivariogram.**    The shrimp example (Example 1) will be reworked using the semivariogram. The semivariogram, obtained from the covariance function [Equation (9)] used in Example 1, is

$$\gamma(x) = 5.1(1 - e^{-0.49x})$$

The semivariance for the two tows 6 nmi. apart is

$$\gamma_{12} = 5.1(1 - e^{-0.49(6)}) = 4.8$$

The semivariances with the new site are $\gamma_{10} = 5.1\{1 - \exp[-0.49(1)]\} = 2.0$ and $\gamma_{20} = 4.7$. The prediction equation [Equation (14)] with the semivariance is

$$\begin{pmatrix} a_1 \\ a_2 \\ m^* \end{pmatrix} = \begin{pmatrix} 0 & 4.8 & 1 \\ 4.8 & 0 & 1 \\ 1 & 1 & 0 \end{pmatrix}^{-1} \begin{pmatrix} 2.0 \\ 4.7 \\ 1 \end{pmatrix}$$

$$= \begin{pmatrix} -0.104 & 0.104 & 0.5 \\ 0.104 & -0.104 & 0.5 \\ 0.5 & 0.5 & -2.4 \end{pmatrix} \begin{pmatrix} 2.0 \\ 4.7 \\ 1 \end{pmatrix} = \begin{pmatrix} 0.78 \\ 0.22 \\ 0.95 \end{pmatrix}$$

so $a_1 = 0.78$, $a_2 = 0.22$, and $m^* = 0.95$.
   The predicted catch [using Equation (5)] is

$$\hat{y}_0 = 0.78(5.526) + 0.22(1.417) = 4.622$$

or 4622 lb at the new location, and the prediction mean square error [using Equation (15)] is

$$\hat{E}(y_0 - \hat{y}_0)^2 = 0.78(2.0) + 0.22(4.7) + 0.95 = 3.5$$

as before.                                                                       □

## 20.4  PREDICTING THE VALUE OVER A REGION

In previous sections, the observed values $y_1, \ldots, y_n$ at $n$ sites were used to predict the $y$-value at a new site. In many spatial sampling situations, however, one wishes to predict the mean (or total) of the $y$-values over a region $A$. Denote this mean by $y_0$, where

$$y_0 = \frac{1}{N} \sum_{i=1}^{N} y_i$$

if $A$ is a study region partitioned into $N$ sites, or

$$y_0 = \frac{1}{|A|} \int_A y_t \, dt$$

where $|A|$ is the area of the study region, if samples can be taken at points $t$ throughout a continuous study region. The population regional mean $y_0$ is a random variable, since the $y$-value at each site is random.

Then the prediction solution [Equation (14)] of Section 20.3 holds, with

$$\gamma_{i0} = \frac{1}{|N|} \sum_{j=1}^{N} \gamma_{ij}$$

in the discrete case, and

$$\gamma_{i0} = \frac{1}{|A|} \int_A \gamma(y_i - y_t) \, dt$$

in the continuous case. In each case, $\gamma_{i0}$ is the average semivariance between site $i$ and sites in the region $A$.

The minimized mean square prediction error is

$$E(y_0 - \hat{y}_0)^2 = \sum_{i=1}^{n} a_i \gamma_{i0} + m^* - \gamma_{00}$$

where the average semivariance $\gamma_{00}$ is

$$\gamma_{00} = \frac{1}{N^2} \sum_{i=1}^{N} \sum_{j=1}^{N} \gamma_{ij}$$

in the discrete case, and

$$\gamma_{00} = \frac{1}{|A|^2} \int_A \int_A \gamma(t - v)\, dt\, dv$$

in the continuous case.

## 20.5 DERIVATIONS AND COMMENTS

The linear prediction equations may be derived using Lagrange's method for finding the minimum of a function subject to a set of constraints. Suppose that a general linear model with $p$ parameters holds for the $n$ random variables $Y_i$ associated with the sample $s$ and for the random variable $Y_0$ that one wishes to predict, that is,

$$E(Y_i) = \sum_{k=1}^{p} \beta_k x_{ik}$$

$$\mathrm{cov}(Y_i, Y_j) = c_{ij}$$

for $i, j = 0, 1, 2, \ldots, n$.

Writing the predictor as $\hat{Y}_0 = \sum_{i=1}^{n} a_i Y_i$, the linear unbiased prediction problem is to find the values of the coefficients $a_1, \ldots, a_n$ that minimize the mean square prediction error $E(\hat{Y}_0 - Y_0)^2$ subject to the unbiasedness condition $E(\hat{Y}_0) = E(Y_0)$.

Note that expectations are in terms of the distribution of the random variables under the model. Implicitly, the expectation is conditional on the sample $s$ selected, so that model unbiasedness is required to hold for the given sample selected, no matter what sample is selected.

The unbiasedness condition implies that

$$\sum_{i=1}^{n} a_i \sum_{k=1}^{p} \beta_k x_{ik} = \sum_{k=1}^{p} \beta_k x_{0k}$$

no matter what the values of the parameters $\beta_k$ may be. Reversing the order of summation on the left-hand side and equating coefficients of $\beta_k$, one obtains the $p$ constraints

$$\sum_{i=1}^{n} a_i x_{ik} = x_{0k}$$

for $k = 1, \ldots, p$.

The mean square prediction error is

$$E(Y_0 - \hat{Y}_0)^2 = \mathrm{var}(Y_0 - \hat{Y}_0)$$

because of the unbiasedness condition $E(Y_0 - \hat{Y}_0) = 0$. This may be written

$$\text{var}(Y_0 - \hat{Y}_0) = \text{var}(\hat{Y}_0) + \text{var}(\hat{Y}_0) - 2\text{cov}(Y_0, \hat{Y}_0)$$

$$= \text{var}(Y_0) + \text{var}\left(\sum_{i=1}^{n} a_i y_i\right) - 2\text{cov}\left(Y_0, \sum_{i=1}^{n} a_i y_i\right)$$

$$= c_{00} + \sum_{i=1}^{n}\sum_{j=1}^{n} a_i a_j c_{ij} - 2\sum_{i=1}^{n} a_i c_{i0} \tag{16}$$

Write

$$F(a_1, \ldots, a_n) = c_{00} + \sum_{i=1}^{n}\sum_{j=1}^{n} a_i a_j c_{ij} - 2\sum_{i=1}^{n} a_i c_{i0}$$

and write

$$G_k(a_1, \ldots, a_n) = \sum_{i=1}^{n} a_i x_{ik} - x_{0k}$$

for $k = 1, \ldots, p$. To minimize $F$ subject to the $p$ constraints $G_k = 0$, define the new function

$$H(a_1, \ldots, a_n, \lambda_1, \ldots, \lambda_p) = F - \lambda_1 G_1 - \cdots - \lambda_p G_p$$

in which the $\lambda_k$ are the Lagrange multipliers.

The partial derivative of $H$ with respect to $a_i$ is

$$\frac{\partial H}{\partial a_i} = 2\sum_{j=1}^{n} a_j c_{ij} - 2c_{i0} - \lambda_1 x_{i1} - \lambda_2 x_{i2} - \cdots - \lambda_p x_{ip}$$

for $i = 1, \ldots, n$. The partial derivative of $H$ with respect to $\lambda_k$ is

$$\frac{\partial H}{\partial \lambda_k} = -\sum_{i=1}^{n} a_i x_{ik} + x_{0k}$$

for $k = 1, \ldots, p$.

Setting each of the partial derivatives to zero gives $(n + p)$ equations in $(n + p)$ unknowns. Defining for simplicity $m_k = -\lambda_k/2$, these equations may be written

$$\sum_{j=1}^{n} a_j c_{ij} + \sum_{k=1}^{p} m_k x_{ik} = c_{i0} \qquad i = 1, \ldots, n$$

$$\sum_{i=1}^{n} a_i x_{ik} = x_{0k} \qquad k = 1, \ldots, p \tag{17}$$

In matrix notation, these equations may be written

$$\mathbf{f}\mathbf{G} = \mathbf{h}$$

where the vectors $\mathbf{f}$ and $\mathbf{h}$ are

$$\mathbf{f} = \begin{pmatrix} a_1 \\ a_2 \\ \vdots \\ a_n \\ m_1 \\ m_2 \\ \vdots \\ m_p \end{pmatrix}, \qquad \mathbf{h} = \begin{pmatrix} c_{10} \\ c_{20} \\ \vdots \\ c_{n0} \\ x_{01} \\ x_{02} \\ \vdots \\ x_{0p} \end{pmatrix}$$

and the matrix $\mathbf{G}$ is

$$\mathbf{G} = \begin{pmatrix} c_{11} & c_{12} & \cdots & c_{1n} & x_{11} & x_{12} & \cdots & x_{1p} \\ c_{21} & c_{22} & \cdots & c_{2n} & x_{21} & x_{22} & \cdots & x_{2p} \\ \vdots & \vdots & \ddots & \vdots & \vdots & \vdots & \ddots & \vdots \\ c_{n1} & c_{n2} & \cdots & c_{nn} & x_{n1} & x_{n2} & \cdots & x_{np} \\ x_{11} & x_{21} & \cdots & x_{n1} & 0 & 0 & \cdots & 0 \\ x_{12} & x_{22} & \cdots & x_{n2} & 0 & 0 & \cdots & 0 \\ \vdots & \vdots & \ddots & \vdots & \vdots & \vdots & \ddots & \vdots \\ x_{1p} & x_{2p} & \cdots & x_{np} & 0 & 0 & \cdots & 0 \end{pmatrix}$$

Assuming that the inverse of $\mathbf{G}$ exists, the solution is given by

$$\mathbf{f} = \mathbf{G}^{-1}\mathbf{h}$$

In partitioned form, the equation may be written

$$\begin{pmatrix} \mathbf{a} \\ \mathbf{m} \end{pmatrix} = \begin{pmatrix} \mathbf{V} & \mathbf{X} \\ \mathbf{X}' & \mathbf{0} \end{pmatrix}^{-1} \begin{pmatrix} \mathbf{c_0} \\ \mathbf{x_0} \end{pmatrix}$$

where $\mathbf{a}$ is the $(n \times 1)$ vector of coefficients, $\mathbf{m}$ the $(p \times 1)$ vector of Lagrange constants, $\mathbf{V}$ is the $(n \times n)$ variance–covariance matrix of the $n$ sample units, $\mathbf{X}$ is the $(n \times p)$ design matrix for the sample under the linear model, $\mathbf{0}$ is a $(p \times p)$ matrix of zeros, $\mathbf{c_0}$ is the $(n \times 1)$ vector of covariances of each sample unit with $Y_0$, and $\mathbf{x_0}$ is the $(n \times 1)$ vector of $x$'s in the linear model for the expected value of $Y_0$.

For computing the solution, the dimension of matrix inversion may be reduced by using the relationship (see C. R. Rao 1973, p. 33)

$$\mathbf{G}^{-1} = \begin{pmatrix} \mathbf{V}^{-1} - \mathbf{V}^{-1}\mathbf{X}(\mathbf{X}'\,\mathbf{V}^{-1}\mathbf{X})^{-1}\mathbf{V}^{-1} & \mathbf{V}^{-1}\mathbf{X}(\mathbf{X}'\,\mathbf{V}^{-1}\mathbf{X})^{-1} \\ (\mathbf{X}'\,\mathbf{V}^{-1}\mathbf{X})^{-1}\mathbf{X}'\,\mathbf{V}^{-1} & -(\mathbf{X}'\,\mathbf{V}^{-1}\mathbf{X})^{-1} \end{pmatrix}$$

The mean square prediction error, using Equations (16) and (17), is

$$
\begin{aligned}
E(Y_0 - \hat{Y}_0)^2 &= c_{00} - 2\sum_{i=1}^{n} a_i c_{i0} + \sum_{i=1}^{n} a_i \left( c_{i0} - \sum_{k=1}^{p} m_k x_{ik} \right) \\
&= c_{00} - \sum_{i=1}^{n} a_i c_{i0} - \sum_{k=1}^{p} m_k \sum_{i=1}^{n} a_i x_{ik} \\
&= c_{00} - \sum_{i=1}^{n} a_i c_{i0} - \sum_{k=1}^{p} m_k x_{0k}
\end{aligned}
$$

If the quantity to be predicted is a linear combination of the $Y$-values of a finite number $N$ of population units, that is, $Y_0 = \mathbf{l}'\mathbf{Y}$, the problem is identical to that considered in the section on the prediction approach to regression estimation in sampling. Using the notation of Chapter 8, in which the subscript $s$ refers to the sample units and $r$ refers to the rest of the units, the covariance of $Y_0$ with the vector $\mathbf{Y}_s$ of sample variables is $\mathbf{c}_0 = \mathbf{V}_{ss}\mathbf{l}_s + \mathbf{V}_{sr}\mathbf{l}_r$, and the predictor $\hat{Y}_0$ can be written as in Section 8.4 as

$$\hat{Y}_0 = \mathbf{a}'\mathbf{Y}_s = \mathbf{l}'_s\mathbf{Y}_s + \mathbf{l}'_r[\mathbf{X}_r\hat{\beta} + \mathbf{V}_{rs}\mathbf{V}_{ss}^{-1}(\mathbf{Y}_s - \mathbf{X}_s\hat{\beta})]$$

where

$$\hat{\beta} = (\mathbf{X}'_s\mathbf{V}_{ss}^{-1}\mathbf{X}_s)^{-1}\mathbf{X}'_s\mathbf{V}_{ss}^{-1}\mathbf{Y}_s$$

Thus, the prediction problem addressed in regression estimation in survey sampling and in 'kriging' of geostatistics are essentially the same, with a historical difference in emphasis. In the sampling literature on the prediction approach to regression, a typical model assumed has the $y$-values uncorrelated (diagonal matrix $\mathbf{V}$) but related to auxiliary variables (the columns of $\mathbf{X}$), and the object is to predict a linear combination of a finite number of units. In the geological literature, the emphasis has been on models with covariances related to distance or relative position of sites, whether or not auxiliary information is available, and on prediction of an integral of the $y$-value over a continuous region.

## EXERCISE

1. The shrimp survey data (Examples 1 and 2) give a covariance function $c(x) = 5.1 \exp(-0.49x)$, where $x$ is distance in nautical miles. The catches at

sites 1 and 2, which are 3 nmi. apart, are 1.200 and 2.400, respectively (units are in thousands of pounds). Predict the catch $y_0$ at a new site:

**(a)** Halfway between sites 1 and 2

**(b)** 5 nmi. from each of sites 1 and 2

**(c)** 1 nmi. from site 2 and 4 miles from site 1

**(d)** 4 nmi. from site 1 and 5 miles from site 2

Along with each prediction, give the prediction mean square error (MSE).

# CHAPTER 21

# Spatial Designs

The mean square prediction error of an unbiased predictor linear $\hat{y}_0$ of a value $y_0$—whether $y_0$ is the value of the variable of interest at a particular site or the mean value over a whole region—depends on the covariances between each sample site and $y_0$ and the covariances among different sites in the sample. One can thus use the information in the covariance function or the variogram to determine what selection of sample sites will give the best prediction, that is, what sampling design is most effective.

Let the covariance $\text{cov}(y_i, y_j)$ between the $y$-values at sites $i$ and $j$ be denoted $c_{ij}$, and let $c_{i0}$ denote the covariance between $y_i$ and $y_0$. Since $E(\hat{y}_0) = E(y_0)$, the mean square prediction error of $\hat{y}_0 = \sum a_i y_i$ is

$$E(y_0 - \hat{y}_0)^2 = \text{var}(y_0 - \hat{y}_0) = \text{var}(\hat{y}_0) + \text{var}(\hat{y}_0) - 2\text{cov}(y_0, \hat{y}_0)$$

Thus,

$$E(y_0 - \hat{y}_0)^2 = c_0 + \sum_{i=1}^{n}\sum_{i=1}^{n} a_i a_j c_{ij} - 2\sum_{i=1}^{n} a_i c_{i0}$$

From the expression above, it is evident that the best predictions will result from a sample of $n$ sites having low covariance with each other and having high covariance with the value to be predicted. Assume that the covariance function is radially symmetric and decreases with distance (equivalently, the variogram is symmetric and increases with distance). High covariance with the value to be predicted is then achieved by locating sample sites near the site (site 0) for which the prediction is to be made, or if the mean value for the whole study region is to be predicted, locating sites in the interior or near the center of the study region. Low covariance among sample sites is achieved by spacing out the sample sites in a systematic fashion or by dividing the study region into many small strata.

## 21.1  DESIGN FOR LOCAL PREDICTION

For predicting $y$-values at individual sites throughout the study region, one may minimize the worst prediction errors by selecting the $n$ sites to minimize the farthest distance of any point in the study region from nearby sample sites. This problem is examined in McBratney et al. (1981a,b) and in Yfantis et al. (1987), in which systematic sample points on square, triangular, and hexagonal grids are compared. For a given number of sample points per unit area, the triangular grid, in which observations are made at the vertices of equilateral triangles, minimized the farthest distance from sample points to nonsample points.

## 21.2  DESIGN FOR PREDICTION OF MEAN OF REGION

Matérn (1960, 1986) investigated the efficiency of various sampling designs for predicting the mean over a region, motivated by problems in forestry surveys. Radially symmetric, decreasing covariance functions were assumed. For systematic samples, he found that the triangular grid was slightly more efficient than the square grid. The square grid could be improved slightly with a modification moving the sample points of every other row by one-half the side of a square. Rectangular grids were less efficient. Boundary effects—influenced by the size and shape of the study region—were ignored.

Matérn also considered the related problem of most efficient stratification for stratified random sampling of such populations. The most efficient design partitioned the region into small strata of compact shape. For a given stratum area, the most efficient shapes in order of efficiency were: circular, hexagonal, square, equilateral triangle. Note, however, that most study regions could not be partitioned into circular strata. The first three of the shapes above were very close in efficiency, while triangular strata were somewhat less efficient. Rectangular strata were considerably less efficient—the less square in shape, the less efficient.

The advantage of a systematic sample derives from spreading the sample locations apart when nearby sample units may be positively correlated. A systematic sample is inefficient, however, when the alignment of units coincides with a periodic pattern in the population. An unaligned systematic sample may be selected as follows (Quenouille 1949; see also Bellhouse 1988a; Cochran 1977, p. 228). A rectangular study region is divided into $m_1 m_2$ squares, in $m_1$ rows and $m_2$ columns. The length of the side of a square is $k$. A random sample of $m_1$ uniform variates from the interval $(0, k)$ gives the horizontal coordinates of sample locations in the first column. A random sample of $m_2$ uniform variates from $(0, k)$ gives vertical coordinates for the first row. The sample point in the first row and first column is thus selected. The remaining sample points are determined by adding multiples of the constant interval $k$ to each chosen horizontal coordinate and to each chosen vertical coordinate for each row and column.

For very general models of spatial covariance, Bellhouse (1977) showed the optimality of systematic sampling within restricted classes of designs. With a model using random lines to approximate the boundaries of regions, Bellhouse (1981) found that a stratified sample with one unit per stratum was more efficient than a systematic grid for estimating the proportion of a study region covered by a resource of interest. Systematic and stratified designs for geological sampling are investigated in Olea (1982, 1984a,b). Barnes (1988) looked at sample size for estimation of extreme values, rather than average values, in a region. Some aspects of spatial sampling for environmental pollution monitoring are summarized in Gilbert (1987). Reviews include Thompson (1997a,b) and a recent general reference to spatial design issues is Muller (2001).

McArthur (1987) compared several kinds of sampling designs and estimators for estimating the mean concentration of a point-source pollutant over a region. Motivated by the problem of estimating radioactive contamination in the vicinity of underground bomb tests, hypothetical pollutant concentration was modeled as bivariate Gaussian density function. Sample designs compared were simple random sampling; stratified random sampling, with higher allocation in strata with higher pollution concentration; systematic sampling; stratified systematic sampling, with a finer grid of sample points in the stratum with higher pollutant concentration; and unequal probability sampling, with selection probabilities higher in regions near the point source. The usual design-unbiased estimators (sample mean, stratified sample mean, Hansen–Hurwitz) were compared with kriging prediction methods. The most efficient strategy was stratified systematic sampling, followed by unstratified systematic sampling, with the usual design-unbiased stratified estimators. The model assumptions under which kriging estimates are best linear unbiased predictors did not hold, since the pollutant concentration was a deterministic function. For this fixed population, the kriging estimates were design-biased downward, and the computed kriging variances underestimated the design-based sampling variances of the predictions.

CHAPTER 22

# Plot Shapes and
# Observational Methods

Populations of plants and animals tend to exhibit characteristic spatial patterns. In many cases, the patterns observed are not consistent with a random distribution, but show evidence of aggregation tendencies caused by behavioral interactions or environmental patchiness, or regularities in spacing caused by such factors as territoriality or other forms of mutual inhibition. The appropriate model for such a population is a stochastic point process (see Cox and Isham 1980; Diggle 1983; Ripley 1981). Matérn (1960, 1986), Schweder (1977), and Thompson and Ramsey (1987) have examined ecological sampling problems in this framework.

A stochastic point process gives rise through some probabilistic mechanism to a pattern of point objects in space—representing, for instance, the locations of animals or plants in a study region. The number of point objects in any region (or subregion) is a random variable. A *realization* of the point process determines a complete spatial pattern of the point objects or locations of plants or animals. For convenience, the point objects will be referred to as animals.

Sampling methods for plant and animal populations include methods with plots, in which every animal or plant within a sample plot is observed, and methods such as line transects or aerial surveys, in which some of the individuals are detected and some are missed, with detection probabilities given by a detectability function. To understand the properties of observations made in sampling such populations—and hence to determine efficient designs and estimation methods to use with such populations—it is useful to work with the mean and covariance density functions of the stochastic point process giving rise to the population.

## 22.1 OBSERVATIONS FROM PLOTS

For any region $A$, let $N(A)$ denote the number of animals in $A$. The region $A$ could, for example, be the whole study region or it could be a sample plot of a certain shape in a given location. The number of animals $N(A)$ is a random variable with expected value

$$E[N(A)] = \int_A \mu(x)\,dx$$

where $\mu(x)$ is the mean or density of the population at location $x$. If the mean (expected number of animals per unit area) is the same everywhere in the study region, then

$$E[N(A)] = \mu \int_A dx = \mu|A|$$

where $|A|$ is the area of the plot or region $A$.

The variance of $N(A)$ is

$$\text{var}[N(A)] = \int_A \mu(x)\,dx + \int_A\int_A c(w,v)\,dw\,dv$$

in which $c(w,v)$ is the covariance density of the process between locations $w$ and $v$. For a second-order stationary process, the covariance density depends only on the relative positions of the locations $w$ and $v$. If the covariance density depends only on the distance between locations, the process is said to be *isotropic*. For a population with a random distribution (Poisson process), the covariance density function $c$ is zero. A positive covariance density function decreasing with distance between locations is characteristic of populations with patchy or clumped spatial patterns.

The number of animals in one plot or region $A_1$ may not be independent of the number of animals in another plot or region $A_2$, especially if the two regions are near each other. The covariance between the number of animals in the first region and the number of animals in the second is

$$\text{cov}[N(A_1), N(A_2)] = \int_{A_1\cap A_2} \mu(x)\,dx + \int_{A_1}\int_{A_2} c(w,v)\,dw\,dv$$

The first integral is taken over the intersection $A_1 \cap A_2$ between the two plots or regions. Thus, the covariance between the number of animals observed at two plots will—not surprisingly—be higher if the two plots overlap.

Consider a covariance density function $c(w,v)$ which is nonnegative and symmetrically decreasing as a function of distance between locations $w$ and $v$. Such covariance functions were thought by Matérn and other writers to be the most common for natural populations. Then a round or square plot—in which locations are on the average close together—will have a higher variance than a long, thin plot of the same area in which some of the locations are far apart. Such variance comparisons have been observed in surveys of natural populations and were confirmed in numerical comparisons by Thompson and Ramsey (1987). Other investigations on the effect of plot size and shape include Matérn (1960, 1986), Starks (1986), and Zhang et al. (1990).

With the assumed nonnegative, symmetrically decreasing covariance density function, the covariance of numbers of animals at two sites will be higher if the sites are closer together (even without overlap). Thus, the spatial design conclusions above would apply, indicating that the most efficient selections of $n$ plots will be systematically configured or stratified into many small strata.

## 22.2  OBSERVATIONS FROM DETECTABILITY UNITS

A plot is associated with a special kind of a detectability function in which detectability is perfect over the plot and zero elsewhere. The detectability function for a plot $A$ can be written $g(x) = 1_A(x)$, where the indicator function $1_A(x)$ is 1 when $x$ is in $A$ and zero otherwise.

Many other observational methods in ecological surveys can be described by detectability functions. With line transects, the detectability function may equal 1 along the line, deceasing laterally with distance to either side. With variable circular plots (see Ramsey and Scott 1979), in which the observer stands at one location and records distances to animals detected, the detectability function may be symmetric and radially decreasing, centered at the observer's location. For trawl surveys of fish or shellfish, the detectability function may be some unknown constant $p$ over the area swept by the net and zero elsewhere. For some aerial surveys, the detectability function may be some constant $p$ over a plot or over the entire study region.

Let $g(x)$ denote whatever detectability function describes the observational method used in a survey of an animal or plant population that is represented as a stochastic point process with mean function $\mu(x)$ and covariance density function $c(w, v)$. Let $y$ be an observation, that is, the number of animals (plants) detected by the observer, whether in a plot, on a transect, at a site, with a trawl, from the air, or by whatever survey method is used.

The expected number of animals seen is

$$E(y) = \int g(x)\mu(x)\,dx$$

If the mean $\mu$ is constant, the expected number is

$$E(y) = \mu \int g(x)\,dx = \mu a$$

where

$$a = \int g(x)\,dx$$

is the effective area observed.

The variance of the observed number of animals $y$ is

$$\mathrm{var}(y) = \int g(x)\mu(x)\,dx + \int\int g(w)g(v)c(w,v)\,dw\,dv$$

Let $y_i$ and $y_j$ be two distinct observations: the numbers of animals observed in two different plots, on transects at two different sites, or with trawls at two different locations, for example. Let $g_i(x)$ and $g_j(x)$ be the detectability functions associated with the two observations: indicator functions for the two plots, line transect detectability functions positioned over two different lines, or functions of height $p$ over two rectangular trawl paths, for example. Then the covariance of $y_i$ and $y_j$ is

$$\mathrm{cov}(y_i, y_j) = \int g_i(x)g_j(x)\mu(x)\,dx + \int\int g_i(w)g_j(v)c(w,v)\,dw\,dv$$

The first term represents the overlap between the two detectability functions, whether true overlap of plots or the possibility of seeing the same animals from nearby transects. The second term depends on both the covariance density of the population and the detectability functions associated with the survey methods used.

Thus, the means, variances, and covariances of the sample observations—and hence the properties of estimators formed from these observations—depend both on the characteristics of the population ($\mu$ and $c$) and on the observational methods employed (the $g$'s). Note that the covariance of an observation $y_i$—say, the number of animals observed on the $i$th transect—with the number of animals $N(A)$ in the whole study region $A$, can be obtained from the formula above by associating with the study area the indicator detectability function $g(x) = 1_A(x)$. Thus the prediction formulas of Chapter 20 can be computed using the covariance expression above.

## 22.3 COMPARISONS OF PLOT SHAPES AND DETECTABILITY METHODS

The density of animals (or other objects) in the study region is $D = N(A)/|A|$. The expected density is $E(D) = \mu$. With a single plot or other detectability unit of effective area $a$, an unbiased estimate of density (assuming that $a$ is known) is

$$\hat{D} = \frac{y}{a}$$

where $y$ is the number of animals detected. The variance $\mathrm{var}(\hat{D})$ and the mean square prediction error $E(\hat{D} - D)^2$ can be computed using the variance and covariance expressions above to compare the effectiveness of any shape of plot or type of detectability function.

Comparisons of the mean square prediction errors obtained with plots of different shapes and with detectability functions associated with different survey methods

**Table 22.1. Comparative Mean Squared Error and Variance of Estimates Using Different Detectability Functions of Equal Effective Area**

| Detectability Function[a] | MSE Term[b] | Variance Term[b] |
|---|---|---|
| Constant (no. 7): $g = 0.25$ | 0 (0.0000) | 0.2091 (0.0027) |
| Line transect, exponential profile (no. 4) | 0.0631 (0.0044) | 0.3179 (0.0101) |
| Square doughnut plot (no. 8) | 0.0744 (0.0094) | 0.1624 (0.0132) |
| Point transect (no. 6) | 0.0778 (0.0050) | 0.3658 (0.0113) |
| Line transect, half-normal profile (no. 5) | 0.1059 (0.0079) | 0.3857 (0.0148) |
| Round doughnut plot (no. 9) | 0.1060 (0.0097) | 0.1706 (0.0145) |
| Rectangular plot (no. 3) | 0.1358 (0.0135) | 0.4284 (0.0212) |
| Square plot (no. 1) | 0.1632 (0.0140) | 0.4863 (0.0222) |
| Round plot (no. 2) | 0.1700 (0.0142) | 0.5010 (0.0226) |

*Source*: Thompson and Ramsey (1987). With permission from the Biometric Society.

[a]Detectability function identification numbers correspond to the descriptions in the text.

[b]Only the double-integral components of MSE and variance are computed. Standard errors in parentheses refer to precision of numerical integrations.

are given in Table 22.1. Each of the detectability functions compared has the same effective area (for the plots, this is the actual area). For the comparisons, the study region is the unit square and each plot or other detectability unit has effective area $a = 0.25$.

The covariance density function assumed for the population is $c(w, v) = \exp(-r^2/\sigma^2)$, where $r$ is the distance between locations $w$ and $v$ and the scale parameter is $\sigma = 0.1$. This covariance function is characteristic of a Poisson cluster process, representing a population with aggregation tendencies.

The nine types of plots and detectability functions compared in the table and illustrated in Figure 22.1 are as follows:

1. A square plot
2. A round plot
3. A rectangular plot
4. A line transect of unit length and exponential profile
5. A line transect of unit length and half-normal profile
6. A variable circular plot (point transect) with half-normal shape
7. A constant detectability over the whole study region
8. A "square doughnut" plot along the perimeter of the region
9. A "round doughnut" plot—actually, a square doughnut with a round hole

The MSE term in the table indicates the relative efficiency in predicting the density of the population in the study region. The mean square prediction error

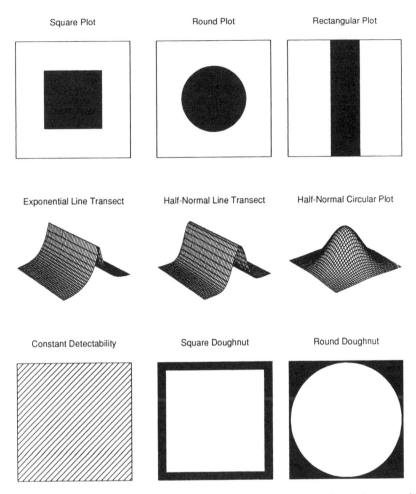

**Figure 22.1.** Plots and other detectability units, each of equal effective area observed, compared for effectiveness in estimating or predicting density in a study region.

$E(\hat{D} - D)^2$ for a given detectability unit equals the MSE term in the table plus a constant that is the same for each type of detectability unit. The variance term in the table pertains to estimating the parameter $\mu$, expected density, an objective with possibly less frequent application. The variance $var(\hat{D})$ equals the variance term in the table plus a constant term which is the same for each unit.

The most efficient method spreads the detectability over the whole study area, as in certain aerial surveys. Long, thin rectangular plots are more efficient than square or round plots. The various line transects, variable circular plots (radial transects), and plots with holes in them give intermediate results. The "doughnut" plots may be little more than a curiosity. They achieve low variance by spreading the parts of the plot away from each other to minimize correlations between points within the plot.

In designing a survey, one cannot at will choose every aspect of the detectability functions involved. Much may depend on the behavior of the animals and on the habitat. Many choices made in the observational methods do affect the detectability functions, however. Certainly, one can choose the size and shape of plots. In line transect and variable circular plot surveys, decisions regarding how far and how fast to walk or how long to remain at a site affect the shape of the detectability function. In trawl surveys of fish, such factors as net dimensions and mesh size and distance towed determine the detectability function. When detectability or effective area is estimated, an additional term of variability is introduced and the estimates are not unbiased.

The efficiencies of different sampling designs, that is, of different spatial selections of $n$ detectability units in a study region, could also be compared using the variance and covariance formulas above. Further research is needed in that direction.

# PART VI

# Adaptive Sampling

CHAPTER 23

# Adaptive Sampling Designs

## 23.1 ADAPTIVE AND CONVENTIONAL DESIGNS AND ESTIMATORS

Conventionally, much of the attention in sampling theory and methodology has been limited to sampling designs in which the selection procedure does not depend in any way on observations made during the survey, so that the entire sample of units may be selected prior to the survey. But in many sampling situations, a researcher may feel inclined to make decisions during a survey, based on what has been observed thus far, as to which sites or how many sites to observe next. For example, in a survey of a rare, spatially clustered animal population, the researcher may wish to add neighboring sites to the sample once high abundance of the species has finally been encountered. *Adaptive sampling* refers to sampling designs in which the procedure for selecting sites or units to be included in the sample may depend on values of the variable of interest observed during the survey.

The primary purpose of adaptive sampling designs is to take advantage of population characteristics to obtain more precise estimates of population abundance or density, for a given sample size or cost, than is possible with conventional designs. For example, many populations of animals and plants have aggregation tendencies due to such factors as schooling, flocking, dispersal patterns, and environmental patchiness. Minerals, fossil fuels, and some human populations can exhibit similar patterns. Often, the location and shape of the aggregations cannot be predicted before a survey so that traditional means of increasing precision such as stratification are not sufficient. For such populations, adaptive sampling strategies may provide a way to increase dramatically the effectiveness of sampling effort. A secondary advantage of adaptive sampling may be the increase in the yield of interesting observations—for example, the increased number of animals observed or amount of mineral obtained during the survey—which may result in better estimates of other parameters of interest.

Adaptive selection procedures may introduce biases into conventional estimators, so that estimators and estimators of variance that are unbiased under the adaptive designs are needed. Design-unbiased estimators are emphasized in the following chapters on adaptive sampling designs; that is, the unbiasedness does not depend on assumptions about the population.

Adaptive sampling methods offer the potential to give large increases in efficiency for some populations. Even after conventional means of increasing precision or practicality, such as stratification and systematic or cluster arrangements, have been applied, adaptive procedures can increase precision still further. For a given population, the ultimate choice of procedure will depend both on the known characteristics of the population and on practical considerations of cost and convenience.

## 23.2   BRIEF SURVEY OF ADAPTIVE SAMPLING

Sampling designs in which the selection procedure depends on observed values of the variable of interest have been of theoretical interest to statisticians for some time. In his paper establishing that the minimal sufficient statistic in finite population sampling is the unordered set of distinct observations together with their unit labels, Basu (1969) expressed the view that the most efficient designs would be ones in which the selection probabilities were conditional on the values observed. Zacks (1969) described an optimal fixed-sample-size adaptive design from a Bayesian perspective; While recognizing the theoretical advantage of designs depending on the values of the variable of interest, Solomon and Zacks (1970) observed that the optimal design as described would be impractically complex to implement and advocated the development of much simpler sequential or two-phase designs. Cassel et al. (1977) summarized the subsequent literature on sampling designs that make use of observed values (*informative designs*, in their terminology), but found little of practical interest there.

In the statistical literature on sequential statistical methods (see Chernoff 1972; Siegmund 1985; Wald 1947; Woodroofe 1982) many results are found showing advantages over nonsequential methods such as increased power, lower expected sample size, and more controllable precision. Sampling designs that depend on the variable of interest are necessarily sequential but go beyond the usual situation considered in sequential statistics in that the unit labels in the sampling data make it possible to choose during a survey not just how much to sample next but which units or group of units to sample next. Although these labels are responsible for many of the complications in the theory of finite population sampling (see discussions in Cassel et al. 1977; Chaudhuri and Vos 1988), estimators that use the labels are in some cases better than estimators that do not use the labels. This is certainly the case with the designs described in the following chapters, in which unbiased estimators utilizing information from the labels in the data have lower variance than the unbiased estimator that does not use the unit labels.

Sampling designs in which the allocation of effort to different strata or primary units are based adaptively on initial observations relate to the wider statistical literature on sequential allocation or *bandit problems* (Barry and Fristedt 1985; Robbins, 1952). But again, the labels identifying the sampling units provide expanded opportunities and problems. Adaptive designs in which the sample size of a simple random sample within primary units or strata depends instead on initial observations within those primary units or strata are discussed in Francis (1984) and Kremers (1987). Adaptive strategies in which the sample size depends instead on observed values in neighboring primary units or strata are presented in Thompson (1988), Thompson and Ramsey (1983), and Thompson et al. (1992).

The concept of unbiased estimation based on the design has had much influence in survey sampling practice since Neyman (1934) and has been the topic of more recent discussions in Cassel et al.(1979), Godambe (1982) and Särndal (1978). By the Rao–Blackwell theorem, any unbiased estimator that is not a function of the minimal sufficient statistic can be improved upon by taking its conditional expectation given the sufficient statistic. Blackwell's (1947) contribution to the topic was motivated by the problem of obtaining an unbiased estimate of the mean following sequential stopping. The method has since been used in a sequential context by Ferebee (1983) for estimating the drift of Brownian motion. Kremers (1987) applied the Rao–Blackwell Theorem to two-stage adaptive sampling of a finite population with sample size depending on the values of initial observations and gave variance and variance estimation expressions for the estimator obtained. Similar use of the method has been made in Kremers (1986) and Kremers and Robson (1987). With finite population sampling, an unbiased estimator obtained by the Rao–Blackwell method is not in general a unique minimum-variance unbiased estimator because the sufficient statistic is not complete, and in the following chapters, more than one distinct estimator is obtained through the Rao–Blackwell method.

Adaptive cluster sampling designs in which the initial sample is selected by simple random sampling, with or without replacement, are described in Thompson (1990). Adaptive cluster sampling designs in which the initial sample may be a strip sample or systematic sample are described in Thompson (1991a). Adaptive cluster sampling designs in which the population is initially stratified (but adaptive follow-up may cross stratum boundaries) are described in Thompson (1991b). The importance of adaptive sampling methods for ecological sampling was discussed by Cormack (1988) and Seber (1986, 1992).

Recent years have seen many new developments and uses of adaptive sampling strategies. These include two-stage adaptive cluster sampling (Salehi and Seber 1997b), adaptive cluster double sampling (Félix Medina 2000b, Félix Medina and Thompson 1999), unequal probability adaptive cluster sampling (Pontius 1997; Roesch 1993; Smith et al. 1995) optimal adaptive allocation (Francis 1984), design-unbiased adaptive allocation (Thompson et al. 1992), adaptive cluster sampling without replacement of networks (Salehi and Seber 1997a), adaptive cluster sampling without replacement of clusters (Dryver 1999), adaptive Latin square sampling (Borkowski 1998; Munholland and Borkowski 1997), adaptive cluster sampling based on order statistics (Thompson 1996), multivariate aspects of adaptive sampling (Thompson 1993), inverse adaptive cluster sampling (Christman and Lan 2001), and restricted adaptive cluster sampling to limit sample size (Brown and Manly 1998; Salehi and Seber 2001). Results to ease the computation of Rao–Blackwell improved estimators are contained in Dryver (1999), Félix Medina

(2000a), and Salehi (1999). Bootstrap confidence intervals for adaptive cluster sampling are discussed in Christman and Pontius (2000). Dealing with imperfect detectability with adaptive designs is addressed in Thompson and Seber (1994). Applications of adaptive sampling to natural and human populations are addressed in Acharyal et al. (2000), Blair (1999), Boomer et al. (2000), Brown (1994), Clausen et al. (1999), Danaher and King (1994), Francis (1984), Khaemba and Stein (2001), Petrucci (1998), Ramsey and Sjamsoe'oed (1994), Roesch (1993), Seber and Thompson (1993), Smith et al. (1995), M. E. Thompson (1997), and Werner et al. (2000). Small sample and asymptotic properties of estimators are examined in Christman (1997, 2000) and Félix Medina (2000b, 2002). Optimal adaptive designs under a spatial model are worked out in Chao (1999, 2002) and Chao and Thompson (2001). A range of topics in the theory and methods of adaptive sampling is described in a monograph by Thompson and Seber (1996).

CHAPTER 24

# Adaptive Cluster Sampling

In a number of sampling situations, field researchers carrying out the survey may feel an inclination to increase sampling effort adaptively in the vicinity of observed values that are high or otherwise interesting. *Adaptive cluster sampling* refers to designs in which an initial set of units is selected by some probability sampling procedure, and whenever the variable of interest of a selected unit satisfies a given criterion, additional units in the neighborhood of that unit are added to the sample. In this chapter, the simplest adaptive cluster sampling designs are considered: namely, those in which the initial sample is selected by simple random sampling with or without replacement.

For the sorts of situations in which field researchers feel the inclination to depart from the preselected sample plan and add nearby or associated units to the sample, adaptive cluster sampling accommodates that inclination almost completely. Consider a survey of a rare and endangered bird species in which observers record the number of individuals of the species seen or heard at sites or units within a study area. At many of the sites selected for observation, zero abundance may be observed. But whenever substantial abundance is encountered, observation of neighboring sites is likely to reveal additional concentrations of individuals of the species. Such patterns of clustering or patchiness are encountered with many types of animals from whales to insects, with vegetation types from trees to lichens, and with mineral and fossil fuel resources. A related pattern is found in epidemiological studies of rare, contagious diseases. Whenever an infected individual is encountered, addition to the sample of closely associated individuals reveals a higher-than-expected incidence rate. The results and examples of this chapter and the next two indicate that for some populations—particularly for rare, clustered populations—the researchers' inclinations are well justified in that manyfold increases in precision of estimates may be obtained with the adaptive strategy, compared to the precision of a conventional design of equivalent sample size. Most results of this chapter were given in Thompson (1990).

The basic idea of the designs in this chapter is illustrated in Figures 24.1 and 24.2, in which the problem is to estimate the mean number of point objects—which could, for example, represent locations of animals or mineral deposits—scattered

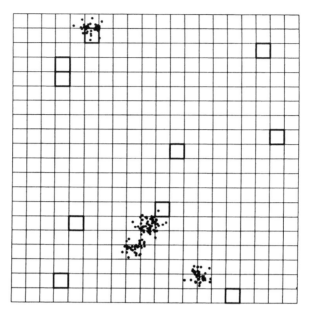

**Figure 24.1.** Adaptive cluster sampling to estimate the number of point objects in a study region of 400 units. An initial random sample of 10 units is shown. Adjacent neighboring units are added to the sample whenever one or more of the objects of the population is observed in a selected unit. [From Thompson (1990). With permission from the American Statistical Association.]

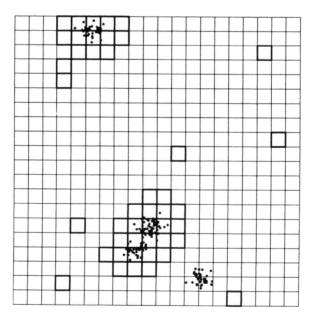

**Figure 24.2.** Sample resulting from method illustrated in Figure 24.1. [From Thompson (1990). With permission from the American Statistical Association.]

unevenly in a study region partitioned into 400 square sampling units. An initial random sample of 10 units is shown in Figure 24.1. Whenever one or more of the objects is observed in a selected unit, the adjacent neighboring units—to the left, right, top, and bottom—are added to the sample. When this process is completed, the sample consists of 45 units, shown in Figure 24.2. Neighborhoods of units may be defined in many ways other than the spatial proximity system of this example. For instance, the neighborhood of a unit could consist of a larger set of contiguous units or a systematic grid pattern of surrounding units.

Since a conventional estimator such as the sample mean may be biased when used with adaptive cluster sampling, estimators that are unbiased for the population mean are given in this chapter, along with unbiased estimators of their variances. The estimators given in this chapter are design-unbiased, that is, the unbiasedness is based on the way the sample is selected rather than on assumptions about the population. The estimators developed for adaptive cluster sampling are related to the Hansen–Hurwitz and Horvitz–Thompson estimators used with unequal probability sampling designs such as network sampling and line-intercept sampling. With adaptive cluster sampling, however, the selection and inclusion probabilities needed for those estimators cannot be determined for every unit in the sample, so modified estimators are used.

## 24.1  DESIGNS

As in the usual finite population sampling situation, the population consists of $N$ units with labels $1, 2, \ldots, N$ and with associated variables of interest $\mathbf{y} = \{y_1, y_2, \ldots, y_n\}$. The sample $s$ is a set of sequence labels identifying the units selected for observation. The data consist of the observed $y$-values together with the associated unit labels. The object of interest is to estimate the population mean

$$\mu = N^{-1} \sum_{i=1}^{N} y_i$$

or total $\tau = N\mu$ of the $y$-values. A sampling *design* is a function $P(s \mid \mathbf{y})$ assigning a probability to every possible sample $s$. In designs such as those described in this chapter, these selection probabilities depend on the population $y$-values.

It is assumed that for every unit $i$ in the population a neighborhood $A_i$ is defined, consisting of a collection of units including $i$. These neighborhoods do not depend on the population $y$-values. In the spatial sampling examples of this chapter, the neighborhood of each unit consists of a set of geographically nearest neighbors, but more elaborate neighborhood patterns are also possible, including a larger contiguous set of units or a noncontiguous set such as a systematic grid pattern around the initial unit. In other sampling situations, neighborhoods may be defined by social or institutional relationships between units. The neighborhood relation is symmetric: If unit $j$ is in the neighborhood of unit $i$, unit $i$ is in the neighborhood of unit $j$.

The condition for additional selection of neighboring units is given by an interval or set $C$ in the range of the variable of interest. The unit $i$ is said to *satisfy the condition* if $y_i \in C$. In the examples of this chapter, a unit satisfies the condition if the variable of interest $y_i$ is greater than or equal to some constant $c$; that is, $C = \{y : y \geq c\}$.

When a selected unit satisfies the condition, all units within its neighborhood are added to the sample and observed. Some of these units may in turn satisfy the condition and some may not. For any of these units that does satisfy the condition, the units in *its* neighborhood are also included in the sample, and so on.

Consider the collection of all the units that are observed under the design as a result of initial selection of unit $i$. Such a collection, which may consist of the union of several neighborhoods, will be termed a *cluster* when it appears in a sample. Within such a cluster is a subcollection of units, termed a *network*, with the property that selection of any unit within the network would lead to inclusion in the sample of every other unit in the network. In the example of Figures 24.1 and 24.2, inside either of the obvious clusters of units in the final sample, the subcollection of units with one or more of the point objects forms a network.

Any unit not satisfying the condition but in the neighborhood of one that does is termed an *edge unit*. While selection of any unit in the network will result in inclusion of all units in the network and all associated edge units, selection of an edge unit will not result in the inclusion of any other units. It is convenient to consider any unit not satisfying the condition of a network of size 1, so that given the $y$-values, the population may be uniquely partitioned into networks.

### Initial Simple Random Sample without Replacement

When the initial sample of $n$ units is selected by simple random sampling without replacement, the $n$ units in the initial sample are distinct due to the without-replacement sampling, but the data may nevertheless contain repeat observations due to selection in the initial sample of more than one unit in a cluster. The unit $i$ will be included in the sample either if any unit of the network to which it belongs (including itself) is selected as part of the initial sample or if any unit of a network of which unit $i$ is an edge unit is selected. Let $m_i$ denote the number of units in the network to which unit $i$ belongs, and let $a_i$ denote the total number of units in networks of which unit $i$ is an edge unit. Note that if unit $i$ satisfies the criterion $C$, then $a_i = 0$, while if unit $i$ does not satisfy the condition, then $m_i = 1$. The probability of selection of unit $i$ on any one of the $n$ draws is

$$p_i = \frac{m_i + a_i}{N} \tag{1}$$

The probability that unit $i$ is included in the sample is

$$\pi_i = 1 - \binom{N - m_i - a_i}{n} \bigg/ \binom{N}{n} \tag{2}$$

## Initial Random Sample with Replacement

When the initial sample is selected by random sampling with replacement, repeat observations in the data may occur due either to repeat selections in the initial sample or to initial selection of more than one unit in a cluster. With this design, the draw-by-draw selection probability is $p_i = (m_i + a_i)/N$, and the inclusion probability is $\pi_i = 1 - (1 - p_i)^n$.

With either initial design, neither the draw-by-draw selection probability $p_i$ nor the inclusion probability $\pi_i$ can be determined from the data for all units in the sample, because some of the $a_i$ may be unknown.

## 24.2 ESTIMATORS

Classical estimators such as the sample mean $\bar{y}$, which is an unbiased estimator of the population mean under a nonadaptive design such as simple random sampling, or the mean of the cluster means $\bar{\bar{y}}$, which is unbiased under cluster sampling with selection probabilities proportional to cluster sizes, are biased when used with the adaptive designs described in this chapter. In this section several estimators are examined which are unbiased for the population mean under the adaptive designs. Derivations of the means and variances of the estimators, as well as of the unbiased estimators of variance, are given later in the chapter.

## Initial Sample Mean

If the initial sample in the adaptive design is selected by simple random sampling, with or without replacement, the mean $\hat{\mu}_0$ of the $n$ initial observations is an unbiased estimator of the population mean. This estimator ignores all observations in the sample other than those selected initially.

## Estimation Using Draw-by-Draw Intersections

For sampling designs in which $n$ units are selected with replacement and the probability $p_i$ of selecting unit $i$ on any draw is known for all units, the Hansen–Hurwitz estimator, in which each $y$-value is divided by the associated selection probability and multiplied by the number of times the unit is selected, is an unbiased estimator of the population mean.

With the adaptive cluster sampling designs of this chapter, the selection probabilities are not known for every unit in the sample. An unbiased estimator can be formed by modifying the Hansen–Hurwitz estimator to make use of observations not satisfying the condition only when they are selected as part of the initial sample. The modified estimator is based on draw-by-draw probabilities that a unit's network is intersected by the initial sample.

Let $\Psi_i$ denote the network that includes unit $i$, and let $m_i$ be the number of units in that network. (Recall that a unit not satisfying the condition is considered a network of size 1.) Let $w_i$ represent the average of the observations in the network that includes the $i$th unit of the initial sample, that is,

$$w_i = \frac{1}{m_i} \sum_{j \in \Psi_i} y_j \tag{3}$$

The modified estimator is

$$\hat{\mu}_1 = \frac{1}{n} \sum_{i=1}^{n} w_i \tag{4}$$

The variance of $\hat{\mu}_1$ is

$$\text{var}(\hat{\mu}_1) = \frac{N-n}{Nn(N-1)} \sum_{i=1}^{N} (w_i - \mu)^2 \tag{5}$$

if the initial sample is selected without replacement, and

$$\text{var}(\hat{\mu}_1) = \frac{1}{n} \sum_{i=1}^{N} \frac{(w_i - \mu)^2}{N} \tag{6}$$

if the initial sample is selected with replacement.

An unbiased estimator of this variance is

$$\widehat{\text{var}}(\hat{\mu}_1) = \frac{N-n}{Nn(n-1)} \sum_{i=1}^{n} (w_i - \hat{\mu}_1)^2 \tag{7}$$

if the initial sample is selected without replacement, and

$$\widehat{\text{var}}(\hat{\mu}_1) = \frac{1}{n(n-1)} \sum_{i=1}^{n} (w_i - \hat{\mu}_1)^2 \tag{8}$$

if the initial sample is selected with replacement.

*Example 1: Draw-by-Draw.* For the sample shown in Figures 24.1 and 24.2, the initial sample of $n = 10$ units was selected by simple random sampling. One of these units, near the top of the study region, intersected a network of $m_1 = 6$ units containing $y_1 = 36$ point objects. Another intersected a network of $m_2 = 11$ units containing $y_2 = 107$ objects. For the other 8 units of the initial sample, $y_i = 0$ and $m_i = 1$.

The estimate based on draw-by-draw intersection probabilities [see Equations (3) and (4)] is

$$\hat{\mu}_1 = \frac{1}{10} \left( \frac{36}{6} + \frac{107}{11} + \frac{0}{1} + \cdots + \frac{0}{1} \right)$$
$$= 0.1(6 + 9.727 + 0 + \cdots + 0) = 1.573$$

objects per unit or $N\hat{\mu}_1 = 400(1.573) = 629$ objects in the population.

The estimated variance [using Equation (5)] is

$$\widehat{\text{var}}(\hat{\mu}_1) = \frac{400 - 10}{400(10)(10 - 1)} [(6 - 1.573)^2 + (9.727 - 1.573)^2$$
$$+ (0 - 1.573)^2 + \cdots + (0 - 1.573)^2]$$
$$= \frac{390}{400(10)} (11.765) = 1.147$$

For the estimate of the total, the estimated variance is $\widehat{\text{var}}(N\hat{\mu}_1) = 400^2$ $(1.147) = 183{,}520$, giving a standard error of $\sqrt{183{,}520} = 428.4$.

Note that the ordinary sample mean of the 45 units in the final sample would have given $\bar{y} = 143/45 = 3.178$ for an estimate of $N\bar{y} = 1271$ objects in the study region. The fact that the adaptive selection procedure produces a high yield of observed objects gives the ordinary sample mean a tendency to overestimate.     □

### Estimation Using Initial Intersection Probabilities

For sampling designs in which the probability $\pi_i$ that unit $i$ is included in the sample is known for every unit, the Horvitz–Thompson estimator, in which each $y$-value is divided by the associated inclusion probability, is an unbiased estimator of the population mean.

With the adaptive designs of this chapter, the inclusion probabilities are not known for all units included in the sample. An unbiased estimator can be formed by modifying the Horvitz–Thompson estimator to make use of observations not satisfying the condition only when they are included in the initial sample. Then the probability that a unit is utilized in the estimator can be computed, even though its actual probability of inclusion in the sample may be unknown. The modified estimator is based on probabilities of the initial sample intersecting networks.

If the initial sample is selected by simple random sampling without replacement, define

$$\pi'_i = 1 - \binom{N - m_i}{n} \bigg/ \binom{N}{n}$$

where $m_i$ is the number of units in the network that includes unit $i$. If the initial selection is made with replacement, define $\pi'_i = 1 - (1 - m_i/N)^n$. For any unit not satisfying the condition, $m_i = 1$. Let the indicator variable $J_i$ be zero if the $i$th unit in the sample does not satisfy the condition and was not selected in the initial sample; otherwise, $J_i = 1$. The modified estimator, written in terms of the individual units in the sample, is

$$\hat{\mu}_2 = \frac{1}{N} \sum_{i=1}^{\nu} \frac{y_i J_i}{\pi_i'} \qquad (9)$$

To obtain the variance of $\hat{\mu}_2$, it will be most convenient to change notation to deal with the networks into which the population is partitioned rather than individual units. Let $K$ denote the number of networks in the population and let $\Psi_k$ be the set of units comprising the $k$th network. Let $x_k$ be the number of units in network $k$. The total of the $y$-values in network $k$ will be denoted $y_k^* = \sum_{i \in \Psi_k} y_i$. The probability $\pi_i'$ that the unit $i$ is utilized in the estimator is the same for all units within a given network $k$; this common probability will be denoted $\alpha_k$. Thus,

$$a_k = 1 - \binom{N - x_k}{n} \bigg/ \binom{N}{n} \qquad (10)$$

with simple random sampling without replacement and $\alpha_k = 1 - (1 - x_k/N)^n$ with replacement.

Let the indicator variable $z_k$ equal 1 if any unit of the $k$th network is in the initial sample, and let $z_k$ equal zero otherwise. With this network notation, the estimator can be written

$$\hat{\mu}_2 = \frac{1}{N} \sum_{k=1}^{K} \frac{y_k^* z_k}{\alpha_k} \qquad (11)$$

For any network not in the sample, $z_k$ will be zero; it will also be zero for any single-unit network in the sample as an edge unit not selected initially.

The probability $\alpha_{kh}$ that the initial sample contains at least one unit in each of the networks $k$ and $h$ is

$$\alpha_{kh} = 1 - \left[ \binom{N - x_k}{n} + \binom{N - x_h}{n} - \binom{N - x_k - x_h}{n} \right] \bigg/ \binom{N}{n}$$

when the initial sample is selected without replacement and $\alpha_{kh} = 1 - \{(1 - x_k/N)^n + (1 - x_h/N)^n - (1 - (x_k + x_h)/N)^n\}$ when the initial selection is with replacement.

With the convention that $\alpha_{kk} = \alpha_k$, the variance of the estimator can be written

$$\text{var}(\hat{\mu}_2) = \frac{1}{N^2} \sum_{k=1}^{K} \sum_{h=1}^{K} \frac{y_k^* y_h^* (\alpha_{kh} - \alpha_k \alpha_h)}{\alpha_k \alpha_h}$$

An unbiased estimator of the variance of $\hat{\mu}_2$ is

$$\widehat{\text{var}}(\hat{\mu}_2) = \frac{1}{N^2} \sum_{k=1}^{K} \sum_{h=1}^{K} \frac{y_k^* y_h^* z_k z_h (\alpha_{kh} - \alpha_k \alpha_h)}{\alpha_k \alpha_h}$$

$$= \frac{1}{N^2} \left[ \sum_{k=1}^{K} \left( \frac{1}{\alpha_k^2} - \frac{1}{\alpha_k} \right) y_k^{*2} z_k + \sum_{k=1}^{K} \sum_{h \neq k} \left( \frac{1}{\alpha_k \alpha_h} - \frac{1}{\alpha_{kh}} \right) y_k^* y_h^* z_k z_h \right]$$

***Example 2: Initial Intersection Probabilities.*** For the sample in Figures 24.1 and 24.2, the intersection probability for the first (top) network [using Equation (10)] is

$$\alpha_1 = 1 - \frac{\binom{400 - 6}{10}}{\binom{400}{10}}$$

$$= 1 - \frac{394!}{10! \, 384!} \frac{10! \, 390!}{400!}$$

$$= 1 - 0.8582 = 0.1418$$

For the other large network intersected by the initial sample,

$$\alpha_2 = 1 - \frac{\binom{400 - 11}{10}}{\binom{400}{10}} = 1 - 0.7542 = 0.2458$$

For the networks of size 1 the probability is $\alpha_k = 10/400 = 0.025$.

The estimate using intersection probabilities [see Equation (11)] is

$$\hat{\mu}_2 = \frac{1}{400} \left( \frac{36}{0.1418} + \frac{107}{0.2458} + \frac{0}{0.025} + \cdots + \frac{0}{0.025} \right) = 1.723$$

objects per unit or $400(1.723) = 689$ total objects in the population. $\qquad\square$

## 24.3 WHEN ADAPTIVE CLUSTER SAMPLING IS BETTER THAN SIMPLE RANDOM SAMPLING

It was pointed out earlier that the unbiasedness of the adaptive designs in this chapter does not depend on the type of population being sampled, because the unbiasedness is design-based. Whether an adaptive design is more efficient or less efficient than a nonadaptive design such as simple random sampling does, however, depend on the type of population being sampled.

Consider adaptive cluster sampling with the initial sample of $n$ units selected by simple random sampling without replacement and with the estimator $\hat{\mu}_1$. With $\sigma^2 = (N-1)^{-1} \sum_{i=1}^{N} (y_i - \mu)^2$ denoting the finite population variance, the variance of the sample mean of a simple random sample of fixed size $n^*$ will have variance $\sigma^2(N - n^*)/Nn^*$. Comparing this quantity with the expression [Equation (5)] for the variance of $\hat{\mu}_1$, gives the following result (see Section 24.7): The adaptive strategy will have lower variance than the sample mean of a simple random sample of size $n^*$ if and only if

$$\left(\frac{1}{n} - \frac{1}{n^*}\right)\sigma^2 < \frac{N-n}{Nn(N-1)} \sum_{k=1}^{K} \sum_{i \in \Psi_k} (y_i - w_i)^2$$

where $\Psi_k$ is the $k$th network in the population. The term on the right contains the within-network variance of the population. Thus, adaptive cluster sampling with the estimator $\hat{\mu}_1$ will be more efficient than simple random sampling if the within-network variance of the population is sufficiently high.

## 24.4   EXPECTED SAMPLE SIZE, COST, AND YIELD

In the examples in this chapter, comparisons of adaptive strategies with simple random sampling are made on the basis of expected (effective) sample size $E(\nu)$, which is the sum of the inclusion probabilities given in Section 24.1 [i.e., $E(\nu) = \sum_{i=1}^{N} \pi_i$]. In classical cluster sampling, comparisons are often made on the basis of cost, since it is often less expensive, in terms of time or money, to sample units within a cluster than to select a new cluster. The same may be true in applications of adaptive cluster sampling. A reasonable cost equation might then be $c = c_0 + c_1 n + c_2 n'$, where $c$ is total cost, $c_0$ is a fixed cost, $c_1$ and $c_2$ are the marginal costs per unit in the initial and subsequent samples, and $n$ and $n'$ are the initial and subsequent sample sizes. In addition, there may in many applications be lower costs associated with observing a unit that does not satisfy the criterion than with observing one that does, in which case the cost equation can be modified accordingly. (For example, if the $y$-variable is biomass of a plant species on sample plots, the measurement is easier on plots with zero.) When the foregoing conditions apply, the relative advantage of the adaptive to the nonadaptive strategy would tend to be greater than in comparisons based solely on sample size. In addition, adaptive cluster sampling tends to increase the "yield" of interesting observations, for example, the number of animals observed or amount of ore assessed.

## 24.5   COMPARATIVE EFFICIENCIES OF ADAPTIVE AND CONVENTIONAL SAMPLING

In this section, adaptive cluster sampling is illustrated using, first, the clustered population of Figures 24.1 and 24.2, and second, the same population with each

$y$-value converted to a zero–one variable indicating the presence or absence of objects in the unit. The population of each example is contained in a square region partitioned into $N = 20 \times 20 = 400$ units. The neighborhood of each unit consists, in addition to itself, of all adjacent units (i.e., that share a common boundary line). A unit satisfies the condition for additional sampling if the $y$-value associated with the unit is greater than or equal to 1.

For each example, variances are computed for the estimators $\hat{\mu}_1$ and $\hat{\mu}_2$ under the design adaptive cluster sampling with the initial sample of $n$ units selected by simple random sampling without replacement. Results are listed in Tables 24.1 and 24.2 for a selection of initial sample sizes, from $n = 1$ to $n = 200$.

For comparison, the variance is also computed for the sample mean $\hat{\mu}_0^*$ of a simple random sample (without replacement) with sample size equal to the expected (effective) sample size $E(\nu)$ under the adaptive design. For each adaptive strategy, the relative efficiency—the variance of the simple random sampling strategy divided by the variance of the adaptive strategy—is also listed. The population $y$-values for the two examples are listed in Table 24.4.

*Population 1*

The population of point objects illustrated in the figures was produced as a realization in the unit square of a Poisson cluster process (see Diggle 1983) with five "parent locations" from a uniform distribution and random Poisson (mean = 40) numbers of point objects dispersed in relation to the parent locations with a symmetric Gaussian distribution having standard deviation 0.02. The population mean is $190/400 = 0.475$.

**Table 24.1. Population 1: Variances of $\hat{\mu}_1$ and $\hat{\mu}_2$ with Adaptive Cluster Sampling and Initial Sample Size $n$ for the Population Illustrated in Figures 24.1 and 24.2** [a]

| $n$ | $E(\nu)$ | var($\hat{\mu}_1$) | var($\hat{\mu}_2$) | var($\hat{\mu}_0^*$) | eff($\hat{\mu}_1$) | eff($\hat{\mu}_2$) |
|---|---|---|---|---|---|---|
| 1 | 1.92 | 4.29705 | 4.29705 | 4.28364 | 1.00 | 1.00 |
| 2 | 3.82 | 2.14314 | 2.12386 | 2.14420 | 1.00 | 1.01 |
| 10 | 18.26 | 0.42001 | 0.38655 | 0.43240 | 1.03 | 1.12 |
| 20 | 34.66 | 0.20462 | 0.17097 | 0.21805 | 1.07 | 1.28 |
| 30 | 49.56 | 0.13282 | 0.10030 | 0.14627 | 1.10 | 1.46 |
| 40 | 63.26 | 0.09693 | 0.06587 | 0.11012 | 1.14 | 1.67 |
| 50 | 76.00 | 0.07539 | 0.04593 | 0.08819 | 1.17 | 1.92 |
| 60 | 87.97 | 0.06103 | 0.03322 | 0.07338 | 1.20 | 2.21 |
| 100 | 130.80 | 0.03231 | 0.01096 | 0.04258 | 1.32 | 3.89 |
| 200 | 223.86 | 0.01077 | 0.00106 | 0.01628 | 1.51 | 15.36 |

*Source*: Thompson (1990). With permission from the American Statistical Association.

[a] The variance of $\bar{y}$ with simple random sampling is calculated for sample size $E(\nu)$, the sample size expected with the adaptive design. Relative efficiencies in the last two columns are eff($\hat{\mu}_1$) = var($\hat{\mu}_0^*$)/var($\hat{\mu}_1$) and eff($\hat{\mu}_2$) = var($\hat{\mu}_0^*$)/var($\hat{\mu}_2$). eff($\hat{\mu}_1$) compares the efficiency of adaptive cluster sampling using the estimator $\hat{\mu}_1$ and simple random sampling with sample size $E(\nu)$. eff($\hat{\mu}_2$) compares adaptive cluster sampling using estimator $\hat{\mu}_2$ and simple random sampling with sample size $E(\nu)$.

**Table 24.2. Population 2: Variance Comparisons with the $y$-Variable Indicating the Presence or Absence of Objects in the Population of Figures 24.1 and 24.2**

| $n$ | $E(\nu)$ | var$(\hat{\mu}_1)$ | var$(\hat{\mu}_2)$ | var$(\hat{\mu}_0^*)$ | eff$(\hat{\mu}_1)$ | eff$(\hat{\mu}_2)$ |
|---|---|---|---|---|---|---|
| 1 | 1.92 | 0.04974 | 0.04974 | 0.02581 | 0.52 | 0.52 |
| 2 | 3.82 | 0.02481 | 0.02459 | 0.01292 | 0.52 | 0.53 |
| 10 | 18.26 | 0.00486 | 0.00448 | 0.00261 | 0.54 | 0.58 |
| 20 | 34.66 | 0.00237 | 0.00198 | 0.00131 | 0.55 | 0.66 |
| 30 | 49.56 | 0.00154 | 0.00116 | 0.00088 | 0.57 | 0.76 |
| 40 | 63.26 | 0.00112 | 0.00076 | 0.00066 | 0.59 | 0.87 |
| 50 | 76.00 | 0.00087 | 0.00053 | 0.00053 | 0.61 | 1.00 |
| 60 | 87.97 | 0.00071 | 0.00038 | 0.00044 | 0.63 | 1.15 |
| 100 | 130.80 | 0.00037 | 0.00012 | 0.00026 | 0.69 | 2.06 |
| 200 | 223.86 | 0.00012 | 0.00001 | 0.00010 | 0.79 | 9.52 |

*Source*: Thompson (1990). With permission from the American Statistical Association.

Table 24.1 lists the expected sample sizes, variances, and relative efficiencies for the different sampling strategies for a selection of initial sample sizes. With an initial sample size of 1, the variances of the adaptive strategies are about equal to that obtained with simple random sampling. The relative advantage of the adaptive strategies increases with increasing $n$. An initial sample of size 10 as illustrated in Figure 24.1 leads to an average final sample of size of about 18 units $[E(\nu)]$ and with the estimator $\hat{\mu}_2$, an efficiency gain of 12% over simple random sampling with equivalent sample size. With an initial sample size of 200, the adaptive strategy leads on average to observing a total of about 224 units and is 15.36 times as efficient as simple random sampling.

*Population 2*
The $y$-values for this population are either zero or 1. The population was obtained from that of Example 41, letting the $y$-value of each unit indicate the presence or absence of point objects in that unit. Thus, the pattern of the "1's" in the population is identical to the pattern of nonzero units in Figure 24.1. For such a population, the within-network variance is zero, since every network in the population consists either of a single unit with $y_i = 0$ or a group of one or more units each with $y_i = 1$. Therefore, by the results of Section 5, the adaptive strategy with the estimator $\hat{\mu}_1$ cannot do better than simple random sampling in this situation. The variance computations in Table 24.2 reveal, however, that the estimator $\hat{\mu}_2$ used with the adaptive design does turn out to be more efficient than simple random sampling for initial sample sizes of 50 or larger.

## 24.6 FURTHER IMPROVEMENT OF ESTIMATORS

Neither $\hat{\mu}_0$, $\hat{\mu}_1$, nor $\hat{\mu}_2$ is a function of the minimal sufficient statistic. Therefore, each of these unbiased estimators can be improved upon using the Rao–Blackwell method of taking their conditional expectations given the minimal sufficient statistic. The minimal sufficient statistic $D$ in the finite population sampling setting is the unordered set of distinct, labeled observations; that is, $D = \{(k, y_k) : k \in s\}$, where $s$ denotes the set of distinct units included in the sample. Each of the estimators $\hat{\mu}_0$, $\hat{\mu}_1$, and $\hat{\mu}_2$ depends on the order of selection; $\hat{\mu}_1$ depends in addition on repeat selections; and when the initial sample is selected with replacement, $\hat{\mu}_0$ also depends on repeat selections.

Letting $\hat{\mu}$ denote any of the three unbiased estimators above, consider the estimator $\hat{\mu}_{RB} = \mathrm{E}(\hat{\mu} \mid D)$, the application of the Rao–Blackwell method to the initial sample mean. Let $\nu$ denote the effective sample size, that is, the number of distinct units included in the sample. When the initial sample is selected by simple random sampling without replacement, define $\eta = \binom{\nu}{n}$, the number of combinations of $n$ distinct units from the $\nu$ in the sample. Let these combinations be indexed in any arbitrary way by the label $g$, and let $\hat{\mu}_g$ denote the value of the estimator $\hat{\mu}$ obtained when the initial sample consists of combination $g$. Similarly, let $\widehat{\mathrm{var}}(\hat{\mu}_g)$ denote the value of the variance estimator obtained with initial sample $g$.

Any initial sample that gives rise through the design to the given value $D$ of the minimal sufficient statistic will be termed *compatible* with $D$. A *sample edge unit* is a unit in the sample that does not satisfy the condition but is in the neighborhood of one or more units *in the sample* that do satisfy the condition. Let $\kappa^*$ denote the number of distinct networks represented in the sample exclusive of sample edge units, that is, the number of distinct networks intersected by the initial sample. Because of the way the sample is selected, an initial sample of $n$ units gives rise to the given value of $D$ if and only if the initial sample contains at least one unit from each of the $\kappa^*$ distinct networks exclusive of sample edge units in $D$. Letting $x_j$ denote the number of units in the initial sample from the $j$th of these networks, an initial sample of $n$ units from the $\nu$ distinct units in $D$ is compatible with $D$ if and only if $x_j \geq 1$ for $j = 1, \ldots, \kappa^*$. Define the indicator variable $I_g$ to be 1 if the $g$th combination of $n$ units from the sample is compatible with $D$ and zero otherwise. The number of compatible combinations is $\xi = \sum_{g=1}^{\eta} I_g$.

With the notation above, the Rao–Blackwell estimator obtained from estimator $\hat{\mu}$ is $\hat{\mu}_{RB} = \xi^{-1} \sum_{g=1}^{\eta} \hat{\mu}_g I_g$. Its variance is $\mathrm{var}(\hat{\mu}_{RB}) = \mathrm{var}(\hat{\mu}) - \mathrm{E}[\mathrm{var}(\hat{\mu} \mid D)]$. An unbiased estimator of this variance is given by $\widehat{\mathrm{var}}(\hat{\mu}_{RB}) = \xi^{-1} \sum_{g=1}^{\eta} [\widehat{\mathrm{var}}(\hat{\mu}_g) - (\hat{\mu}_g - \hat{\mu}_{RB})^2] I_g$. If the initial sample is selected with replacement, the formulas above hold with $\eta = \nu^n$, the number of sequences of $n$ units, distinguishing order and allowing repeats, from the $\nu$ distinct units in the sample and with the label $g$ identifying a sequence in the collection. Although unbiased, the estimator of variance above can, with some sets of data, take on negative values. Computational aspects of the Rao–Blackwell estimators deserve further study, since the numbers of terms in the expressions above are potentially large.

The estimator obtained when the Rao–Blackwell method is applied to $\hat{\mu}_0$ is shown in Section 24.7 to be identical to that obtained when the Rao–Blackwell method is applied to $\hat{\mu}_1$. A different estimator is obtained, however, when the Rao–Blackwell method is applied to $\hat{\mu}_2$, as demonstrated in the small population example (Example 3). The reason that a unique minimum variance unbiased estimator is not obtained is that the minimal sufficient statistic in the finite population sampling setting is not complete. The incompleteness of $D$ in the finite population sampling situation is due basically to the presence of the unit labels in $D$. Yet the labels cannot be regarded as just a nuisance, as good use is made of these labels in constructing estimators for use with the adaptive designs in this chapter. Of the five unbiased estimators in this section, all but the initial sample mean depend on the labels in the data.

*Example 3: Small Population.* In this section, the sampling strategies are applied to a very small population in order to shed light on the computations and properties of the adaptive strategies in relation to each other and to conventional strategies. The population consists of just five units, the $y$-values of which are $\{1, 0, 2, 10, 1000\}$. The neighborhood of each unit includes all adjacent units (of which there are either one or two). The condition is defined by $C = \{x : x \geq 5\}$. The initial sample size is $n = 2$.

With the adaptive design in which the initial sample is selected by simple random sampling without replacement, there are $\binom{5}{2} = 10$ possible samples, each having probability $1/10$. The resulting observations and the values of each estimator are listed in Table 24.3.

**Table 24.3. All Possible Outcomes of Adaptive Cluster Sampling for a Population of Five Units with $y$-Values 1, 0, 2, 10, 1000, in Which the Neighborhood of Each Unit Consists of Itself Plus Adjacent Units** [a]

| Observations | $\hat{\mu}_0$ | $\hat{\mu}_1$ | $\hat{\mu}_{RB1}$ | $\hat{\mu}_2$ | $\hat{\mu}_{RB2}$ | $\bar{y}$ | $\bar{\bar{y}}$ |
|---|---|---|---|---|---|---|---|
| 1,0 | 0.50 | 0.50 | 0.50 | 0.50 | 0.50 | 0.50 | 0.50 |
| 1,2 | 1.50 | 1.50 | 1.50 | 1.50 | 1.50 | 1.50 | 1.50 |
| 1,10;2,1000 | 5.50 | 253.00 | 253.00 | 289.07 | 289.07 | 253.25 | 169.67 |
| 1,1000;10,2 | 500.50 | 253.00 | 253.00 | 289.07 | 289.07 | 253.25 | 169.67 |
| 0,2 | 1.00 | 1.00 | 1.00 | 1.00 | 1.00 | 1.00 | 1.00 |
| 0,10;2,1000 | 5.00 | 252.50 | 252.50 | 288.57 | 288.57 | 253.00 | 168.67 |
| 0,1000;10,2 | 500.00 | 252.50 | 252.50 | 288.57 | 288.57 | 253.00 | 168.67 |
| 2,10;1000 | 6.00 | 253.50 | 337.33 | 289.57 | 289.24 | 337.33 | 337.33 |
| 2,1000;10 | 501.00 | 253.50 | 337.33 | 289.57 | 289.24 | 337.33 | 337.33 |
| 10,1000;2 | 505.00 | 505.00 | 337.33 | 288.57 | 289.24 | 337.33 | 337.33 |
| Mean: | 202.6 | 202.6 | 202.6 | 202.60 | 202.60 | 202.75 | 169.17 |
| Bias: | 0 | 0 | 0 | 0 | 0 | 0.15 | −33.43 |
| MSE: | 59,615 | 22,862 | 18,645 | 17,418.4 | 17,418.3 | 18,660 | 18,086 |

*Source*: Thompson (1990). With permission from the American Statistical Association.

[a] The initial sample of two units is selected by simple random sampling without replacement. Whenever an observed $y$-value exceeds 5, the neighboring units are added to the sample. Initial observations are separated from subsequent observations in the table by a semicolon. For each possible sample, the value of each estimator is given. The bottom line of the table gives the mean square error for each estimator. The sample mean of a simple random sample of equivalent sample size has variance 24,359.

In this population, the fourth and fifth units, with the $y$-values 10 and 1000, respectively, form a network, while the third, fourth, and fifth units, with $y$-values 2, 10, and 1000, form a cluster. In the fourth row of the table, the first and fifth units, with $y$-values 1 and 1000, were selected initially; since $1000 \geq 5$, the single neighbor of the fifth unit, having $y$-value 10, was added to the sample. Since 10 also exceeds 5, the neighboring unit with $y$-value 2 is also added to the sample. The computations for the estimators [Equations (4) and (11)] are

$$\hat{\mu}_1 = \frac{1 + (10 + 1000)/2}{2} = 253$$

$$\hat{\mu}_2 = \frac{1/0.4 + 10/0.7 + 1000/0.7}{5} = 289.07$$

in which $\alpha_1 = 1 - \binom{4}{2} / \binom{5}{2} = 0.4$ and $\alpha_2 = \alpha_3 = 1 - \binom{3}{2} / \binom{5}{2} = 0.7$. The classical estimator $\bar{y}$ (the sample mean) $= 253.25$ is obtained by averaging all four observations in the sample, while the mean of cluster means

$$\bar{\bar{y}} = \frac{1 + (10 + 2 + 1000)/3}{2} = 169.17$$

The six distinct values of the minimal sufficient statistic $D$ are indicated by the distinct values of the Rao–Blackwell estimators $\hat{\mu}_{RB1}$ and $\hat{\mu}_{RB2}$, which are obtained by averaging $\hat{\mu}_1$ and $\hat{\mu}_2$, respectively, over all samples with the same value of $D$. In the last row of Table 24.3, initial selection of the units with $y$-values 10 and 1000 leads to addition of the adjacent unit with value 2, which receives no weight in the estimators $\hat{\mu}_1$ and $\hat{\mu}_2$ since it does not satisfy the condition and was not included in the initial sample. But the Rao–Blackwell estimates based on this sample do not utilize the value 2 by averaging over the last three rows of the table.

The population mean is 202.6, and the population variance (defined with $N - 1$ in the denominator) is 198,718. One sees from Table 24.3 that the unbiased adaptive strategies indeed have expectation 202.6, while the estimators $\bar{y}$ and $\bar{\bar{y}}$, used with the adaptive design, are biased.

With the adaptive design, the effective sample size $\nu$ varies from sample to sample, with expected sample size 3.1. For comparison, the sample mean with a simple random sampling design (without replacement) and a sample size of 3.1 has (inserting the noninteger value 3.1 into the standard formula) variance $[(198,718)(5 - 3.1)]/[5(3.1)] = 24,359$.

From the variances and mean square error given in the last row of Table 24.3, one sees that for this population, the adaptive design with the estimator $\hat{\mu}_{RB2}$ has the lowest variance among the unbiased strategies [note, however, the extra digit of reporting precision necessary in the table to show that $\text{var}(\hat{\mu}_{RB2})$ is slightly less than $\text{var}(\hat{\mu}_2)$], and that all of the adaptive strategies are more efficient than simple random sampling. Among the five unbiased adaptive strategies, the four that make use of labels in the data have lower variance than the one ($\hat{\mu}_0$) that does not.    □

## 24.7 DERIVATIONS

The expected value of an estimator $\hat{\mu}$ is defined in the design sense, that is, $E(\hat{\mu}) = \sum \hat{\mu}_s P(s \mid \mathbf{y})$, where $\hat{\mu}_s$ is the value of the estimate computed when sample $s$ is selected, $P(s \mid \mathbf{y})$ is the probability under the design of selecting the sample $s$ given the population values $\mathbf{y} = \{y_1, \ldots, y_N\}$, and the summation is over all possible samples $s$. The sampling strategy—the estimator together with the design—is *design-unbiased* for the population mean if $E(\hat{\mu}) = N^{-1} \sum_{i=1}^{N} y_i$ for all population vectors $\mathbf{y}$.

*Mean and Variance of $\hat{\mu}$.*    To see that $\hat{\mu}_1$ is unbiased, let $r_i$ indicate the number of times the $i$th unit of the population appears in the estimator, which is exactly the number of units in the initial sample that intersect the network including unit $i$. Note that $r_i$ may be less than the number of times that unit $i$ appears in the sample, which includes selections of unit $i$ as an edge unit. The random variable $r_i$ has a hypergeometric distribution when the initial sample is selected by simple random sampling without replacement and a binomial distribution when the initial sample is selected by simple random sampling with replacement. With either design, $r_i$ has expected value $E(r_i) = nm_i/N$. Writing the estimator in the form $\hat{\mu}_1 = n^{-1} \sum_{i=1}^{N} r_i y_i / m_i$, it follows that $E(\hat{\mu}_1) = N^{-1} \sum_{i=1}^{N} y_i$, so that $\hat{\mu}_1$ is a design-unbiased estimator of the population mean.

The estimator $\hat{\mu}_1 = n^{-1} \sum_{k=1}^{n} w_i$ can be viewed as a sample mean, based on a simple random sample, in which the variable of interest associated with the $i$th unit in the population is $w_i$, the mean of the $y$-values in the network that includes unit $i$. The expressions for the variance and estimator of variance then follow from classical results on the sample mean of a simple random sample.    □

*Mean and Variance of $\hat{\mu}_2$.*    To see that $\hat{\mu}_2$ is unbiased, let $J_i = 1$ if unit $i$ is utilized in the estimator and $J_i = 0$ otherwise. For any $i$, $J_i$ is a Bernoulli random variable with expected value $\pi_i'$. Writing the estimator as $\hat{\mu}_2 = N^{-1} \sum_{i=1}^{N} J_i y_i / \pi_i'$ it follows that $E(\hat{\mu}_2) = N^{-1} \sum_{i=1}^{N} y_i$, the population mean.

Define the indicator variable $z_k$ to be 1 if the initial sample contains one or more units from the $k$th network and zero otherwise, so that $\hat{\mu}_2 = N^{-1} \sum_{k=1}^{N} z_k y_k^* / \alpha_k$. For any network $k$, $z_k$ is a Bernoulli random variable with expected value $E(z_k) = \alpha_k$ and $\mathrm{var}(z_k) = \alpha_k(1 - \alpha_k)$. For $k \neq h$, the covariance of the indicator variables is $(z_k, z_h) = E(z_k z_h) - E(z_k)E(z_h) = \alpha_{kh} - \alpha_k \alpha_h$. Thus,

$$\mathrm{var}(\hat{\mu}_2) = N^{-2} \sum_{k=1}^{N} \sum_{h=1}^{N} \frac{y_k^* y_h^* \mathrm{cov}(z_j, z_h)}{\alpha_k \alpha_h}$$

$$= N^{-2} \sum_{k=1}^{N} \sum_{h=1}^{N} \frac{y_k^* y_h^* (\alpha_{kh} - \alpha_k \alpha_h)}{\alpha_k \alpha_h}$$

To see that $\widehat{\mathrm{var}}(\hat{\mu}_2)$ is unbiased, let $z_{kh}$ be 1 if units from both networks $k$ and $h$ are selected in the initial sample and zero otherwise. Then $\widehat{\mathrm{var}}(\hat{\mu}_2) = N^{-2}$

$\sum_{k=1}^{K} \sum_{h=1}^{K} y_k^* y_h^* z_{kh} (\alpha_{kh} - \alpha_k \alpha_h) / \alpha_k \alpha_h \alpha_{kh}$,   and   unbiasedness   follows   since
$E(z_{kh}) = \alpha_{kh}$.                                                                          □

*Variances of Rao–Blackwell Estimators.*   When the initial sample is selected
by simple random sampling without replacement, each of the $\binom{M}{n}$ possible combinations of $n$ distinct units form the $N$ units in the population has equal probability
of being selected as the initial sample. When the initial sample is selected by simple
random sampling with replacement, each of the $N^n$ possible sequences, which distinguish order and can include repeat selections, of $n$ units chosen from the $N$ units
in the population is equally probable. Conditional on the minimal sufficient statistic
$D$, all initial samples of $n$ units that give rise through the design to the given value of
$D$ have equal selection probability; all other initial samples have conditional probability zero. Since the units of the initial sample are included in the $\nu$ distinct units
in $D$, only $\binom{\nu}{n}$ combinations or $\nu^n$ sequences need be considered conditional on $D$.

The conditional expectation of $\hat{\mu}$, given $D$, is therefore the average of the values
of $\hat{\mu}$ over all initial samples that are compatible with $D$. Since $\hat{\mu}_{RB} = E(\hat{\mu} \mid D)$,
where $D$ is the minimal sufficient statistic, the variance of $\hat{\mu}$ can be decomposed
as $\text{var}(\hat{\mu}) = E[\text{var}(\hat{\mu} \mid D)] + \text{var}[E(\hat{\mu} \mid D)]$, so that the variance of $\hat{\mu}_{RB}$ can be written $\text{var}(\hat{\mu}_{RB}) = \text{var}(\hat{\mu}) - E[\text{var}(\hat{\mu} \mid D)]$. An unbiased estimator of $\text{var}(\hat{\mu})$ is $\widehat{\text{var}}(\hat{\mu})$.
But by the Rao–Blackwell theorem, a better unbiased estimator of $\text{var}(\hat{\mu})$ is
$E[\widehat{\text{var}}(\hat{\mu}) \mid D]$, which is the average, over all compatible selections of $n$ observations
from $D$, of the variance estimates $\widehat{\text{var}}(\hat{\mu}_g)$. An unbiased estimator of the variance of
$\hat{\mu}_{RB}$ is thus provided by $\widehat{\text{var}}(\hat{\mu}_{RB}) = E[\widehat{\text{var}}(\hat{\mu}) \mid D] - \text{var}(\hat{\mu} \mid D)$. The second term on
the right is computed from the sample as $\text{var}(\hat{\mu} \mid D) = \xi^{-1} \sum (\hat{\mu}_g - \hat{\mu}_{RB})^2$, where
$\hat{\mu}_g$ denotes the value of $\hat{\mu}$ obtained from the $g$th compatible selection, and the summation is over the $\xi$ compatible selections of $n$ observations from $D$.                      □

*Equivalence of Rao–Blackwell Estimators from $\hat{\mu}_0$ and $\hat{\mu}_1$.*   To establish that
$E(\hat{\mu}_0 \mid D) = E(\hat{\mu}_1 \mid D)$, it is helpful to consider the statistic $D^*$ consisting of the
unordered set of labeled observations together with information about the number
of times the network of each unit is intersected by the sample; that is,
$D^* = \{(i, y_i, f_i), i \in s\}$, where $s$ is the set of distinct units included in the sample
and $f_i$ is the number of times the network of unit $i$ was intersected by the sample.
The statistic $D^*$ is sufficient but not minimally sufficient.

An initial selection of $n$ units giving rise to the statistic $D^*$ determines $n$
networks, some of which may be repeats, contained in $D^*$. Let $\kappa$ be the number
of distinct networks among these. (Note that a sample edge unit forms one of these
groups only if it was included in the initial selection.) Because of the way the sample is selected, the same value of the statistic $D^*$ will arise from any initial sample
of $n$ units having exactly the given numbers of units in each of the $\kappa$ groups.

Let $\Psi_i$ denote the network that includes unit $i$, $m_i$ the number of units in it, and
$w_i$ the average of the observations in it. Let $r_i$ be the number of times $\Psi_i$ is represented in the initial sample. (If the unit $i$ in the sample is not a sample edge unit,

$r_i = f_i$. If unit $i$ is a sample edge unit, $r_i$ equals $f_i$ less the number of times the networks of which it is an edge unit are included in the sample.) Let $u_i$ be the number of times unit $i$ is in the initial sample.

Conditional on $D^*$ (which fixes $r_i$), the distribution of $u_i$ for any unit $i$ included in the $\kappa$ networks above is Bernoulli with expected value $r_i/m_i$ if the initial sample is selected by simple random sampling without replacement. If the sampling is with replacement, the distribution is binomial with expectation $r_i/m_i$. For any unit $i$ not included in the $\kappa$ networks with initial representation in $D^*$, $u_i = 0$.

Writing $\hat{\mu}_0 = n^{-1} \sum_{i=1}^{N} u_i y_i$, the conditional expectation is $\mathrm{E}(\hat{\mu}_0 \mid D^*) = n^{-1} \sum_{k=1}^{\kappa} \sum_{j \in \Psi_k} r_j y_j / m_j = n^{-1} \sum_{i=1}^{n} w_i$, since $r_j$ is constant for $j \in \Psi_k$. Thus, $\mathrm{E}(\hat{\mu}_0 \mid D^*) = \hat{\mu}_1$. Since $D$ is a function of $D^*$, $\mathrm{E}(\hat{\mu}_0 \mid D^*) = \mathrm{E}(\hat{\mu}_0 \mid D^*, D)$. Therefore, $\mathrm{E}(\hat{\mu}_1 \mid D) = \mathrm{E}[\mathrm{E}(\hat{\mu}_0 \mid D^*, D) \mid D] = \mathrm{E}(\hat{\mu}_0 \mid D)$. □

***When Adaptive Cluster Sampling Is Better.*** Consider adaptive cluster sampling with the initial sample of $n$ units selected by simple random sampling without replacement and with the estimator $\hat{\mu}_1$. Since $\hat{\mu}_1 = \mathrm{E}(\hat{\mu}_0 \mid D^*)$, where $D^*$ is the unordered collection of labeled observations with repeat frequencies, the variance of $\hat{\mu}_1$ can be written $\mathrm{var}(\hat{\mu}_1) = \mathrm{var}(\hat{\mu}_0) - \mathrm{E}[\mathrm{var}(\hat{\mu}_0 \mid D^*)]$. Thus, the variance of $\hat{\mu}_1$ will always be less than or equal to the variance of $\hat{\mu}_0$, which is $\sigma^2(N - n)/(Nn)$. The variance of the sample mean of a simple random sample of fixed size $n^*$ will, by comparison, have variance $\sigma^2(N - n^*)/Nn^*$. Comparing this quantity with the expression above for the variance of $\hat{\mu}_1$ gives the result: The adaptive strategy will have lower variance than the sample mean of a simple random sample of size $n$ if and only if $(n^{-1} - n^{*-1})\sigma^2 < \mathrm{E}[\mathrm{var}(\hat{\mu}_0 \mid D^*)]$.

The expression for the variance of $\hat{\mu}_1$ given in Section 24.2 can be rewritten in terms of the $K$ distinct networks in the population as follows: $\mathrm{var}(\hat{\mu}_1) = b \sum_{i=1}^{N} (w_i - \mu)^2 = b \sum_{k=1}^{K} \sum_{i \in \Psi_k} (w_i - \mu)^2$, where $b$ is the constant term $(N - n)/[Nn(N - 1)]$ and $\Psi_k$ is the $k$th network in the population. Similarly, the variance of $\hat{\mu}_0$ can be written $\mathrm{var}(\hat{\mu}_0) = b \sum_{k=1}^{K} \sum_{i \in \Psi_k} (y_i - \mu)^2$. Decomposition of the total sum of squares into terms between and within networks then shows that $\mathrm{E}[\mathrm{var}(\hat{\mu}_0 \mid D^*)]$ is the within-network variance, that is, $\mathrm{E}[\mathrm{var}(\hat{\mu}_0 \mid D^*)] = (N - n)/(Nn)^{-1} \sum_{k=1}^{K} \sum_{i \in \Psi_k} (y_i - w_i)^2/(N - 1)$.

With adaptive cluster sampling, the improved estimator $\hat{\mu}_{RB1}$, obtained from $\hat{\mu}_1$ by the Rao–Blackwell method, is more efficient than $\hat{\mu}_1$. Since $\hat{\mu}_{RB1} = \mathrm{E}(\hat{\mu}_0 \mid D)$, the variance of $\hat{\mu}_{RB1}$ can be written $\mathrm{var}(\hat{\mu}_{RB1}) = \mathrm{var}(\hat{\mu}_0) - \mathrm{E}[\mathrm{var}(\hat{\mu}_0 \mid D)]$ and a corresponding result obtained: The adaptive strategy with $\hat{\mu}_{RB1}$ will have lower variance than simple random sampling with $\bar{y}$ if and only if $(n^{-1} - n^{*-1})\sigma^2 < \mathrm{E}[\mathrm{var}(\hat{\mu}_0 \mid D)]$. □

## 24.8 DATA FOR EXAMPLES AND FIGURES

The $20 \times 20$ matrix of $y$-values for population 1 and for the figures and examples is listed in Table 24.4. The data for population 2 are obtained by replacing all positive entries with 1's.

**Table 24.4. 20 × 20 Matrix of Values for Population 1**

| | | | | | | | | | | | | | | | | | | | |
|---|---|---|---|---|---|---|---|---|---|---|---|---|---|---|---|---|---|---|---|
| 0 | 0 | 0 | 0 | 5 | 13 | 3 | 0 | 0 | 0 | 0 | 0 | 0 | 0 | 0 | 0 | 0 | 0 | 0 | 0 |
| 0 | 0 | 0 | 0 | 2 | 11 | 2 | 0 | 0 | 0 | 0 | 0 | 0 | 0 | 0 | 0 | 0 | 0 | 0 | 0 |
| 0 | 0 | 0 | 0 | 0 | 0 | 0 | 0 | 0 | 0 | 0 | 0 | 0 | 0 | 0 | 0 | 0 | 0 | 0 | 0 |
| 0 | 0 | 0 | 0 | 0 | 0 | 0 | 0 | 0 | 0 | 0 | 0 | 0 | 0 | 0 | 0 | 0 | 0 | 0 | 0 |
| 0 | 0 | 0 | 0 | 0 | 0 | 0 | 0 | 0 | 0 | 0 | 0 | 0 | 0 | 0 | 0 | 0 | 0 | 0 | 0 |
| 0 | 0 | 0 | 0 | 0 | 0 | 0 | 0 | 0 | 0 | 0 | 0 | 0 | 0 | 0 | 0 | 0 | 0 | 0 | 0 |
| 0 | 0 | 0 | 0 | 0 | 0 | 0 | 0 | 0 | 0 | 0 | 0 | 0 | 0 | 0 | 0 | 0 | 0 | 0 | 0 |
| 0 | 0 | 0 | 0 | 0 | 0 | 0 | 0 | 0 | 0 | 0 | 0 | 0 | 0 | 0 | 0 | 0 | 0 | 0 | 0 |
| 0 | 0 | 0 | 0 | 0 | 0 | 0 | 0 | 0 | 0 | 0 | 0 | 0 | 0 | 0 | 0 | 0 | 0 | 0 | 0 |
| 0 | 0 | 0 | 0 | 0 | 0 | 0 | 0 | 0 | 0 | 0 | 0 | 0 | 0 | 0 | 0 | 0 | 0 | 0 | 0 |
| 0 | 0 | 0 | 0 | 0 | 0 | 0 | 0 | 0 | 0 | 0 | 0 | 0 | 0 | 0 | 0 | 0 | 0 | 0 | 0 |
| 0 | 0 | 0 | 0 | 0 | 0 | 0 | 0 | 0 | 0 | 0 | 0 | 0 | 0 | 0 | 0 | 0 | 0 | 0 | 0 |
| 0 | 0 | 0 | 0 | 0 | 0 | 0 | 0 | 0 | 0 | 0 | 0 | 0 | 0 | 0 | 0 | 0 | 0 | 0 | 0 |
| 0 | 0 | 0 | 0 | 0 | 0 | 0 | 0 | 0 | 0 | 0 | 0 | 0 | 0 | 0 | 0 | 0 | 0 | 0 | 0 |
| 0 | 0 | 0 | 0 | 0 | 0 | 0 | 0 | 0 | 3 | 1 | 0 | 0 | 0 | 0 | 0 | 0 | 0 | 0 | 0 |
| 0 | 0 | 0 | 0 | 0 | 0 | 0 | 0 | 5 | 39 | 10 | 0 | 0 | 0 | 0 | 0 | 0 | 0 | 0 | 0 |
| 0 | 0 | 0 | 0 | 0 | 0 | 0 | 0 | 5 | 13 | 4 | 0 | 0 | 0 | 0 | 0 | 0 | 0 | 0 | 0 |
| 0 | 0 | 0 | 0 | 0 | 0 | 0 | 2 | 22 | 3 | 0 | 0 | 0 | 0 | 0 | 0 | 0 | 0 | 0 | 0 |
| 0 | 0 | 0 | 0 | 0 | 0 | 0 | 0 | 0 | 0 | 0 | 0 | 10 | 8 | 0 | 0 | 0 | 0 | 0 | 0 |
| 0 | 0 | 0 | 0 | 0 | 0 | 0 | 0 | 0 | 0 | 0 | 0 | 7 | 22 | 0 | 0 | 0 | 0 | 0 | 0 |
| 0 | 0 | 0 | 0 | 0 | 0 | 0 | 0 | 0 | 0 | 0 | 0 | 0 | 0 | 0 | 0 | 0 | 0 | 0 | 0 |

## EXERCISES

1. In a survey to estimate the abundance of a rare animal species in a study region divided into $N = 1000$ units, an initial simple random sample of $n = 100$ units is selected. An adaptive cluster sampling design is used with adjacent units added to the sample whenever one or more of the animals is encountered. One of the units in the initial sample intersected a network of two units containing three of the animals, and another intersected a network of three units containing six animals. The other 98 units of the initial sample contained none of the animals. Estimate the total number of the animals in the population using draw-by-draw intersection probabilities and estimate the variance of the estimator.

2. For the survey in Exercise 1, estimate the population total using overall intersection probabilities.

CHAPTER 25

# Systematic and Strip Adaptive Cluster Sampling

In this chapter, adaptive cluster sampling designs are considered in which the initial sample is selected in terms of primary units and subsequent additions to the sample are in terms of secondary units. For example, in an aerial survey of walruses or polar bears or in a ship survey of whales sighted by their spouts, the strip observed in each selected transect forms a primary unit. If, whenever animals are sighted, the area to the side of the transect is searched—with still further searching if additional animals are sighted while on this search—the searching pattern defines neighborhoods of secondary units added to the sample. In surveys of bird and fish species, the selection of sites at which to make observations is often done systematically and a single systematic selection forms a primary unit. If additional observations are made in the neighborhood of any site at which abundance is observed, the subsequent observations would not in general follow the initial systematic pattern. With such survey situations, one can think of the study region as partitioned into secondary units representing all possible sites at which observations may be made, while the primary units from which the initial sample is selected consist of clusters—such as long, thin strips or systematic arrangements—of the secondary units.

Examples of the types of designs described in this chapter are illustrated in Figures 25.1 and 25.2, in which the object is to estimate the mean number of point objects—representing the locations of animals (or plants, mineral deposits, or other objects) in a clumped population—in the study region. In Figure 25.1, the initial sample consists of five randomly selected strips (primary units). The secondary units are small, square plots. Whenever a plot in the sample contains one or more of the animals, adjacent plots are added to the sample. If, in turn, any of the new plots in the sample contain any animals, additional adjacent plots—to the left, right, top, or bottom—are added. The final sample resulting from this procedure is shown in the figure.

In Figure 25.2, the initial sample is a spatial systematic sample with two randomly selected starting points. Whenever animals of the species are observed

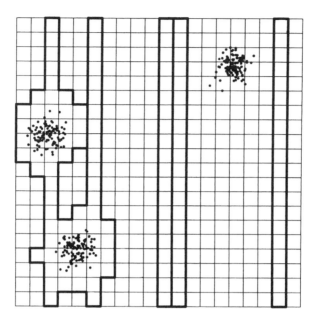

**Figure 25.1.** Adaptive cluster sample with initial random selection of five strip plots. The final sample obtained is outlined. [From Thompson (1991). With permission from the Biometric Society.]

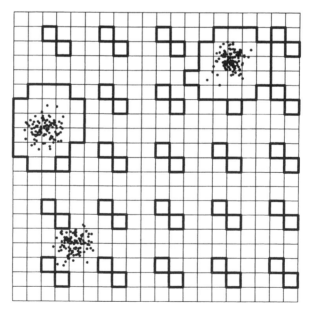

**Figure 25.2.** Adaptive cluster sample with initial random selection of two systematic samples. The final sample obtained is outlined. [From Thompson. With permission from the Biometric Society.]

in any plot of the sample, adjacent plots are added to the sample. The final sample is illustrated in the figure. Most results of this chapter were given in Thompson (1991a).

## 25.1 DESIGNS

For the adaptive cluster sampling designs considered in this chapter, the population is composed of $N$ primary units. Each primary unit contains $M$ secondary units (which may be referred to simply as *units*). The $MN$ units of the population are denoted $u_{ij}$, for $i = 1, \ldots, N$ and $j = 1, \ldots, M$. Associated with the $j$th secondary unit of the $i$th primary unit is a variable of interest $y_{ij}$. The object of inference is estimation of the population mean $\mu = (MN)^{-1} \sum_{j=1}^{N} \sum_{j=1}^{M} y_{ij}$ or, equivalently, of the population total $\tau = MN\mu$.

For every (secondary) unit of the population, a collection of units called the *neighborhood* of that unit is defined. The neighborhood of unit $u_{ij}$ includes unit $u_{ij}$, and if unit $u_{ij}$ belongs to the neighborhood of unit $u_{i'j'}$, then unit $u_{i'j'}$ belongs to the neighborhood of unit $u_{ij}$. In applications, the neighborhood of a unit will typically be defined as a contiguous set of surrounding units or a systematic pattern of surrounding units. In the examples of Figures 25.1 and 25.2, the secondary units are square plots. The neighborhood of one of these plots consists of itself plus its adjacent units to the left, right, top, and bottom, so that for a plot not on the boundary of the study area, the neighborhood consists of five plots in a cross shape. Many other neighborhood configurations are possible. For example, only the units to the right and left (but not top and bottom) might be included, or the neighborhood could be defined to include a larger set of contiguous units than the five in the examples. A neighborhood could, in fact, consist of a set of noncontiguous units, spread out, for example, in a systematic grid pattern about the original unit.

The unit $u_{ij}$ is said to satisfy the *condition of interest* if the associated value $y_{ij}$ is in a specified set $C$. For problems in the estimation of animal abundance, the condition may commonly be defined so that a unit satisfies the condition if its $y$-value equals or exceeds some constant $c$. In the examples of Figures 25.1 and 25.2, a plot satisfies the condition if the number of animals in it equals 1 or more. It would also be possible to set the criterion $c$ at a higher level, so that only units with $y$-values of, say, 50 or more would satisfy the condition.

In the adaptive cluster sampling designs of this chapter, an initial sample of $n$ primary units is selected by simple random sampling without replacement. Whenever the observed value of a (secondary) unit in the sample satisfies the condition of interest, all units in its neighborhood are added to the sample. If, in turn, any of these subsequently added units satisfies the condition, the units of its neighborhood are also added to the sample, so that finally, the sample contains every unit in the neighborhood of any sample unit satisfying the condition.

A population with a given set of $y$-values can be uniquely partitioned into $K$ sets called *networks* so that whenever a unit $u_{ij}$ satisfying the condition is in the neighborhood of unit $u_{i'j'}$, also satisfying the condition, units $u_{ij}$ and $u_{i'j'}$ belong to the same network. Thus, if an initially selected primary unit intersects a given network, every unit in that network will be included in the sample. A unit that does not satisfy the condition belongs to a network consisting just of itself. The population of Figures 25.1 and 25.2 has three networks that are larger than single-plot size. Every plot within one of these large networks contains at least one animal.

A unit $u_{ij}$ which does not satisfy the condition will be included in the sample if either the primary unit that includes it is initially selected or if any primary unit of the initial selection intersects the network of one or more units satisfying the condition in the neighborhood of unit $u_{ij}$. A unit not satisfying the condition but in the neighborhood of one or more units that do satisfy the condition is called an *edge unit*. In Figures 25.1 and 25.2, a number of such plots containing no animals have been added to the sample at the edges of the large networks. Thus, a unit will be included in the sample if the initial sample intersects either the network to which it belongs or a network of which it is an edge unit.

Note that while neighborhoods are defined by such relationships as physical proximity and do not depend on the $y$-values of the population, networks do depend on the population $y$-values, corresponding roughly to the natural aggregations of animals, plants, or other individuals in the population.

If each initial primary unit consists of a set of units evenly spaced in some arrangement throughout the population, the initial sample will be termed a *systematic* initial sample (Figure 25.2). The initial primary units will be called *strips* if each initial primary unit consists of a row of units arranged in a straight line (Figure 25.1). Many other arrangements of primary units are, of course, possible.

The draw-by-draw selection probability $p_{ij}$ for unit $u_{ij}$ is the probability in any initial draw of selecting any one of the primary units that intersects the network containing unit $u_{ij}$, or if unit $u_{ij}$ does not satisfy the condition, selecting a primary unit that intersects the network of any unit satisfying the condition in the neighborhood of unit $u_{ij}$. That is,

$$p_{ij} = \frac{m_{ij} + a_{ij}}{N}$$

where $m_{ij}$ is the number of primary units that intersect the network containing unit $u_{ij}$, and $a_{ij}$ is the number of primary units that do not intersect the network of unit $u_{ij}$ but intersect the network of one or more units satisfying the condition in the neighborhood of unit $u_{ij}$. For a unit satisfying the condition, $a_{ij} = 0$, and for a unit not satisfying the condition, $m_{ij} = 1$.

The probability $\pi_{ij}$ that unit $u_{ij}$ is included in the sample is the probability that one or more primary units of the initial sample either intersects the network that includes unit $u_{ij}$ or intersects the network of any unit satisfying the condition in the neighborhood of unit $u_{ij}$. That is,

$$\pi_{ij} = 1 - \left( \begin{array}{c} N - m_{ij} - a_{ij} \\ n \end{array} \right) \bigg/ \left( \begin{array}{c} N \\ n \end{array} \right)$$

The expected sample size, that is, the expected number of distinct secondary units in the final sample, is the sum of the inclusion probabilities (Godambe 1955; see also Cassel et al. 1977, p. 11), so that the expected sample size $\nu$ expressed in terms of the equivalent number of primary units in the final sample is

$$E(\nu) = \frac{1}{M} \sum_{i=1}^{N} \sum_{j=1}^{M} \pi_{ij}$$

## 25.2 ESTIMATORS

With the adaptive cluster sampling designs described in this chapter, standard estimators of the population mean and total are biased. With spatially aggregated populations, for example, if additional units are added to the sample whenever high abundance is observed, the final sample tends to contain units with higher-than-average abundance, and the sample mean will overestimate the population mean. If, on the other hand, the estimator is formed by averaging first all $y$-values associated with the selection of a primary unit—that is, the units of the primary unit together with all units adaptively added to the sample as a result of initial selection of that primary unit—the mean of these averages may tend to underestimate the population mean, due to the fact that whenever units with higher-than-average $y$-values are selected, additional sampling begins until low values are obtained, while when units with low values are selected, no such compensatory procedure begins.

In this section, therefore, estimators are given which are unbiased with the adaptive cluster sampling designs. Since these estimators are, in fact, design-unbiased, the unbiasedness does not depend on any assumptions about the population itself.

### Initial Sample Mean

One way to obtain an unbiased estimator of the population mean $\mu$ is to ignore all units adaptively added to the sample and use the sample mean of the initial sample. For notational simplicity let $Y_i$ denote the total of the $y$-values in the $i$th primary unit, that is, $Y_i = \sum_{j=1}^{M} y_{ij}$. The estimator of $\mu$ based on the initial sample mean is

$$\hat{\mu}_0 = \frac{1}{Mn} \sum_{i=1}^{n} Y_i$$

This estimator does not make use of the observations adaptively added to the sample. It is of interest in this chapter because it offers the basis for nonadaptive alternatives with which the adaptive strategies may be compared.

From classical results on simple random sampling without replacement, $\hat{\mu}_0$ is unbiased for $\mu$ and has variance

$$\text{var}(\hat{\mu}_0) = \frac{N-n}{M^2 Nn} \sigma_0^2$$

where

$$\sigma_0^2 = \frac{1}{N-1}\sum_{i=1}^{N}(Y_i - M\mu)^2$$

An unbiased estimator of variance is

$$\widehat{\text{var}}(\hat{\mu}_0) = \frac{N-n}{M^2 Nn}s_0^2$$

where

$$s_0^2 = \frac{1}{n-1}\sum_{i=1}^{n}(Y_i - M\hat{\mu}_0)^2$$

An unbiased estimate of variance is, of course, not available for systematic samples with only one starting point [but see Wolter (1984) for methods useful in practice].

## Estimator Based on Partial Selection Probabilities

It is also possible to obtain unbiased estimators that do make use of observations in addition to those selected initially. Estimators such as the Hansen–Hurwitz estimator, the multiplicity estimator of network sampling, and related estimators used in line-intercept sampling achieve unbiasedness by dividing each observation by its selection probability and multiplying by the number of times the unit was selected. Jolly (1979) gives an estimator of this form for aerial surveys of animals in large herds. With the adaptive cluster sampling designs, however, not all of the selection probabilities as given in Section 25.1 can be determined from the sample data. When a unit not satisfying the condition appears in the sample, one may not know whether its selection probability is influenced by the presence of units in its neighborhood that do satisfy the condition. Thus, the constants $a_{ij}$ on which the selection probabilities of Section 25.1 depend may not be known.

The unbiased estimator of this section therefore depends only on aspects of the selection probabilities that are known. While the Hansen–Hurwitz estimator would achieve unbiasedness by dividing each observed $y$-value by the selection probability $p_{ij}$ as given in Section 25.1, the estimator of this section will effectively divide instead by $m_{ij}/N$, the "known part" of the selection probability.

For the estimator of this section and the next, it will be convenient to relabel the variables in terms of the networks of the population rather than in terms of the individual units. Let the $K$ networks of the population be labeled $1, \ldots, K$, and let $y_k$ denote the total of the $y$-values in the $k$th network. Define the indicator variable $I_{ik}$ to be 1 if the $i$th primary unit intersects the $k$th network and zero otherwise. Let

$x_k$ be the number of primary units in the population that intersect the $k$th network, that is, $x_k = \sum_{i=1}^{N} I_{ik}$. (For any unit $u_{ij}$ in the $k$th network, $m_{ij} = x_k$, where $m_{ij}$ was defined in Section 25.1.) The draw-by-draw probability that the selected primary unit will intersect the $k$th network is $x_k/N$.

For the $i$th primary unit, define the new variable $w_i$ as

$$w_i = \frac{1}{M} \sum_{k=1}^{K} \frac{y_k I_{ik}}{x_k} \tag{1}$$

The estimator based on partial selection probabilities is

$$\hat{\mu}_1 = \frac{1}{n} \sum_{i=1}^{n} w_i \tag{2}$$

Note that variables only for those networks that are intersected by selected primary units enter into the estimator.

A network $y$-value is utilized in the estimator as many times as there are primary units in the initial sample that intersect it. Some observations in the data— associated with units not satisfying the condition and not included in the initial sample—are not utilized at all in the estimator. The actual selection probability for network $k$ is related to $x_k$ but may also depend, in a manner not known from the data at hand, on other networks.

It is shown in the derivations that $\hat{\mu}_1$ is an unbiased estimator of the population mean with variance

$$\text{var}(\hat{\mu}_1) = \frac{N-n}{Nn} \sigma_w^2 \tag{3}$$

where

$$\sigma_w^2 = \frac{1}{N-1} \sum_{i=1}^{N} (w_i - \mu)^2$$

An unbiased estimator of the variance of $y_2$ is given by

$$\widehat{\text{var}}(\hat{\mu}_1) = \frac{N-n}{Nn} s_w^2 \tag{4}$$

where

$$s_w^2 = \frac{1}{n-1} \sum_{i=1}^{n} (w_i - \hat{\mu}_1)^2$$

**Estimator Based on Partial Inclusion Probabilities**

The Horvitz–Thompson estimator achieves unbiasedness by dividing the $y$-value for each unit in the sample by the probability that the unit is included in the sample. With the adaptive cluster sampling designs, not all of these inclusion probabilities, as given in Section 25.1, are known from the sample data. In particular, the constants $a_{ij}$ as defined in Section 25.1 may not be known because the sample may not reveal units that satisfy the condition in the neighborhoods of sample units that do not satisfy the condition. In this section, an unbiased estimator is given based on the partial knowledge of inclusion probabilities obtainable from the data.

Let $\alpha_k$ denote the probability that one or more of the primary units which intersects network $k$ is included in the initial sample. With the adaptive cluster sampling designs, this probability is given by

$$\alpha_k = 1 - \binom{N - x_k}{n} \Big/ \binom{N}{n} \tag{5}$$

Let $\alpha_{kj}$ denote the probability that one or more of the primary units which intersect both networks $k$ and $j$ is included in the initial sample. With the designs of this chapter,

$$\alpha_{kj} = 1 - \left[ \binom{N - x_k}{n} + \binom{N - x_j}{n} - \binom{N - x_k - x_j + x_{kj}}{n} \right] \Big/ \binom{N}{n} \tag{6}$$

where $x_k$ is the number of primary units that intersect network $k$ and $x_{kj}$ is the number of primary units that intersect both networks $k$ and $j$. It is emphasized that the $\alpha$'s are not the actual network inclusion probabilities, but unlike the inclusion probabilities, are computable from the sample data.

Define the indicator variable $z_k$ to be 1 if one or more of the primary units that intersect network $k$ are included in the initial sample and zero otherwise. Consider the estimator $\hat{\mu}_2$ given by

$$\hat{\mu}_2 = \frac{1}{MN} \sum_{k=1}^{K} \frac{y_k z_k}{\alpha_k} \tag{7}$$

so that the summation is over the distinct networks in the sample that intersect one or more primary units of the initial sample. The weight an observation receives in the estimator does not depend, as it does with $\hat{\mu}_1$, on the number of intersecting primary units selected, as long as at least one of them is included in the initial sample. Also, some observations in the data may receive zero weight.

The estimator $\hat{\mu}_2$ is unbiased for the population mean (see derivations) and has variance

$$\operatorname{var}(\hat{\mu}_2) = \frac{1}{M^2 N^2} \sum_{k=1}^{K} \sum_{j=1}^{K} y_k y_j \left( \frac{\alpha_{kj}}{\alpha_k \alpha_j} - 1 \right) \tag{8}$$

with the convention that $\alpha_{kk} = \alpha_k$.

An unbiased estimator of this variance is given by

$$\widehat{\text{var}}(\hat{\mu}_2) = \frac{1}{N^2 N^2} \sum_{k=1}^{K} \sum_{j=1}^{K} \frac{y_k y_j z_k z_j}{\alpha_{kj}} \left( \frac{\alpha_{kj}}{\alpha_k \alpha_j} - 1 \right) \tag{9}$$

provided that none of the joint probabilities $\alpha_{kj}$ is zero. For any design involving an initial simple random sample of at least two primary units, all of these joint inclusion probabilities would be greater than zero. For an initial systematic sample with only one starting point (i.e., only one primary unit is selected), some of the joint inclusion probabilities are zero, underscoring the fact that an unbiased estimator of variance is not available for such a design. Small-sample coverage probabilities of confidence limits based on the variance estimators $\widehat{\text{var}}(\hat{\mu}_1)$ and $\widehat{\text{var}}(\hat{\mu}_2)$ require further investigation.

The estimator $\hat{\mu}_2$ differs from the Horvitz–Thompson estimator in that it is based on intersection probabilities rather than inclusion probabilities. The actual inclusion probabilities, although not known completely from the data, do determine relevant properties of the adaptive designs, such as the expected sample size as given in Section 25.1.

Just as the Horvitz–Thompson estimator has small variance when the $y$-values are approximately proportional to the inclusion probabilities, the estimator $\hat{\mu}_2$ should have low variance when the network totals $y_k$ are approximately proportional to the intersection probabilities $\alpha_k$. With an animal population with aggregation tendencies, if the largest concentrations of animals occurred in sizable clusters, each intersecting several primary units, high $y_k$-values would be associated with high intersection probabilities. For cases in which the $y$-values are approximately proportional to some other variable $b$, Hájek (1971) suggested a ratio weighting of the Horvitz–Thompson estimator. The corresponding ratio modification of $\hat{\mu}_2$ would be $\hat{\mu}_2 \sum b_k / (\sum b_k z_k / \alpha_k)$, where the summations are over the $K$ networks in the population. An estimator of this form would not, however, be unbiased.

## 25.3 CALCULATIONS FOR ADAPTIVE CLUSTER SAMPLING STRATEGIES

Sample calculations for the adaptive cluster sampling strategies of this chapter will be illustrated with two types of designs. In one, the primary units consists of long, thin strips. The other has an initial systematic sampling design, with starting points chosen at random in a $4 \times 4$ square and the positions repeated throughout the study area. The neighborhood of a unit is defined to consist of itself together with all adjacent units (those sharing a full edge). Thus, for a unit not on the boundary of the study region, the neighborhood consists of five units in a cross shape.

The square study region of 400 units is depicted in Figures 25.1 and 25.2. The locations of individuals or objects in the study region were produced with a realization of a Poisson cluster process (cf. Diggle 1983), with three parent locations selected at random and Poisson (mean $= 100$) numbers of offspring distributed about each parent with a bivariate Gaussian distribution (with standard derivation 0.03 in the unit square). The object of sampling is to estimate the number of objects in the study region (the correct answer: 326) or, equivalently, the mean number per unit (0.815). The 400 population $y$-values for the example are listed at the end of this chapter (Table 25.3).

A unit is considered to satisfy the condition if it contains at least one individual of the population, so that any time a selected unit contains one or more individuals, the remaining units in its neighborhood are added to the sample. Figure 25.1 shows the sample obtained as the result of the initial selection of five of the strips. Figure 25.2 shows the sample obtained with an initial selection of two systematic starting points. Sample calculations will be carried out for the samples illustrated.

***Example 1: Initial Strip Plots.*** In the initial strip plot sample (Figure 25.1), the first (leftmost) of the primary units in the sample intersects two collections of units that satisfy the condition, leading to the additional clusters of units added to the sample. The network of units satisfying the condition within the uppermost of these clusters has total $y$-value 106; it can be determined from the sample that this network intersects four of the primary units. The lower cluster has total $y$-value 105 and also intersects four primary units. All other observations in the data are zero. The variable $w_i$ associated with the first sample primary unit is $w_1 = (1/120)$ $[(106/4) + (105/4)] = 2.6375$. For the second sample primary unit, the term is $w_2 = (1/20)105/4 = 1.3125$, and for the other three primary units in the sample, $w_i = 0$. The estimate $\hat{\mu}_1$ [Equation (2)] is $(1/5)(2.6375 + 1.3125 + 0 + 0 + 0) = 0.79$. The variance estimate [Equation (4)] is $\widehat{\text{var}}(\hat{\mu}_1) = (20 - 5)/[(20)(5)]$ $(1.38964) = 0.2084$, in which 1.38964 is the sample variance of the $w$-values 2.6375, 1.3125, 0, 0, and 0.

The intersection probability [Equation (5)] for each of the two nonzero networks in the sample is $1 - \left(\begin{array}{c} 20 - 4 \\ 5 \end{array}\right) \Big/ \left(\begin{array}{c} 20 \\ 5 \end{array}\right) = 0.7183$, and the estimator $\hat{\mu}_2$ [Equation (7)] is $(1/20)(1/20)[(106/0.7183) + (105/0.7183)] = 0.7344$. Since two primary units intersect both networks, the joint intersection probability [Equation (6)] for the two sample networks is

$$1 - \left[\left(\begin{array}{c} 20 - 4 \\ 5 \end{array}\right) + \left(\begin{array}{c} 20 - 4 \\ 5 \end{array}\right) - \left(\begin{array}{c} 20 - 4 - 4 + 2 \\ 5 \end{array}\right)\right] \Big/ \left(\begin{array}{c} 20 \\ 5 \end{array}\right) = 0.5657$$

The sample estimate of variance [Equation (9)] is

$$\widehat{\text{var}}(\hat{\mu}_2) = (1/20^2)(1/20^2)\{(106^2/0.7183)[(1/0.7183) - 1]$$
$$+ (105^2/0.7183)[(1/0.7183) - 1]$$
$$+ 2(106)(105/0.5676)[(0.5676/0.7183^2) - 1]\} = 0.09963 \quad \square$$

*Example 2: Systematic Initial Sample.* For the sample with the initial systematic design (Figure 25.2), the first primary unit (based on the starting position in the third column of the second row) intersects the central left network, with $y$-total 106, while the second primary unit (starting position in fourth column of third row) intersects both that network and the top right network, which has total $y$-value 115. By dividing the study area into $4 \times 4$ squares, it can be determined that 10 primary units (i.e., 10 of the 16 possible systematic samples) intersect the top right network and 13 intersect the central left network, while nine intersect both. For this sample, $\hat{\mu}_1 = (1/2)[0.3262 + 0.7862] = 0.5562$, and $\widehat{\text{var}}(\hat{\mu}_1) = [(16 - 2)/(16)(2)] (0.1058) = 0.0463$. The intersection probability for the top network is 0.875 and for the left network is 0.975, and their joint intersection probability is 0.8583. For this sample, $\hat{\mu}_2 = (1/25)(1/16)(115/0.875 + 106/0.975) = 0.6004$, and

$$\widehat{\text{var}}(\hat{\mu}_2) = (1/25^2)(1/16^2)\{(115^2/0.875)[(1/0.875) - 1] + (106^2/0.975)$$
$$\times [(1/0.975) - 1] + 2(115)(106/0.8583)[(0.8583/((0.875)(0.975))) - 1]\}$$
$$= 0.01684 \qquad \square$$

## 25.4 COMPARISONS WITH CONVENTIONAL SYSTEMATIC AND CLUSTER SAMPLING

The actual variances for each of the unbiased estimators are given in Table 25.1 for the design with the initial strips and in Table 25.2 for the initially systematic design. In addition to the estimators $\hat{\mu}_0$, $\hat{\mu}_1$, and $\hat{\mu}_2$, the variance has been computed for the

**Table 25.1. Variance with Initial Long, Thin Strip Plots**[a]

| $n$ | $\text{E}(\nu)$ | $\text{var}(\hat{\mu}_0)$ | $\text{var}(\hat{\mu}_0^*)$ | $\text{var}(\hat{\mu}_1)$ | $\text{var}(\hat{\mu}_2)$ |
|---|---|---|---|---|---|
| 1 | 1.57 | 1.30628 | 0.80706 | 0.79253 | 0.79253 |
| 2 | 3.01 | 0.61876 | 0.38751 | 0.37541 | 0.34713 |
| 3 | 4.35 | 0.38959 | 0.24758 | 0.23637 | 0.19944 |
| 4 | 5.58 | 0.27501 | 0.17749 | 0.16685 | 0.12651 |
| 5 | 6.74 | 0.20625 | 0.13530 | 0.12514 | 0.08378 |
| 6 | 7.82 | 0.16042 | 0.10702 | 0.09733 | 0.05636 |
| 7 | 8.85 | 0.12768 | 0.08666 | 0.07746 | 0.03788 |
| 8 | 9.82 | 0.10313 | 0.07123 | 0.06257 | 0.02510 |
| 9 | 10.76 | 0.08403 | 0.05907 | 0.05098 | 0.01621 |
| 10 | 11.66 | 0.06875 | 0.04917 | 0.04171 | 0.01008 |

*Source*: Thompson (1991). With permission from the Biometric Society.

[a] $n$ is the initial sample size and $\text{E}(\nu)$ is the expected sample size with the adaptive design. Nonadaptive strategies are represented by $\hat{\mu}_0$ and $\hat{\mu}_0^*$, with sample sizes $n$ and $\text{E}(\nu)$, respectively. Adaptive strategies are represented by $\hat{\mu}_1$ and $\hat{\mu}_2$.

**Table 25.2. Variances with Initial Systematic Samples**

| $n$ | $E(\nu)$ | $\text{var}(\hat{\mu}_0)$ | $\text{var}(\hat{\mu}_0^*)$ | $\text{var}(\hat{\mu}_1)$ | $\text{var}(\hat{\mu}_2)$ |
|---|---|---|---|---|---|
| 1 | 2.98 | 0.44078 | 0.12825 | 0.08441 | 0.08441 |
| 2 | 4.36 | 0.20570 | 0.07846 | 0.03939 | 0.01684 |
| 3 | 5.31 | 0.12734 | 0.05919 | 0.02439 | 0.00363 |
| 4 | 6.15 | 0.08816 | 0.04701 | 0.01688 | 0.00072 |
| 5 | 6.98 | 0.06465 | 0.03798 | 0.01238 | 0.00011 |
| 6 | 7.80 | 0.04898 | 0.03089 | 0.00938 | 0.00001 |
| 7 | 8.62 | 0.03778 | 0.02516 | 0.00724 | 0.00000 |
| 8 | 9.44 | 0.02939 | 0.02042 | 0.00563 | 0.00000 |

*Source*: Thompson (1991). With permission from the Biometric Society.

sample mean $\hat{\mu}_0^*$ of a simple random sample of primary units with sample size equal to the expected sample size under the adaptive designs. The variance of $\hat{\mu}_0^*$ is computed using the formula of Section 25.2 with sample size $E(\nu)$, even if fractional, in place of $n$. Thus, $\hat{\mu}_0^*$ offers one way to compare the adaptive strategies with nonadaptive counterparts of equivalent sample size. Sample sizes in the tables are expressed in terms of primary units. One primary unit consists of 20 secondary units in the strip design and 25 secondary units in the systematic design. The tables give variances obtained with initial sample sizes ranging from 1 up to a sampling fraction of $1/2$.

With the initial strip design, the adaptive strategies with an initial sample size of one primary unit are slightly more efficient than the comparable nonadaptive strategy for the example population. The relative advantage of the adaptive strategies increases with increasing initial sample size, and also the efficiency of $\hat{\mu}_2$ relative to $\hat{\mu}_1$ increases. With an initial sample size of 10 (initial sampling fraction of $1/2$), the adaptive cluster sampling strategy increases the expected sample size by 16.6% [$E(\nu) = 11.66$], but is almost five times as efficient as the equivalent nonadaptive strategy [$\text{var}(\hat{\mu}_0^*)/\text{var}(\hat{\mu}_2) = 0.04917/0.01008 = 4.88$].

With the initial systematic sampling design, the adaptive strategies are dramatically more efficient than their nonadaptive counterparts for the example population. Also, comparing Tables 25.1 and 25.2, one sees that even with conventional systematic strategies ($\hat{\mu}_0$ and $\hat{\mu}_0^*$), variances are considerably lower than with the conventional strategies using strips. [This result would be expected due to the positive, monotonically decreasing covariance density function of the Poisson cluster process (see, e.g., Matérn 1986; Thompson and Ramsey, 1987).] The efficiency of the adaptive strategy with $\hat{\mu}_2$ relative to the comparable nonadaptive systematic strategy with $\hat{\mu}_0^*$ ranges from 152% for a single initial systematic sample $(0.12852/0.08441 = 1.52)$ to infinity—the adaptive strategy has zero variance for initial selections of more than six, as the intersection probability for each of the three networks in the population becomes one with such a design.

## 25.5  DERIVATIONS

***Estimator*** $\hat{\mu}_2$. The estimator $\hat{\mu}_1$ can be written

$$\hat{\mu}_1 = \frac{1}{Mn} \sum_{k=1}^{K} \frac{y_k r_k}{x_k}$$

where the random variable $r_k$ denotes the number of primary units in the initial sample that intersect the $k$th network of the population. Under the design, $r_k$ has a hypergeometric distribution with expected value $nx_k/N$. The expected value of $\hat{\mu}_1$ is thus

$$\mathrm{E}(\hat{\mu}_1) = \frac{1}{Mn} \sum_{k=1}^{K} \frac{y_k n x_k}{x_k N} = \frac{1}{MN} \sum_{k=1}^{K} y_k = \mu$$

so $\hat{\mu}_1$ is an unbiased estimator of the population mean. Since $\hat{\mu}_1$ is the sample mean of the $w_i$ for a simple random sample of size $n$, the formulas for the variance and the estimator of variance for $\hat{\mu}_1$ follow from the usual results on simple random sampling. $\qquad\square$

***Estimator*** $\hat{\mu}_2$. In the expression for the estimator $\hat{\mu}_2$, the random variable $z_k$ equals 1 if one or more primary units of the initial sample intersect network $k$,

**Table 25.3. $y$-Values for the Population Used in Examples and Figures**

| | | | | | | | | | | | | | | | | | | | |
|---|---|---|---|---|---|---|---|---|---|---|---|---|---|---|---|---|---|---|---|
| 0 | 0 | 0 | 0 | 0 | 0 | 0 | 0 | 0 | 0 | 0 | 0 | 0 | 0 | 0 | 0 | 0 | 0 | 0 | 0 |
| 0 | 0 | 0 | 0 | 0 | 0 | 0 | 0 | 0 | 0 | 0 | 0 | 0 | 0 | 0 | 0 | 0 | 0 | 0 | 0 |
| 0 | 0 | 0 | 0 | 0 | 0 | 0 | 0 | 0 | 0 | 0 | 0 | 0 | 0 | 14 | 25 | 2 | 0 | 0 | 0 |
| 0 | 0 | 0 | 0 | 0 | 0 | 0 | 0 | 0 | 0 | 0 | 0 | 0 | 0 | 22 | 38 | 3 | 0 | 0 | 0 |
| 0 | 0 | 0 | 0 | 0 | 0 | 0 | 0 | 0 | 0 | 0 | 0 | 0 | 2 | 2 | 6 | 1 | 0 | 0 | 0 |
| 0 | 0 | 0 | 0 | 0 | 0 | 0 | 0 | 0 | 0 | 0 | 0 | 0 | 0 | 0 | 0 | 0 | 0 | 0 | 0 |
| 0 | 1 | 1 | 1 | 0 | 0 | 0 | 0 | 0 | 0 | 0 | 0 | 0 | 0 | 0 | 0 | 0 | 0 | 0 | 0 |
| 2 | 11 | 26 | 5 | 0 | 0 | 0 | 0 | 0 | 0 | 0 | 0 | 0 | 0 | 0 | 0 | 0 | 0 | 0 | 0 |
| 2 | 22 | 19 | 8 | 0 | 0 | 0 | 0 | 0 | 0 | 0 | 0 | 0 | 0 | 0 | 0 | 0 | 0 | 0 | 0 |
| 0 | 3 | 5 | 0 | 0 | 0 | 0 | 0 | 0 | 0 | 0 | 0 | 0 | 0 | 0 | 0 | 0 | 0 | 0 | 0 |
| 0 | 0 | 0 | 0 | 0 | 0 | 0 | 0 | 0 | 0 | 0 | 0 | 0 | 0 | 0 | 0 | 0 | 0 | 0 | 0 |
| 0 | 0 | 0 | 0 | 0 | 0 | 0 | 0 | 0 | 0 | 0 | 0 | 0 | 0 | 0 | 0 | 0 | 0 | 0 | 0 |
| 0 | 0 | 0 | 0 | 0 | 0 | 0 | 0 | 0 | 0 | 0 | 0 | 0 | 0 | 0 | 0 | 0 | 0 | 0 | 0 |
| 0 | 0 | 0 | 0 | 0 | 0 | 0 | 0 | 0 | 0 | 0 | 0 | 0 | 0 | 0 | 0 | 0 | 0 | 0 | 0 |
| 0 | 0 | 0 | 0 | 1 | 1 | 0 | 0 | 0 | 0 | 0 | 0 | 0 | 0 | 0 | 0 | 0 | 0 | 0 | 0 |
| 0 | 0 | 0 | 17 | 26 | 9 | 0 | 0 | 0 | 0 | 0 | 0 | 0 | 0 | 0 | 0 | 0 | 0 | 0 | 0 |
| 0 | 0 | 1 | 10 | 26 | 6 | 0 | 0 | 0 | 0 | 0 | 0 | 0 | 0 | 0 | 0 | 0 | 0 | 0 | 0 |
| 0 | 0 | 0 | 5 | 2 | 1 | 0 | 0 | 0 | 0 | 0 | 0 | 0 | 0 | 0 | 0 | 0 | 0 | 0 | 0 |
| 0 | 0 | 0 | 0 | 0 | 0 | 0 | 0 | 0 | 0 | 0 | 0 | 0 | 0 | 0 | 0 | 0 | 0 | 0 | 0 |
| 0 | 0 | 0 | 0 | 0 | 0 | 0 | 0 | 0 | 0 | 0 | 0 | 0 | 0 | 0 | 0 | 0 | 0 | 0 | 0 |

and $z_k$ equals zero otherwise. Under the design, $z_k$ is a Bernoulli random variable with expected value $\alpha_k$. Also, $\text{var}(z_i) = \alpha_i(1 - \alpha_i) = \alpha_i - \alpha_i^2$, and $\text{cov}(z_i, z_j) = \alpha_{ij} - \alpha_i \alpha_j$ for $i \neq j$. The unbiasedness of $\hat{\mu}_2$ for $\mu$ and the expression in Section 2 for $\text{var}(\hat{\mu}_2)$ then follow according to the usual derivations for the Horvitz–Thompson estimator, although the $\alpha_k$ give intersection probabilities, not inclusion probabilities. The expression in Section 25.2 for $\widehat{\text{var}}(\hat{\mu}_2)$ is unbiased for $\text{var}(\hat{\mu}_2)$ because $E(z_i z_j) = \alpha_{ij}$.                                                           □

## 25.6 EXAMPLE DATA

The 400 $y$-values for the population in the examples and figures are listed in Table 25.3.

## EXERCISES

1. An adaptive cluster sample initially selects a random sample of $n = 4$ strips from the $N = 25$ in the study region. Each strip contains $M = 12$ secondary units. Neighborhoods are defined to include adjacent secondary units as in the example in the text. The first sample unit intersects a network that intersects $m_1 = 3$ strips in the population and has $y_1 = 50$ animals. Another network, with $y_2 = 30$ animals, intersects both the first and second sample strips, and intersects $m_2 = 5$ strips altogether; $m_{12} = 2$ strips in the population intersect both these networks. No other animals are observed. Estimate the number of animals in the study area using draw-by-draw initial probabilities. Estimate the variance of the estimator.

2. For Exercise 1, use the estimator based on overall intersection probabilities.

# CHAPTER 26

# Stratified Adaptive Cluster Sampling

In stratified adaptive cluster sampling, an initial stratified sample is selected from a population, and whenever the value of the variable of interest for any unit is observed to satisfy a specified condition, additional units from the neighborhood of that unit are added to the sample. Still more units may be added to the sample if in turn any of the units added subsequently satisfies the condition.

Stratified adaptive cluster sampling designs are important from a practical point of view because for many populations prior information exists on which an initial stratification can be based and yet the exact distribution or patterns of concentration of the population cannot be predicted. In conventional stratified sampling, units that are thought to be similar are grouped a priori into strata, based on prior information about the population or simple proximity of the units. Adaptive cluster sampling, on the other hand, provides a means of taking advantage of clustering tendencies in a population, when the locations and shapes of the clusters cannot be predicted prior to the survey. The sampling designs described in this chapter combine the two methods.

Conventional estimators such as the stratified sample mean are not unbiased with the adaptive designs, so estimators that are unbiased under the designs are given in this chapter. A complication that arises in stratified adaptive cluster sampling is that a selection in one stratum may result in the addition of units from other strata to the sample, so that observations in separate strata are not independent as in conventional stratified sampling. The different unbiased estimators given in this chapter handle such crossing of stratum boundaries in slightly different ways. Most results of this chapter were given in Thompson (1991b).

## 26.1  DESIGNS

For the adaptive cluster sampling designs of this chapter, the population is partitioned into $L$ strata, of which stratum $h$ composed of $N_h$ units, and the total number of units in the population is denoted $N$. Associated with unit $u_{hi}$, the $i$th unit of stratum $h$, is a variable of interest $y_{hi}$. For any unit $u_{hi}$ of the population, the

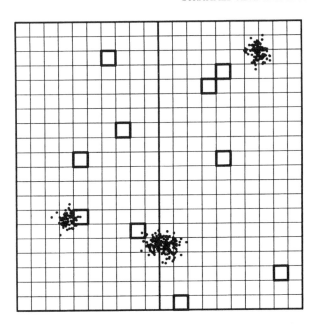

**Figure 26.1.** Initial stratified random sample of five units in each of two strata. Whenever a unit in the sample contains one or more of the point objects, the adjacent units are added to the sample. [From Thompson (1991). With permission from the Biometrika Trustees.]

*neighborhood* of unit $u_{hi}$ is defined as a collection of units that includes $u_{hi}$ and with the property that if unit $u_{h'i'}$ is in the neighborhood of unit $u_{hi}$, then unit $u_{hi}$ is in the neighborhood of unit $u_{h'i'}$. The neighborhood of a unit may include units from more than one stratum. A unit $u_{hi}$ is said to *satisfy the condition of interest* if the $y$-value associated with that unit is in a specified set $C$.

In the designs considered in this chapter, an initial sample of units is selected from a population using stratified random sampling; that is, within stratum $h$, a simple random sample of $n_h$ units is selected without replacement, the selections for separate strata being made independently. Whenever a selected unit satisfies the condition, all units in its neighborhood not already in the sample are added to the sample. Still more units may be added to the sample whenever any of the additionally added units satisfies the condition, so that the final sample contains every unit in the neighborhood of any sample unit satisfying the condition.

An example is illustrated in Figure 26.1, in which the object is to estimate the abundance of a clustered population, that is, the total across-area units of the numbers $y$ of point objects within each unit. The point object could, for example, represent the location of a plant or animal. A unit satisfies the condition here if it contains one or more point objects; that is, $y \geq 1$. The population is divided into two strata, and a simple random sample of five units selected from each stratum is shown in Figure 26.1. The neighborhood of a unit consists of that unit together with all the adjacent units to the north, south, east, and west. Applying the stratified adaptive cluster sampling design gives the final sample shown in Figure 26.2.

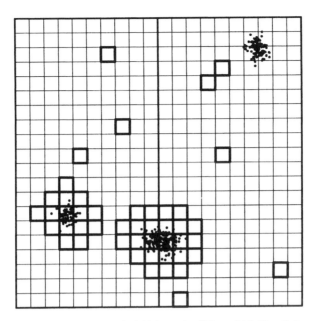

**Figure 26.2.** Final sample resulting from the initial sample of Figure 26.1. Note that some units in stratum 2 (right) were included in the sample as a result of an initial selection in stratum 1. [From Thompson (1991). With permission from the Biometrika Trustees.]

The population may be partitioned into $K$ sets of units, termed *networks*, such that selection in the initial sample of any unit in a network will result in inclusion in the final sample of all units in that network. A unit not satisfying the condition belongs to a network consisting just of itself. Initial selection of a unit satisfying the condition will typically result in the addition to the sample not only of all the other units in its network, but also of units not in its network, that is, units not satisfying the condition but in the neighborhood of one or more members of the network. In Figure 26.3, the networks intersected by the initial sample are outlined in thick lines. The other units in the sample, the edge units, do not satisfy the condition and were not in the initial sample but are each in the neighborhood of one ore more units satisfying the condition in the networks intersecting the initial sample.

The number of times a unit is selected equals the number of units from its network or from a network intersecting its neighborhood that are selected in the initial sample. Let $r_{hi}$ represent the number of times that unit $u_{hi}$ is selected. Let $m_{khi}$ denote the number of units in the intersection of stratum $k$ with the network that contains unit $u_{hi}$. For a unit $u_{hi}$ not satisfying the condition, let $a_{khi}$ be the total number of units in the intersection of stratum $k$ with the collection of distinct networks, exclusive of $u_{hi}$ itself, which intersect the neighborhood of unit $u_{hi}$. Initial selection of any of these $a_{khi}$ units will result in the addition of unit $u_{hi}$ to the sample. Define $a_{khi}$ to be zero for any unit $u_{hi}$ satisfying the condition. The expected number of

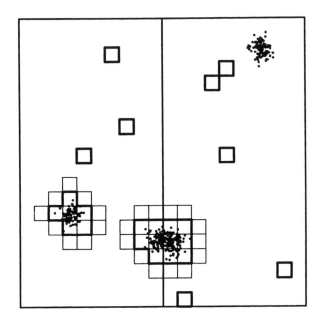

**Figure 26.3.** Distinct networks intersected by the initial sample are outlined with bold lines.

times unit $u_{hi}$ is selected is

$$E(r_{hi}) = \sum_{k=1}^{L} n_k \frac{m_{khi} + a_{khi}}{N_k}$$

The unit $u_{hi}$ will be included in the sample if one or more units from the network to which $u_{hi}$ belongs is included in the initial selection or, for a unit $u_{hi}$ not satisfying the condition, if one or more units from any network that intersects the neighborhood of unit $u_{hi}$ is included in the initial sample. Because of the initial stratified random sampling, the inclusion probability $\pi_{hi}$ for unit $u_{hi}$ is

$$\pi_{hi} = 1 - \prod_{k=1}^{L} \binom{N_k - m_{khi} - a_{khi}}{n_k} \bigg/ \binom{N_k}{n_k}$$

The expected sample size $\nu$, that is, the expected number of distinct units in the final sample, is the sum of the $N$ inclusion probabilities in the population (see Cassel et al. 1977, pp. 11; Godambe 1955).

## 26.2 ESTIMATORS

Conventional estimators such as the stratified sample mean, although unbiased for the population mean with classical stratified random sampling, are not unbiased with the adaptive designs (see Example 1 below). An unbiased, if inefficient estimator $\hat{\mu}_0$ of the population mean can be obtained, however, simply by using the conventional stratified estimator of the mean based on the initial sample, ignoring all subsequent observations.

### Estimators Using Expected Numbers of Initial Intersections

For sampling with replacement with known selection probabilities, the Hansen–Hurwitz estimator achieves unbiasedness by dividing the $y$-value of each unit by the draw-by-draw selection probability of that unit. More precisely, each observation is divided by the expected number of times that it is selected in the sample and multiplied by the number of times it is selected. With stratified adaptive cluster sampling, the selection probabilities and hence the expected number of times selected are not known for every unit in the sample, so that an unbiased estimator must be based only on the aspects of the expected selection numbers that can be determined from the data.

For the unit $u_{hi}$, define the new variable $w_{hi}$ to be the total of the $y$-values of the network to which $u_{hi}$ belongs, weighted by the stratum sampling fraction and divided by a weighted sum of the network–stratum intersection sizes as follows:

$$w_{hi} = \frac{n_h}{N_h} \sum_{k=1}^{L} \xi_{khi} \Bigg/ \sum_{k=1}^{L} \frac{n_k}{N_k} m_{khi}$$

where $\xi_{khi}$ is the total of the $y$-values in the intersection of stratum $k$ with the network that includes unit $u_{hi}$ and $m_{khi}$ is the number of units in this intersection. The estimator of the population mean is

$$\hat{\mu}_1 = \frac{1}{N} \sum_{h=1}^{L} \frac{N_h}{n_h} \sum_{i=1}^{n_h} w_{hi} \tag{1}$$

Letting the random variable $r_{khi}$ represent the number of units in the initial sample that are in the intersection of stratum $k$ with the network to which unit $u_{hi}$ belongs, the estimator can be written in the alternative form

$$\hat{\mu}_1 = \frac{1}{N} \sum_{h=1}^{L} \sum_{i=1}^{N_h} \left( y_{hi} \sum_{k=1}^{L} r_{khi} \Bigg/ \sum_{k=1}^{L} \frac{n_k}{N_k} m_{khi} \right)$$

Since $E(r_{khi}) = n_k m_{khi}/N_k$, it follows that $\hat{\mu}_1$ is an unbiased estimator of the population mean.

With $w_{hi}$ as the variable of interest for unit $u_{hi}$ for each unit in the population, $\hat{\mu}_1$ is the stratified sample mean of a stratified random sample and hence has variance

$$\text{var}(\hat{\mu}_1) = \frac{1}{N^2} \sum_{h=1}^{L} N_h (N_h - n_h) \frac{\sigma_h^2}{n_h} \tag{2}$$

in which the stratum population variance term is

$$\sigma_h^2 = \frac{1}{N_h - 1} \sum_{i=1}^{N_h} (w_{hi} - \bar{W}_h)^2 \tag{3}$$

and the stratum population mean is $\bar{W}_h = (1/n_h) \sum w_{hi}$.

An unbiased estimator $v(\hat{\mu}_1)$ of the variance $\hat{\mu}_1$ is obtained by replacing $\sigma_h^2$ in formula (3) with the sample variance

$$s_h^2 = \frac{1}{n_h - 1} \sum_{i=1}^{n_h} (w_{hi} - \bar{w}_h)^2 \tag{4}$$

using the sample mean $\bar{w}_h = (1/n_h) \sum w_{hi}$.

A variation $\hat{\mu}_1'$ on the estimator $\hat{\mu}_1$ may be constructed that is related to the stratified "multiplicity" estimator of network sampling (Birnbaum and Sirken 1965; Levy 1977; Sirken 1972a), in which the weight an observation receives depends on the stratum in which the initial sample intersects the network of that unit. For unit $u_{hi}$, define the new variable $w_{hi}'$ to be the total of the $y$-values in the entire network to which unit $u_{hi}$ belongs, divided by the total number of units in that network; that is,

$$w_{hi}' = \sum_{k=1}^{L} \xi_{khi} \Big/ \sum_{k=1}^{L} m_{khi} \tag{5}$$

The modified stratified multiplicity estimator is given by Equation (1) with $w'$ replacing $w$.

For every time any unit of a network is selected in the initial sample, the estimator includes a term with the total of the $y$-values for that network, divided by the network size and weighted by $N_k/n_k$ for the stratum from which the unit was selected. Thus, each individual $y$-value occurs in the estimator every time any unit from the network to which it belongs is selected in the initial sample, but with weightings depending on the strata from which the initial selections came. Thus, the estimator $\hat{\mu}_1'$ can be written in the alternative form

$$\hat{\mu}_1' = \frac{1}{N} \sum_{h=1}^{L} \sum_{i=1}^{N_h} \left( y_{hi} \sum_{k=1}^{L} \frac{N_k}{n_k} r_{khi} \Big/ \sum_{k=1}^{L} m_{khi} \right) \tag{6}$$

Unbiasedness of $\hat{\mu}_1'$ for the population mean follows from the fact that $E(r_{khi}) = n_k m_{khi}/N_k$.

Associating the variable $w_{hi}'$ with unit $u_{hi}$, the estimator $\hat{\mu}_1'$ is a stratified sample mean of a stratified random sample. Hence, the variance and estimated variance of $\hat{\mu}_1'$ are given by Equations (2), (3), and (4) with $w'$ replacing $w$.

It is also possible to use an estimator $\hat{\mu}_1''$, which ignores all units added through crossing stratum boundaries. For this estimator, let $w_{hi}''$ be the total of the $y$-values in the intersection of the stratum and network of unit $u_{hi}$, divided by the number of units in that intersection. The estimator and its variance expressions are then given by Equations (1) through (4) with $w''$ replacing $w$. Unbiasedness and other properties follow from the unstratified case, since components in different strata are independent.

### Estimator Using Initial Intersection Probabilities

For any design in which inclusion probabilities are known, the Horvitz–Thompson estimator achieves unbiasedness by dividing the $y$-value for each unit in the sample by the probability that unit is included in the sample. With adaptive cluster sampling, these inclusion probabilities cannot be determined from the data for every unit in the sample. However, an estimator can be formed using for each unit the probability that the initial sample intersects the network to which that unit belongs, and giving zero weight to any observation not satisfying the condition that was not included in the initial sample.

Let the $K$ distinct networks of the population be labeled $1, 2, \ldots, K$, without regard to stratum boundaries. Let $y_i$ denote the total of the $y$-values in the $i$th network of the population. Let $x_{hi}$ be the number of units in stratum $h$ that intersect network $i$. The probability $\alpha_i$ that the initial sample intersects network $i$ is

$$\alpha_i = 1 - \prod_{k=1}^{L} \binom{N_k - x_{ki}}{n_k} \Big/ \binom{N_k}{n_k} \tag{7}$$

Letting $q_i = 1 - \alpha_i$, the probability $\alpha_{ij}$ that the initial sample intersects both networks $i$ and $j$ is

$$\alpha_{ij} = 1 - q_i - q_j + \prod_{k=1}^{L} \binom{N_k - x_{ki} - x_{kj}}{n_k} \Big/ \binom{N_k}{n_k} \tag{8}$$

Let the indicator variable $z_i$ be 1 if the initial sample intersects network $i$ and zero otherwise. The stratified estimator of modified Horvitz–Thompson type is

$$\hat{\mu}_2 = \frac{1}{N} \sum_{i=1}^{K} \frac{y_i z_i}{\alpha_i} \tag{9}$$

For $i = 1, \ldots, K$, $z_i$ is a Bernoulli random variable with $E(z_i) = \alpha_i$, $var(z_i) = \alpha_i(1 - \alpha_i)$, and $cov(z_i, z_j) = \alpha_{ij} - \alpha_i \alpha_j$, for $i \neq j$. It follows that $\hat{\mu}_2$ is an unbiased estimator of the population mean, and with the convention that $\alpha_{ii} = \alpha_i$,

$$var(\hat{\mu}_2) = \frac{1}{N^2} \sum_{i=1}^{K} \sum_{j=1}^{K} y_i y_j \left( \frac{\alpha_{ij}}{\alpha_i \alpha_j} - 1 \right) \tag{10}$$

An unbiased estimator of this variance, since $\mathrm{E}(z_i z_j) = \alpha_{ij}$, is

$$\widehat{\mathrm{var}}(\hat{\mu}_2) = \frac{1}{N^2} \sum_{i=1}^{K} \sum_{j=1}^{K} \frac{y_i y_j z_i z_j}{\alpha_{ij}} \left( \frac{\alpha_{ij}}{\alpha_i \alpha_j} - 1 \right) \tag{11}$$

provided that the joint intersection probability $\alpha_{ij}$ is not zero for any pair of networks.

The estimator $\hat{\mu}_2$ is not a true Horvitz–Thompson estimator because the intersection probabilities $\alpha_i$ are not identical to the inclusion probabilities under the adaptive cluster sampling design. Expected sample size and other properties of the sampling strategy depend on the actual inclusion probabilities given in Section 26.1.

***Example 1: Stratified Adaptive Cluster Sampling of a Clumped Population.*** The spatially clumped population of Figures 26.1, 26.2, and 26.3 was produced as a realization of a Poisson cluster process (cf. Diggle 1983). Four "parent" locations were randomly located in the study region, and "offspring" locations were distributed about each parent location according to a symmetric Gaussian distribution with dispersion parameter $\sigma = 0.02$. The numbers of offspring were Poisson random variables, each with mean 100. The $y$-values for each of the 400 units (plots) in the population are listed in Section 26.5. The actual number of point objects in the region is 397, so that the true population mean is $\mu = 397/400 = 0.9925$.

For the design, the study region is divided into two strata, and initial samples are selected by stratified random sampling with equal sample sizes in each stratum. A unit satisfies the condition if it contains one or more of the point objects. The neighborhood of a unit includes all adjacent units, so that a typical neighborhood away from the boundary consists of five plots in a cross shape.

Consider the design with initial sample sizes of five units in each stratum. An outcome of the initial sample selection is shown in Figure 26.1, and Figure 26.2 shows the final sample that results. Sample computations are illustrated using the illustrated sample (Figure 26.2). In stratum 1 (on the left), the initial sample has intersected two networks of larger than single-unit size. The first network (on the left) consists of six units, the total $y$-value of which is 96. The second network has five units within the first stratum and six units within the second stratum. The total of the $y$-values in the intersection of this network with the first stratum is 78, while the total of the $y$-values in the intersection of the network with the second stratum is 114. Thus, the second network has a total of 11 units and a total $y$-value of 192. In the second stratum, none of the five units of the initial sample (Figure 26.1) satisfied the condition.

Using the data of this sample (Figure 26.2), the value of the variable $w''_{hi}$ for the estimator $\hat{\mu}''_1$, which ignores crossover between strata, is zero for all units not satisfying the condition. In the first network intersected in stratum 1, the value is $w''_{11} = 96/6 = 16$. For the second network intersected, the value is $w''_{12} = 78/5 = 15.6$, based only on units within stratum 1. The estimate of the population mean is $\hat{\mu}_1'' = (1/400)$ $[(200/5)(16 + 15.6 + 0 + 0 + 0) + (200/5)$ $(0 + 0 + 0 + 0 + 0)] = 3.16$. The estimated variance is $\widehat{\text{var}}(\hat{\mu}_1'') = (1/400^2)$ $[200(200 - 5)(74.9)/5 + 0] = 3.65$, in which 74.9 is the sample variance of the five numbers 16, 15.6, 0, 0, and 0.

For the estimator $\hat{\mu}_1$, the variable $w_{hi}$ for the first network of the sample is $w_{11} = 96/6 = 16$. For the second network intersected by the sample, the value is $w_{12} = 192/11 = 17.45$. The estimate is [using Equation (1)] $\hat{\mu}_1 = (1/400)$ $[(200/5)(16 + 17.45 + 0 + 0 + 0) + 0] = 3.35$. The variance estimate [using Equation (2) with Equation (4) is $\widehat{\text{var}}(\hat{\mu}_1) = (1/400^2)[200(200 - 5)(84.2)/$ $5 + 0] = 4.10$, in which 84.2 is the sample variance of the five sample values of $w_{1i}$ in the first stratum. The estimator $\hat{\mu}_1'$ and its estimated variance assume the same values as $\hat{\mu}_1$ because of the equal stratum and sample sizes.

For the estimator $\hat{\mu}_2$, the intersection probabilities must first be calculated. For every unit in the initial sample not satisfying the condition, the intersection probability is $\alpha_0 = n_h/N_h = 5/200 = 0.025$ and is the same in each stratum, because of the equal sample and stratum sizes. For the first of the large networks intersected (the one on the left in Figure 26.3), the inclusion probability [using Equation (7)] is
$$\alpha_1 = 1 - \binom{200 - 6}{5} \Big/ \binom{200}{5} = 0.14261.$$
For the second network, since it intersects both strata, the intersection probability is $\alpha_2 = 1 - \binom{200 - 5}{5}$ $\binom{200 - 6}{5} \Big/ \binom{200}{5}\binom{200}{5} = 0.24554$. The joint inclusion probability for both networks [using Equation (8)] is

$$\alpha_{12} = 1 - (1 - 0.14261) - (1 - 0.24554) + \binom{200 - 6 - 5}{5}\binom{200 - 0 - 6}{5} \Big/$$
$$\binom{200}{5}\binom{200}{5} = 0.03240$$

The stratified estimator [using Equation (9)] is $\hat{\mu}_2 = (1/400)[(96/0.14261) +$ $(192/0.24554) + (0/0.025) + \cdots + (0/0.025)] = 3.64$. The estimated variance [using Equation (11)] is

$$\widehat{\text{var}}(\hat{\mu}_2) = (1/400^2)\{(96^2/0.14261)[(1/0.14261) - 1] + (192^2/0.24554)$$
$$\times [(1/0.24554) - 1] + 2(96)(192)(0.0324^{-1})[0.0324/(0.14261)$$
$$\times (0.24554) - 1] + 0 + \cdots + 0\} = 4.78 \qquad \square$$

***Example 2: When Stratum Sizes and Initial Sample Sizes Are Unequal.*** The estimators $\hat{\mu}_1$ and $\hat{\mu}_1'$ are identical when stratum sizes and initial sample sizes are equal. To illustrate the computations when they are not equal, consider an initial sample of five units in the first stratum, as illustrated in Figure 26.1 but with an initial sample of only three units in the second stratum, and suppose again that none of the sample units in the second stratum contain any point objects. For the estimator $\hat{\mu}_1$ [Equation (1)] the values are $w_{11} = [(5/200)(96)]/[(5/200)(6)] = 16$ and $w_{12} = [(5/200)(192)]/[(5/200)(5) + (3/200)(6)] = 22.33$. The estimate is $\hat{\mu}_1 = (1/400)[(200/5)(16 + 22.33 + 0 + 0 + 0) + 0] = 3.83$. For the estimator $\hat{\mu}_1'$, $w_{11}' = 96/6 = 16$ and $w_{12}' = 192/11 = 17.45$ as in Example 1, and the estimate is $\hat{\mu}_1' = 3.35$ with $\widehat{\text{var}}(\hat{\mu}_1') = 4.10$, as in that example. $\qquad\square$

## 26.3  COMPARISONS WITH CONVENTIONAL STRATIFIED SAMPLING

The actual variances of the estimators for the population shown in Figures 26.1 through 26.4 with the design with equal initial sample sizes in each of the two strata are given in Table 26.1 for a selection of initial sample sizes, from samples of size 1 in each stratum to initial samples of 100 units in each stratum—an initial sampling fraction of $1/2$. The first column of the table gives the total initial sample size. The second column gives $E(\nu)$, the total expected number of distinct units in the final sample. The third column gives the variance of the initial stratified sample mean $\hat{\mu}_0$. For comparisons with adaptive strategies, the fourth column gives the variance of the stratified sample mean $\hat{\mu}_0^*$ of the stratified mean of a conventional stratified random sample with total sample size equal to $E(\nu)$, the expected sample size under the adaptive design. The last four columns of the table give the variances of the basic four types of unbiased estimators used with stratified adaptive cluster sampling.

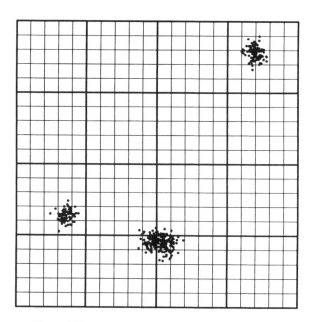

**Figure 26.4.** Sixteen strata for adaptive cluster sampling.

**Table 26.1. Two Strata, Equal Sample Sizes** [a]

| $n$ | $E(\nu)$ | $\text{var}(\hat{\mu}_0)$ | $\text{var}(\hat{\mu}_0^*)$ | $\text{var}(\hat{\mu}_1'')$ | $\text{var}(\hat{\mu}_1')$ | $\text{var}(\hat{\mu}_1)$ | $\text{var}(\hat{\mu}_2)$ |
|---|---|---|---|---|---|---|---|
| 2 | 3.99 | 19.28371 | 9.61712 | 8.63371 | 8.58974 | 8.58974 | 8.53095 |
| 4 | 7.89 | 9.59340 | 4.81846 | 4.29516 | 4.27329 | 4.27329 | 4.17070 |
| 6 | 11.69 | 6.36330 | 3.21874 | 2.84898 | 2.83447 | 2.83477 | 2.71809 |
| 8 | 15.41 | 4.74825 | 2.41875 | 2.12589 | 2.11506 | 2.11506 | 1.99240 |
| 10 | 19.04 | 3.77922 | 1.93866 | 1.69203 | 1.68342 | 1.68342 | 1.55748 |
| 20 | 36.08 | 1.84116 | 0.97749 | 0.82432 | 0.82013 | 0.82013 | 0.69143 |
| 40 | 65.59 | 0.87213 | 0.49402 | 0.39047 | 0.38848 | 0.38848 | 0.26809 |
| 50 | 78.64 | 0.67832 | 0.39598 | 0.30370 | 0.30215 | 0.30215 | 0.18734 |
| 60 | 90.84 | 0.54912 | 0.32979 | 0.24585 | 0.24460 | 0.24460 | 0.13564 |
| 80 | 113.31 | 0.38761 | 0.24517 | 0.17354 | 0.17266 | 0.17266 | 0.07564 |
| 100 | 133.99 | 0.29071 | 0.19239 | 0.13016 | 0.12949 | 0.12949 | 0.04413 |
| 200 | 226.06 | 0.09690 | 0.07456 | 0.04339 | 0.04316 | 0.04316 | 0.00322 |

*Source*: Thompson (1991). With permission from the Biometrika Trustees.

[a] The first column gives the total initial sample size. The second column gives $E(\nu)$, the total expected number of distinct units in the final sample. The third column gives the variance of the initial stratified sample mean $\hat{\mu}_0$. For comparisons with adaptive strategies, the fourth column gives the variance of the stratified sample mean $\hat{\mu}_0^*$ of the stratified mean of a conventional stratified random sample with total sample size equal to $E(\nu)$, the expected sample size under the adaptive design. The last four columns of the table give the variances of the basic four types of unbiased estimators used with stratified adaptive cluster sampling.

For this population, efficiency increases from left to right across Table 26.1. In addition, there is a tendency for the relative advantage of the adaptive strategies, compared to the nonadaptive strategies $\hat{\mu}_0$ and $\hat{\mu}_0^*$, to increase with increasing sample size. For example, with the design depicted in Figures 26.1 and 26.2, having initial sample sizes of 5 in each stratum and total initial sample size $n = 10$, the relative efficiency of the adaptive design with the estimator $\hat{\mu}_2$ compared to the nonadaptive strategy $\hat{\mu}_0^*$ with equivalent sample size is $1.93866/1.55748 = 1.24$, so that the adaptive strategy is 24% more efficient than the comparable nonadaptive one. For the comparison, the fractional sample size of 19.04 is used in the variance formula for the variance of the conventional stratified random sample mean $\hat{\mu}_0^*$, with exactly $19.04/2 = 9.52$ units allocated to each stratum, although in practice integer sample sizes would be necessary. With initial sample sizes of 100 (total sample size 200), the relative efficiency is $0.07456/0.00322 = 23.16$, so that the adaptive strategy is about 23 times more efficient; that is, conventional stratified sampling with initial sample sizes of 113 in each stratum (total sample size 226) would have a variance 23 times as large.

Table 26.2 summarizes variances for the strategies with initial sample sizes in the ratio 2 : 1 (but rounded to whole numbers) in the two strata, so that the initial sample size in the first stratum is as near as possible to twice the sample size in the second. For example, when the total initial sample size $n$ is 6, the two stratum initial sample sizes are $n_1 = 4$ and $n_2 = 2$. When $n = 8$, rounding gives initial sample sizes of $n_1 = 5$ and $n_2 = 3$, as in Example 2. When $n = 200$, $n_1 = 133$ and

**Table 26.2. Two Strata, Sample Sizes in Ratio 2 : 1**[a]

| $n$ | $E(\nu)$ | $var(\hat{\mu}_0)$ | $var(\hat{\mu}_0^*)$ | $var(\hat{\mu}_1'')$ | $var(\hat{\mu}_1')$ | $var(\hat{\mu}_1)$ | $var(\hat{\mu}_2)$ |
|-----|----------|--------------------|----------------------|----------------------|---------------------|--------------------|--------------------|
| 4   | 7.86     | 12.90630           | 6.52072              | 6.44492              | 6.16427             | 5.51233            | 5.41609            |
| 6   | 11.67    | 7.20188            | 3.65693              | 3.47436              | 3.36374             | 3.20237            | 3.09003            |
| 8   | 15.39    | 5.08782            | 2.59904              | 2.41122              | 2.34938             | 2.28465            | 2.16485            |
| 10  | 18.98    | 4.54119            | 2.34637              | 2.22361              | 2.14148             | 1.99178            | 1.86911            |
| 20  | 35.98    | 2.04102            | 1.09135              | 0.97973              | 0.95023             | 0.91060            | 0.78354            |
| 40  | 65.25    | 1.01236            | 0.58305              | 0.49313              | 0.47579             | 0.44640            | 0.32721            |
| 50  | 78.24    | 0.77030            | 0.45726              | 0.37339              | 0.36089             | 0.34123            | 0.22723            |
| 60  | 90.28    | 0.63297            | 0.38816              | 0.30839              | 0.29753             | 0.27916            | 0.17094            |
| 80  | 112.55   | 0.44709            | 0.28977              | 0.21834              | 0.21047             | 0.19687            | 0.10031            |
| 100 | 132.95   | 0.34328            | 0.23418              | 0.16907              | 0.16249             | 0.14994            | 0.06498            |
| 200 | 225.27   | 0.12151            | 0.09701              | 0.06181              | 0.05874             | 0.05166            | 0.01092            |

*Source*: Thompson (1991). With permission from the Biometrika Trustees.

[a] The first column gives the total initial sample size. The second column gives $E(\nu)$, the total expected number of distinct units in the final sample. The third column gives the variance of the initial stratified sample mean $\hat{\mu}_0$. For comparisons with adaptive strategies, the fourth column gives the variance of the stratified sample mean $\hat{\mu}_0^*$ of the stratified mean of a conventional stratified random sample with total sample size equal to $E(\nu)$, the expected sample size under the adaptive design. The last four columns of the table give the variances of the basic four types of unbiased estimators used with stratified adaptive cluster sampling.

$n_2 = 67$. With the unequal sample sizes, the estimator $\hat{\mu}_1$ performs slightly better than $\hat{\mu}_1'$ for this population.

With more than two strata, networks may intersect more than two strata—an initial sample unit in any of these strata may then result in units of the others being added to the sample. Figure 26.4 shows the study area with the clumped population divided into 16 strata, with 4 of these strata intersecting the middle network. Table 26.3 summarizes the variances obtained for the various strategies with equal sample sizes in each of the 16 strata. With this stratification, the efficiency of the strategies again shows a tendency to increase from left to right in the table. The relative efficiency of the most efficient adaptive strategy $\hat{\mu}_2$ compared to the nonadaptive strategy $\hat{\mu}_0^*$ ranges from $1.25567/0.86451 = 1.45$ with initial samples of size 1 in each stratum (a total sample size of 16) to $0.07360/0.00147 = 50.07$ with sample sizes of 13 in each stratum (total sample size 208). Thus, with an initial sampling fraction of just over $1/2$, the adaptive strategy has 50 times the precision of conventional stratified random sampling for estimating the mean of the clumped population illustrated.

## 26.4  FURTHER IMPROVEMENT OF ESTIMATORS

None of the five unbiased estimators above is a function of the minimal sufficient statistic, so each may be improved by the Rao–Blackwell method of taking its conditional expectation, given the minimal sufficient statistic. The minimal sufficient

**Table 26.3. Sixteen Strata, Equal Sample Sizes** [a]

| $n$ | $E(\nu)$ | $\text{var}(\hat{\mu}_0)$ | $\text{var}(\hat{\mu}_0^*)$ | $\text{var}(\hat{\mu}_1'')$ | $\text{var}(\hat{\mu}_1')$ | $\text{var}(\hat{\mu}_1)$ | $\text{var}(\hat{\mu}_2)$ |
|-----|---------|--------|--------|--------|--------|--------|--------|
| 16  | 30.31   | 2.47109 | 1.25567 | 1.25161 | 1.13492 | 1.13492 | 0.86451 |
| 32  | 55.65   | 1.18407 | 0.63716 | 0.59973 | 0.54381 | 0.54381 | 0.34784 |
| 48  | 77.48   | 0.75506 | 0.42859 | 0.38244 | 0.34678 | 0.34678 | 0.18356 |
| 64  | 96.87   | 0.54055 | 0.32220 | 0.27379 | 0.24826 | 0.24826 | 0.10721 |
| 80  | 114.55  | 0.41185 | 0.25658 | 0.20860 | 0.18915 | 0.18915 | 0.06573 |
| 96  | 131.04  | 0.32605 | 0.21133 | 0.16514 | 0.14975 | 0.14975 | 0.04131 |
| 112 | 146.70  | 0.26476 | 0.17777 | 0.13410 | 0.12160 | 0.12160 | 0.02628 |
| 128 | 161.80  | 0.21879 | 0.15158 | 0.11082 | 0.10049 | 0.10049 | 0.01678 |
| 144 | 176.49  | 0.18304 | 0.13039 | 0.09271 | 0.08407 | 0.08407 | 0.01068 |
| 160 | 190.91  | 0.15444 | 0.11277 | 0.07823 | 0.07093 | 0.07093 | 0.00673 |
| 176 | 205.14  | 0.13104 | 0.09780 | 0.06637 | 0.06019 | 0.06019 | 0.00417 |
| 192 | 219.25  | 0.11154 | 0.08488 | 0.05650 | 0.05123 | 0.05123 | 0.00252 |
| 208 | 233.27  | 0.09504 | 0.07360 | 0.04814 | 0.04365 | 0.04365 | 0.00147 |

*Source*: Thompson (1991). With permission from the Biometrika Trustees.

[a] The first column gives the total initial sample size. The second column gives $E(\nu)$, the total expected number of distinct units in the final sample. The third column gives the variance of the initial stratified sample mean $\hat{\mu}_0$. For comparisons with adaptive strategies, the fourth column gives the variance of the stratified sample mean $\hat{\mu}_0^*$ of the stratified mean of a conventional stratified random sample with total sample size equal to $E(\nu)$, the expected sample size under the adaptive design. The last four columns of the table give the variances of the basic four types of unbiased estimators used with stratified adaptive cluster sampling.

statistic $D$ in the finite population setting is the unordered set of distinct, labeled observations (Basu 1969). Starting with any of the unbiased estimators $\hat{\mu}$, one may obtain the Rao–Blackwell version $\hat{\mu}_{RB} = E(\hat{\mu} \mid D)$. With the stratified adaptive cluster sampling design, the estimator $\hat{\mu}_{RB}$ is the average of the values of $\hat{\mu}$ obtained over all samples with the same value of the minimal sufficient statistic.

The Rao–Blackwell version $\hat{\mu}_{RB0}$ of the initial stratified sample mean $\hat{\mu}_0$ is identical, based on the result for the unstratified case, with the Rao–Blackwell version $\hat{\mu}_{RB1}''$ of $\hat{\mu}_1''$. The Rao–Blackwell versions $\hat{\mu}_{RB1}$, $\hat{\mu}_{RB1}'$, $\hat{\mu}_{RB1}''$, and $\hat{\mu}_{RB2}$ are, however, distinct estimators, as demonstrated in the small example of the following section.

***Example 3: A Small Population.*** In the first example, the computational differences between the estimators of Section 26.2 are illustrated with a very small stratified population. In addition, the example demonstrates the bias of the conventional stratified mean with adaptive cluster sampling and shows that the four Rao–Blackwell estimators are distinct, even though each is a function of the minimal sufficient statistic.

Consider a population of 5 units with $y$-values $\{1, 2, 10, 1000, 3\}$ divided into two strata so that the first stratum contains the units with the values $\{1, 2, 10\}$ and the second stratum contains the units with the values $\{100, 3\}$. Let the condition of interest be specified by $C = \{y : y \geq 5\}$, so that whenever a value greater than or equal to 5 is observed, the units in the neighborhood of that observation are added to the sample. The neighborhood of each unit is defined to include its immediately adjacent units. Thus, for example, if the unit with value 10 is observed, the adjacent units, having values 2 and 1000, are added to the sample; then since 1000 also exceeds 5, the adjacent unit with value 3 is also added to the sample. The two units with values $\{10, 1000\}$, the only units in the population that satisfy the condition, form a network that crosses the boundary between strata.

Consider a stratified adaptive cluster sampling design with an initial sample size of 1 in each stratum (i.e., $n_1 = n_2 = 1$). The six possible samples obtainable under this design, each with equal probability, are listed in Table 26.4. In column 1, the initial observations for each of the six possible samples is followed, after the semi-colon, by the observations subsequently added to the sample with the adaptive pro-cedure. For each possible sample, the value of each of the unbiased estimators, other than the initial stratified sample mean, is computed. At the bottom of the table are given the means (equal in each case to 203.2, the population mean) and var-iances of the estimators under the adaptive design.

In the third row of Table 26.4, for illustration, the initial sample selected the unit with value 2 from the first stratum and the unit with value 1000 from the second stratum, resulting in the addition to the sample of the units with values 10 and 3. The computations for each of the estimators are as follows: The intersections of the network of $\{10, 1000\}$ with each stratum have only one unit each, so $\hat{\mu}_1'' = (1/5)[3(2) + 2(1000)] = 401.2$. The sample unit with value 3 does not satisfy the condition and was not intersected by the initial sample, so $\hat{\mu}_1' = (1/5)[3(2) + 2(10 + 1000)/2] = 203.2$. The expected number of times the unit with value 2 is intersected by an initial sample is $1/3$. The expected number for the unit with value 10, as well as for the unit with value 1000, is $1/3 + 1/2 = 5/6$. Thus, $\hat{\mu}_1 = (1/5)[2/(1/3) + 10/(5/6) + 1000/(5/6)] = 243.6$. The intersection probability $\alpha_i$ for the unit with value 2 is $1/3$. For the units with values 10 and 1000, the intersection probability [Equation (7)] is $1 - (2/3)(1/2) = 2/3$. Thus, $\hat{\mu}_2 = (1/5)[2/(1/3) + 10/(2/3) + 1000/(2/3)] = 304.2$.

**Table 26.4. Values of Estimators for the Six Possible Samples in the Example**[a]

| Observations | $\hat{\mu}_1''$ | $\hat{\mu}_1'$ | $\hat{\mu}_1$ | $\hat{\mu}_2$ |
|---|---|---|---|---|
| 1,1000; 10,2,3 | 400.6 | 202.6 | 243.0 | 303.6 |
| 1,3 | 1.8 | 1.8 | 1.8 | 1.8 |
| 2,1000; 10,3 | 401.2 | 203.2 | 243.6 | 304.2 |
| 2,3 | 2.4 | 2.4 | 2.4 | 2.4 |
| 10,1000; 2,3 | 406.0 | 505.0 | 484.8 | 303.0 |
| 10,3; 2,1000 | 7.2 | 304.2 | 243.6 | 304.2 |
| Mean | 203.2 | 203.2 | 203.2 | 203.2 |
| Variance | 39,766.2 | 30,361.2 | 27,504.9 | 20,220.8 |

*Source*: Thompson (1991). With permission from the Biometrika Trustees.

[a] Values of estimators with the small population of 5 units. Any unit with $y \geq 5$ satisfies the condition, and the neighborhood of a unit contains the adjacent units. The observations (col.1) obtained in the initial sample are followed, after the semicolon, by observations obtained subsequently.

**Table 26.5. Values of Estimators Improved by the Rao–Blackwell Method[a]**

| Observations | $\hat{\mu}''_{RB2}$ | $\hat{\mu}'_{RB2}$ | $\hat{\mu}_{RB2}$ | $\hat{\mu}_{RB3}$ |
|---|---|---|---|---|
| 1,1000; 10,2,3 | 400.6 | 202.6 | 243.0 | 303.6 |
| 1,3 | 1.8 | 1.8 | 1.8 | 1.8 |
| 2,1000; 10,3 | 271.47 | 337.47 | 324.0 | 303.8 |
| 2,3 | 2.4 | 2.4 | 2.4 | 2.4 |
| 10,1000; 2,3 | 271.47 | 337.47 | 324.0 | 303.8 |
| 10,3; 2,1000 | 271.47 | 337.47 | 324.0 | 303.8 |
| Mean | 203.2 | 203.2 | 203.2 | 203.2 |
| Variance | 22,305.1 | 22,494.3 | 21,040.8 | 20,220.6 |

[a] Values of estimators with the small population of 5 units. Any unit with $y \geq 5$ satisfies the condition, and the neighborhood of a unit contains the adjacent units. The observations (col.1) obtained in the initial sample are followed, after the semicolon, by observations obtained subsequently.

*Source*: Thompson (1991). With permission from the Biometrika Trustees.

The conventional stratified sample mean for the sample of the third row would be $(1/5)[3(2+10)/2 + 2(1000+3)/2] = 204.2$. The mean of these estimates, over the six possible samples, is 136.67. Hence, the conventional stratified sample mean is biased when used with the adaptive design.

**Table 26.6. Array of y-Values for Figures and Examples**

```
⎛ 0  0  0   0  0  0  0  0  0   0   0    0  0  0  0  0    0    0  0  0 ⎞
⎜ 0  0  0   0  0  0  0  0  0   0   0    0  0  0  0  0   23   14  0  0 ⎟
⎜ 0  0  0   0  0  0  0  0  0   0   0    0  0  0  0  0   36   34  0  0 ⎟
⎜ 0  0  0   0  0  0  0  0  0   0   0    0  0  0  0  0    2    0  0  0 ⎟
⎜ 0  0  0   0  0  0  0  0  0   0   0    0  0  0  0  0    0    0  0  0 ⎟
⎜ 0  0  0   0  0  0  0  0  0   0   0    0  0  0  0  0    0    0  0  0 ⎟
⎜ 0  0  0   0  0  0  0  0  0   0   0    0  0  0  0  0    0    0  0  0 ⎟
⎜ 0  0  0   0  0  0  0  0  0   0   0    0  0  0  0  0    0    0  0  0 ⎟
⎜ 0  0  0   0  0  0  0  0  0   0   0    0  0  0  0  0    0    0  0  0 ⎟
⎜ 0  0  0   0  0  0  0  0  0   0   0    0  0  0  0  0    0    0  0  0 ⎟
⎜ 0  0  0   0  0  0  0  0  0   0   0    0  0  0  0  0    0    0  0  0 ⎟
⎜ 0  0  0   0  0  0  0  0  0   0   0    0  0  0  0  0    0    0  0  0 ⎟
⎜ 0  0  0   0  0  0  0  0  0   0   0    0  0  0  0  0    0    0  0  0 ⎟
⎜ 0  0  0   2  0  0  0  0  0   0   0    0  0  0  0  0    0    0  0  0 ⎟
⎜ 0  0  3  63  9  0  0  0  0   0   0    0  0  0  0  0    0    0  0  0 ⎟
⎜ 0  0  0  16  3  0  0  0  2  12  12    1  0  0  0  0    0    0  0  0 ⎟
⎜ 0  0  0   0  0  0  0  0  2  57  65   17  0  0  0  0    0    0  0  0 ⎟
⎜ 0  0  0   0  0  0  0  0  0   5  14    5  0  0  0  0    0    0  0  0 ⎟
⎜ 0  0  0   0  0  0  0  0  0   0   0    0  0  0  0  0    0    0  0  0 ⎟
⎝ 0  0  0   0  0  0  0  0  0   0   0    0  0  0  0  0    0    0  0  0 ⎠
```

Three of the possible samples in Table 26.4—those in the third, fifth, and sixth rows—have the same set of four distinct observations, and hence have the same value of the minimal sufficient statistic. The Rao–Blackwell version of any of the estimators for each of these samples is obtained by averaging the value of the corresponding estimator over the three samples. The values of the Rao–Blackwell versions of each of these estimators is listed in Table 26.5 for each of the six possible samples, and the variances of these improved unbiased estimators are given at the bottom of the table.                                                  □

## 26.5 EXAMPLE DATA

The array of y-values for the clumped population example and the figures is given in Table 26.6 for each of the 400 units in the study region.

## EXERCISES

1. In a survey of a rare clumped species, the study area is divided into two strata, with $N_1 = 60$ units in the first stratum and $N_2 = 50$ units in the other. Initial simple random samples of sizes $n_1 = 3$ and $n_2 = 2$ are used. Whenever one or more of the animals are found, adjacent units are added as in text examples. In stratum 1, the first initial sample unit had one animal in it; no further animals were found in adjacent units. The second sample unit had no animals. The third sample unit had two animals; of the adjacent units added within that stratum, the only one with animals had three. That unit was adjacent to the stratum boundary, so adding its adjacent units resulted in crossing the stratum boundary and discovering three more units, with two, one, and one animals, respectively. The two initial units selected in stratum 2 each had no animals. Estimate the number of animals in the study region using $\hat{\mu}_0$ and $\hat{\mu}_1$. Estimate the variance of the estimator.

2. For the data in Exercise 1, use $\hat{\mu}_2$.

# Answers to Selected Exercises

## Chapter 2

2. (a) $N\bar{y} = 310$. $s^2 = 1.88$. $\widehat{\text{var}}(N\bar{y}) = 1690$.

   (b) $\bar{y} = 3.1$. $\widehat{\text{var}}(\bar{y}) = 0.169$.

3. (a) $\mu = 2$, $\tau = 10$, $\sigma^2 = 4$. $P(s) = 1/10$ for each sample.

   (b) $\text{E}(\bar{y}) = (1.33 + 1.67 + 3 + 1.33 + 2.67 + 3 + 0.67 + 2 + 2.33 + 2)/$
      $10 = 2$.
      $\text{E}(median) = (1 + 1 + 3 + 1 + 3 + 3 + 1 + 1 + 1 + 1)/10 = 1.6$, while
      the population median is 1, so the sample median is not unbiased.

## Chapter 3

2. $t_9(0.05) = 1.833$.

   (a) $310 \pm 75 = (235, 385)$.

   (b) $3.10 \pm 0.75 = (2.35, 3.85)$.

3. (a) $\text{var}(\bar{y}) = [(1.33 - 2)^2 + (1.67 - 2)^2 + \cdots + ((2 - 2)^2]/10 = 0.533$.
      $\text{var}(median) = [(1 - 1.6)^2 + (1 - 1.6)^2 + \cdots + (1 - 1.6)^2]/10 = 0.84$.

   (b) The 10 values of the sample variance are 2.33, 2.33, 4, 2.33, 6.33, 4, 0.33, 7, 5.33, 7. The 10 values of $\widehat{\text{var}}(\bar{y})$ are 0.31, 0.18, 0.53, 0.31, 0.84, 0.53, 0.04, 0.93, 0.71, 0.93.

   (c) The $t$-value for the confidence intervals is 4.303. Of the 10 confidence intervals corresponding to the 10 possible samples, only the one with $y$-values 1, 0, 1, giving confidence interval $0.67 \pm 0.86$, misses the true mean of 2, so the actual coverage probability is 0.90.

## Chapter 4

1. To within 500: $n = 409$. To within 1000: $n = 148$. To within 2000: $n = 42$.

2. To within 500: $n = 692$. To within 1000: $n = 173$. To within 2000: $n = 44$.

## Chapter 5

1. $\hat{p} = 0.46$. $\widehat{\text{var}}(\hat{p}) = 0.000207$. 95% confidence interval is $0.46 \pm 1.96$ $(0.01439) = 0.46 \pm 0.028$ or approximately $(0.43, 0.49)$.

3. $n = 924$. Ignoring the finite population correction factor gives $n = 2401$.

4. $n^* = 1020$, using $n = 3184$.

## Chapter 6

1. (a) $\hat{\tau}_p = 56.57$.
   (b) $\widehat{\text{var}}(\hat{\tau}_p) = 44.44$.

2. (a) $\pi_i = 1 - (1 - p_i)^3$. $\pi_1 = 0.1694$, $\pi_2 = 0.4880$, $\pi_3 = 0.2710$. $\hat{\tau}_\pi = 64.03$.
   (b) $\pi_{ij} = \pi_i + \pi_j - [1 - (1 - p_i - p_j)^3]$. $\pi_{12} = 0.0626$, $\pi_{13} = 0.0331$, $\pi_{23} = 0.1020$. $\widehat{\text{var}}(\hat{\tau}_\pi) = 62.46$.

3. (a) PPS (with replacement).
   (b) $p_1 = 1.2/80 = 0.015$, $p_2 = 0.0025$, $p_3 = 0.00625$. $\hat{\mu}_p = 3.02$ (one term repeats).
   (c) $\widehat{\text{var}}(\hat{\mu}_p) = 2.3$.

## Chapter 7

1. (a) $r = 6$. $\hat{\mu}_r = 30$.
   (b) $\widehat{\text{var}}(\hat{\mu}_r) = 75$.

2. (a) Units are households, the variable of interest is cost of food, the auxiliary variable is number of people in the household.
   (b) $\bar{y}$ is unbiased, $\hat{\mu}_r$ is biased.
   (c) $\bar{y} = 147.5$. $s^2 = 1691.7$. $\widehat{\text{var}}(\bar{y}) = 422.9$.
   (d) $r = 45.48$, $\mu_x = 2.9$, $\hat{\mu}_r = 131.6$. $\widehat{\text{var}}(\hat{\mu}_r) = 119.6$.
   (e) $\hat{\mu}_r$ is better based on the estimated variance or on a scatterplot of the data.

3. (a) There are six possible samples, each with probability $1/6$. The six values of $N\bar{y}$ are 10, 4, 6, 6, 8, 2. The six values of $\hat{\tau}_r$ are 6.67, 6, 5.14, 7.2, 6, 4.
   (b) $E(N\bar{y}) = 6$, $\text{var}(N\bar{y}) = 6.67$.
   (c) $E(\hat{\tau}_r) = (6.67 + 6 + \cdots + 4)/6 = 5.8$.
   $E(\hat{\tau}_r - \tau)^2 = [(6.67 - 6)^2 + \cdots + (4 - 6)^2]/6 = 1.1$.

## Chapter 8

1. (a) $b = -2$, $a = 22$, $\hat{\mu}_L = 12$.
   (b) $\widehat{\text{var}}(\hat{\mu}_L) = 0.218$.
   (c) When $x_i = 4$, $\hat{y}_i = 14$.

## Chapter 11

1. (a) $\hat{\mu}_{\text{st}} = 24.44$.

   (b) $\widehat{\text{var}}(\hat{\mu}_{\text{st}}) = 5.827$. 95% confidence interval is $24.44 \pm 4.73 = (19.71, 29.17)$.

2. (a) $n_1 = 40$, $n_2 = 60$.

   (b) $n_1 = 60$, $n_2 = 40$.

## Chapter 12

1. (a) $\hat{\tau} = 76.7$.

   (b) $s_u^2 = 16.33$, $\widehat{\text{var}}(\hat{\tau}) = 381.1$.

2. (a) $r = 0.657$, $\hat{\tau}_r = 65.7$.

   (b) $s_r^2 = 1.00$, $\widehat{\text{var}}(\hat{\tau}_r) = 23.33$.

3. (a) $\hat{\tau}_{\text{pps}} = 70$.

   (b) $\widehat{\text{var}}(\hat{\tau}_{\text{pps}}) = 33.33$.

5. (a) $\hat{r} = (10/1)(1 + 0 + 2 + 3 + 0 + 1) = 70$; simple random sample of one primary unit.

   (b) ratio (or PPS) estimator. Not unbiased.

   (c) A design-unbiased estimator of variance is not available with a sample size of one primary unit.

## Chapter 13

1. (a) $\hat{\mu} = 4$.

   (b) $s_u^2 = 72$, $s_1^2 = 4$, $s_2^2 = 1$, $\widehat{\text{var}}(\hat{\mu}) = 0.84$.

## Chapter 14

1. (a) $r = 1.29$, $\hat{r}_x = 297.5$, $\hat{r}_r = 384.3$. $s^2 = 1250$, $s_r^2 = 1.3889$, $\widehat{\text{var}}(\hat{r}_r) = 18784.7$.

   (b) $n/n' = 0.024$.

## Chapter 16

2. The estimated abundance is $\hat{\tau} = 60/0.25 = 240$ animals in the region. The estimated density is $\hat{D} = 240/100 = 2.4$ animals per square kilometer. The variance estimates are $\widehat{\text{var}}(\hat{\tau}) = 240(1 - 0.25)/(0.25) = 720$ and $\widehat{\text{var}}(\hat{D}) = 720/100^2 = 0.072$.

4. The sample mean of the four observations is $\bar{y} = 4.4$, and their sample variance is $s^2 = 19.3$. The estimate of the total number of animals in the study region is $\hat{\tau} = 100(4.4/0.80) = 550$. The estimated variance is $\widehat{\text{var}}(\hat{\tau}) = 100^2\{(95/100)(19.3)/[0.80^2(5)] + 0.20(4.4)/[0.80^2(100)](4.4)]\} = 57{,}434$.

## Chapter 17

1. (a) Using a 5-m-wide strip, $\hat{D} = 0.014$ bird/m$^2$ or 140 birds/ha.
   (b) With $\hat{f}(0) = 0.12$ by eye, $\hat{D} = 156$ birds/ha.
   (c) $\bar{x} = 5.538$, $\hat{D} = 235$ birds/ha.

## Chapter 18

1. (a) $\hat{\tau} = 150,000$.
   (b) $\widehat{\text{var}}(\hat{\tau}) = 50,400,000$. 95% confidence interval is $150,000 \pm 13,915 = (136,085, 163,915)$.
   (c) Overestimation.
   (d) Overestimation.

## Chapter 20

1. (a) $\gamma_{01} = \gamma_{02} = 2.64$. $a_1 = a_2 = 0.5$. $\hat{y}_0 = 1.800$. $m^* = 0.685$. $\hat{E}(\hat{y}_0 - y_0)^2 = 3.325$.
   (b) $\gamma_{01} = \gamma_{02} = 4.62$. $a_1 = a_2 = 0.5$. $\hat{y}_0 = 1.800$. $m^* = 2.665$. $\hat{E}(\hat{y}_0 - y_0)^2 = 7.285$.
   (c) $\gamma_{01} = 4.37$, $\gamma_{02} = 1.98$. $a_1 = 0.1943$, $a_2 = 0.8057$. $\hat{y}_0 = 2.167$. $m^* = 1.22$. $\hat{E}(\hat{y}_0 - y_0)^2 = 3.664$.
   (d) $\gamma_{01} = 4.37$, $\gamma_{02} = 4.62$. $a_1 = 0.5320$, $a_2 = 0.4680$. $\hat{y}_0 = 1.762$. $m^* = 2.54$. $\hat{E}(\hat{y}_0 - y_0)^2 = 7.027$.

## Chapter 24

1. $\hat{\tau}_1 = 35$. $\widehat{\text{var}}(\hat{\tau}_1) = 557$.

2. $\alpha_1 = 0.19$, $\alpha_2 = 0.27$. $\hat{\tau}_2 = 38$. $\alpha_{12} = 0.0511$. $\widehat{\text{var}}(\hat{\tau}_2) = 552$.

## Chapter 26

1. $\hat{\tau}_0 = N\hat{\mu}_0 = 60$.   $N\hat{\mu}_1 = 1/[(3/60)(1)] + (2 + 3 + 2 + 1 + 1)/[(3/60)(2) + (2/50)(3)] + 0 + 0 + 0 = 61$.

2. $\alpha_1 = 0.05$, $\alpha_2 = 0.20$. $\hat{\tau}_2 = N\hat{\mu}_2 = (1/0.05) + (9/0.20) + 0 + 0 + 0 = 65$.

# References

Acharya1, B., Bhattarai, G., de Gier, A., and Stein, A. (2000). Systematic adaptive cluster sampling for the assessment of rare tree species in Nepal. *Forest Ecology and Management*, **137**, 65–73.

Agresti, A. (1984). *Analysis of Ordinal Categorical Data*. New York: Wiley.

Agresti, A. (1990). *Categorical Data Analysis*, New York: Wiley.

Ahlo, J. M. (1990). Logistic regression in capture–recapture models. *Journal of the American Statistical Association*, **46**, 623–635.

Ames, M. H., and Webster, J. T. (1991). On estimating approximate degrees of freedom. *The American Statistician*, **45**, 45–50.

Anthony, R. M. (1990). Personal communication Anchorage, AK: Alaska Fish and Wildlife Research Center, U.S. Fish and Wildlife Service.

Barnes, R. (1988). Bounding the required sample size for geologic site characterization. *Mathematical Geology*, **20**, 477–490.

Barnett, V. (1991) *Sample Survey Principles and Methods* New York: Oxford University Press.

Barry, D. A., and Fristedt, B. (1985). *Bandit Problems: Sequential Allocation of Experiments*. London: Chapman & Hall.

Barry, S. C., and Welsh, A. H. (2001). Distance sampling methodology. *Journal of the Royal Statistical Society B*, **63**, 31–51.

Bart, J., Fligner, M. A., and Notz, W. I. (1998). *Sampling and Statistical Methods for Behavioral Ecologists*. Cambridge: Cambridge University Press.

Basu, D. (1958). On sampling with and without replacement. *Sankhyā*, **20**, 287–294.

Basu, D. (1969). Role of the sufficiency and likelihood principles in sample survey theory. *Sankhyā A*, **31**, 441–454.

Basu, D. (1971). An essay on the logical foundations of survey sampling, part one. In V. P. Godambe and D. A. Sprott (eds.), *Foundations of Statistical Inference*. Toronto, Ontario, Canada: Holt, Rinehart, and Winston, P. 236.

Basu, D., and Ghosh, J. K. (1967). Sufficient statistics in sampling from a finite universe. *Proceedings of the 36th Session of the International Statistical Institute*, pp. 850–859.

Becker, E. F. (1990). Personal communication. Anchorage, AK: Alaska Department of Fish and Game.

Becker, E. F. (1991). A terrestrial furbearer estimator based on probability sampling. *Journal of Wildlife Management*, **55**, 730–737.

Becker, E. F., and Gardner, C. (1990). Wolf and wolverine density estimation techniques. *Federal Aid in Wildlife Restoration Research Report W–23–3*. Juneau, Alaska Department of Fish and Game, Division of Wildlife Conservation.

Becker, E. F., and Reed, D. J. (1990). A modification of a moose population estimator. *Alces*, **26**, 73–79.

Bellhouse, D. R. (1977). Some optimal designs for sampling in two dimensions. *Biometrika*, **64**, 605–611.

Bellhouse, D. R. (1981). Area estimation by point-counting techniques. *Biometrics*, **37**, 303–312.

Bellhouse, D. R. (1988a). Systematic sampling. In P. R. Krishnaiah and C. R. Rao (eds.), *Handbook of Statistics*, Vol. 6, Sampling. Amsterdam: Elsevier Science Publishers, pp. 125–145.

Bellhouse, D. R. (1988b). A brief history of random sampling methods. In P. R. Krishnaiah and C. R. Rao (eds.), *Handbook of Statistics*, Vol. 6, *Sampling*. Amsterdam: Elsevier Science Publishers, pp. 1–14.

Biemer, P. P., Groves, R. M., Lyberg, L. E., Mathiowetz, N. A., and Sudman, S. (1991). *Measurement Errors in Surveys*. New York: Wiley.

Binder, D. A., and Patak, Z. (1994). Use of estimating functions for estimation from complex surveys. *Journal of the American Statistical Association*, **89**, 1035–1043.

Birnbaum, Z. W., and Sirken, M. G. (1965). Design of sample surveys to estimate the prevalence of rare diseases: Three unbiased estimates. *Vital and Health Statistics*, Ser. 2, No. 11. Washington, DC: U.S. Government Printing Office.

Bishop, Y. M. M., Fienberg, S. E., and Holland, P. W. (1975). *Discrete Multivariate Analysis*. Cambridge, MA: MIT Press.

Bjørnstad, J. F. (1990). Predictive likelihood: A review. *Statistical Science*, **5**, 242–265.

Blackwell, D. (1947). Conditional expectation and unbiased sequential estimation. *Annals of Mathematical Statistics*, **18**, 105–110.

Blair, J. (1999). A probability sample of gay urban males: The use of two-phase adaptive sampling. *Journal of Sex Research*, **36**, 25–38.

Bloemena, A. R. (1969). *Sampling from a Graph*. Mathematical Centre Tracts 2. Amsterdam: Mathematisch Centrum.

Bolfarine, H., and Zacks, S. (1992). *Prediction Theory for Finite Populations*. New York: Springer-Verlag.

Boomer, K., Werner, C., and Brantley, S. (2000). $CO_2$ emissions related to the Yellowstone volcanic system: 1. Developing a stratified adaptive cluster sampling plan. *Journal of Geophysycal Research*, **105**, 817–830.

Borkowski, J. J. (1999). Network inclusion probabilities and Horvitz–Thompson estimation for adaptive simple Latin square sampling. *Environmental and Ecological Statistics*, **6**, 291–311.

Brewer, K. R. W. (1963). Ratio estimation and finite populations: Some results deducible from the assumption of an underlying stochastic process. *Australian Journal of Statistics*, **5**, 93–105.

Brewer, K. R. W. (1979). A class of robust sampling designs for large-scale surveys. *Journal of the American Statistical Association*, **74**, 911–915.

Brewer, K. R. W., and Hanif, M. (1983). *Sampling with Unequal Probabilities*. New York: Springer-Verlag.

Brewer, K. R. W., Hanif, M., and Tam, S. M. (1988). How nearly can model-based prediction and design-based estimation be reconciled? *Journal of the American Statistical Association*, **83**, 128–132.

Brown, J. A. (1994). The application of adaptive cluster sampling to ecological studies. In D. J. Fletcher and B. F. J. Manly (eds.), *Statistics in Ecology and Environmental Monitoring*, Otago Conference Series 2. Dunedin, New Zealand: University of Otago Press, pp. 86–97.

Brown, J. A., and Manly, B. F. J. (1998). Restricted adaptive cluster sampling. *Environmental and Ecological Statistics*, **5**, 47–62.

Buckland, S. T. (1980). A modified analysis of the Jolly–Seber capture–recapture model. *Biometrics*, **36**, 419–435.

Buckland, S. T. (1982). A note on the Fourier series model for analysing line transect data. *Biometrics*, **38**, 469–477.

Buckland, S. T. (1984). Monte Carlo confidence intervals. *Biometrics*, **40**, 811–817.

Buckland, S. T. (1985). Perpendicular distance models for line transect sampling. *Biometrics*, **41**, 177–195.

Buckland, S. T. (1987). On the variable circular plot method of estimating animal density. *Biometrics*, **43**, 363–384.

Buckland, S. T. (1988). Quantifying precision of mark–recapture estimates using the bootstrap. Unpublished ms.

Buckland, S. T., Anderson, D. R., Burnham, K. P., and Laake, J. L. (1992). *Distance Sampling: Estimating Abundance of Biological Populations*. London: Chapman & Hall.

Burnham, K. P. (1979). A parametric generalization of the Hayne estimator for line transect sampling. *Biometrics*, **35**, 587–595.

Burnham, K. P., and Anderson, D. R. (1976). Mathematical models for nonparametric inferences from line transect data. *Biometrics*, **32**, 325–336.

Burnham, K. P., and Overton, W. S. (1978). Estimation of the size of a closed population when capture probabilities vary among animals. *Biometrika*, **65**, 625–633.

Burnham, K. P., Anderson, D. R., and Laake, J. L. (1980). Estimation of density from line transect sampling of biological populations. *Wildlife Monograph*, *72*, supplement to *Journal of Wildlife Management*.

Burnham, K. P., Anderson, D. R., and Laake, J. L. (1981). Line transect estimation of bird population density using a Fourier series. *Studies in Avian Biology*, **6**, 466–482.

Cassel, C. M., Särndal, C. E., and Wretman, J. H. (1976). Some results on generalized difference estimation and generalized regression estimation for finite populations. *Biometrika*, **63**, 615–620.

Cassel, C. M., Särndal, C. E., and Wretman, J. H. (1977). *Foundations of Inference in Survey Sampling*. New York: Wiley.

Cassel, C. M., Särndal, C. E., and Wretman, J. H. (1979). Prediction theory for finite populations when model-based and design-based principles are combined. *Scandinavian Journal of Statistics*, **6**, 97–106.

Caughley, G. (1974). Bias in aerial survey. *Journal of Wildlife Management*, **38**, 921–933.

Caughley, G., Sinclair, R., and Scott-Kemmis, D. (1976). Experiments in aerial survey. *Journal of Wildlife Management*, **40**, 290–300.

Chao, T.-C. (1999). Adaptive sampling designs. Ph.D. dissertation. University Park, PA: Pennsylvania State University.

Chao, T.-C. (2002). Markov chain Monte Carlo on optimal adaptive sampling selections. *Environmental and Ecological Statistics*, to appear.

Chao, T.-C., and Thompson, S. K. (2001). Optimal adaptive selection of sampling sites. *Environmetrics*, **12**, 517–538.

Chaudhuri, A., and Stenger, H. (1992). *Survey Sampling: Theory and Methods*. New York: Marcel Dekker.

Chaudhuri, A., and Vos, J. W. E. (1988). *Unified Theory and Strategies of Survey Sampling*. Amsterdam: North-Holland.

Chernoff, H. (1972). *Sequential Analysis and Optimal Design*. Philadelphia: Society for Industrial and Applied Mathematics.

Christensen, R. (1987). The analysis of two-stage sampling data by ordinary least squares. *Journal of the American Statistical Association*, **82**, 492–498.

Christman, M. (1997). Efficiency of some sampling designs for spatially clustered populations. *Environmetrics*, **8**, 145–166.

Christman, M. C. (2000). A review of quadrat-based sampling of rare, geographically-clustered populations. *Journal of Agricultural, Biological and Environmental Statistics*, **5**, 168–201.

Christman, M. C., and Lan, F. (2001). Inverse adaptive cluster sampling. *Biometrics*, **57**, 1096–1105.

Christman, M. C., and Pontius, J. R. (2000). Bootstrap confidence intervals for adaptive cluster sampling. *Biometrics*, **56**, 503–610.

Clausen, D., Hanselman, D., Lunsford, C., Quinn, T., and Heifetz, J. (1999). Rockfish adaptive sampling experiment in the central Gulf of Alaska, 1998. *AFSC Processed Report 99–04.*, Juneau, AK: Alaska Fisheries Science Center.

Cochran, W. G. (1977). *Sampling Techniques*, 3rd ed. New York: Wiley.

Cormack, R. M. (1979). Models for capture–recapture. In R. M. Cormack, G. P. Patil, and D. S. Robson (eds.), *Sampling Biological Populations*. Fairland, MD: International Co-operative Publishing House, pp. 217–255.

Cormack, R. M. (1980). Model selection in capture–recapture experiments. *GLIM Newsletter*, Dec., pp. 27–29.

Cormack, R. M. (1981). Loglinear models for capture–recapture experiments on open populations. In R. W. Hiorns and D. Cooke (eds.), *The Mathematical Theory of the Dynamics of Biological Populations II*. London: Academic Press, pp. 217–235.

Cormack, R. M. (1985). Examples of the use of GLIM to analyse capture–recapture studies. In B. J. T. Morgan and P. M. North (eds.), *Statistics in Ornithology*. Lecture Notes in Statistics, No. 29. New York: Springer-Verlag, pp. 243–273.

Cormack, R. M. (1988). Statistical challenges in the environmental sciences: A personal view. *Journal of the Royal Statistical Society A*, **151**, 201–210.

Cormack, R. M. (1989). Log-linear models for capture–recapture. *Biometrics*, **45**, 395–413.

Cowan, C.D., and Malec, D. (1986). Capture–recapture models when both sources have clustered observations. *Journal of the American Statistical Association*, **81**, 347–353.

Cox, D. R., and Isham, V. (1980). *Point Processes*. London: Chapman & Hall.

Crain, B. R., Burnham, K. P., Anderson, D. R., and Laake, J. L. (1979). Nonparametric estimation of population density for line transect sampling using Fourier series. *Biometrical Journal*, **21**, 731–748.

Cressie, N. (1986). Kriging nonstationary data. *Journal of the American Statistical Association*, **81**, 625–634.

Cressie, N. (1989). Geostatistics. *The American Statistician*, **43**, 197–202.

Cressie, N. (1991). *Statistics for Spatial Data*. New York: Wiley.

Cumberland, W. G., and Royall, R. M. (1981). Prediction models in unequal probability sampling. *Journal of the Royal Statistical Society B*, **43**, 353–367.

Cumberland, W. G., and Royall, R. M. (1988). Does simple random sampling provide adequate balance? *Journal of the Royal Statistical Society B*, **50**, 118–124.

Czaja, R. F., Snowdon, C. B., and Casady, R. J. (1986). Reporting bias and sampling errors in a survey of a rare population using multiplicity counting rules. *Journal of the American Statistical Association*, **81**, 411–419.

Danaher, P. J., and King, M. (1994). Estimating rare household characteristics using adaptive sampling. *The New Zealand Statistician*, **29**, 14–23.

Das, A. C. (1951). On two phase sampling and sampling with varying probabilities. *Bulletin of the International Statistical Institute*, **33**(2), 105–112.

David, I. P., and Sukhatme, B. V. (1974). On the bias and mean square error of the ratio estimator. *Journal of the American Statistical Association*, **69**, 464–466.

Davis, J. L., Valkenburg, P., and Harbo, S. (1979). Refinement of the aerial photo-direct count-extrapolation caribou census technique. *Final Research Report W-17-11*. Fairbanks, AK: Alaska Department of Fish and Game Wildlife Conservation Division.

Dawid, A. P., and Dickey, J. M. (1977). Likelihood and Bayesian inference from selectively reported data. *Journal of the American Statistical Association*, **72**, 845–850.

Deng, L. Y., and Wu, C. F. J. (1987). Estimation of variance of the regression estimator. *Journal of the American Statistical Association*, **82**, 568–576.

DeVries, P. G. (1979). Line intersect sampling: Statistical theory, applications, and suggestions for extended use in ecological inventory. In R. M. Cormack, G. P. Patil, and D. S. Robson (eds.), *Sampling Biological Populations*. Fairland, MD: International Co-operative Publishing House, pp. 1–70.

Diggle, P. J. (1983). *Statistical Analysis of Spatial Point Patterns*. New York: Academic Press.

Drummer, T. D., and McDonald, L. L. (1987) Size bias in line transect sampling. *Biometrics*, **43**, 13–21.

Dryver, A. L. (1999). Adaptive sampling designs and associated estimators. Ph.D. dissertation. University Park, PA: Pennsylvania State University.

Eberhardt, L. L. (1978a). Transect methods for population studies. *Journal of Wildlife Management*, **42**, 1–31.

Eberhardt, L. L. (1978b). Appraising variability in population studies. *Journal of Wildlife Management*, **42**, 207–238.

Eberhardt, L. L., and Simmons, M. A. (1987). Calibrating population indices by double sampling. *Journal of Wildlife Management*, **51**, 665–675.

Efron, B. (1982). The jackknife, the bootstrap, and other resampling plans. *CBMS-NSF Monograph 38*. Philadelphia: Society for Industrial and Applied Mathematics.

Efron, B., and Gong, G. (1983). A leisurely look at the bootstrap, the jackknife, and cross-validation. *The American Statistician*, **37**, 36–48.

Erdös, P., and Rényi, A. (1959). On a central limit theorem for samples from a finite population. *Publications of the Mathematical Institute of the Hungarian Academy of Science*, **4**, 49–61.

Erickson, B. (1979). Some problems of inference from chain data. *Sociological Methodology*, **10**, 276–302.

Ericson, W. A. (1969). Subjective Bayesian models in sampling finite populations: I. *Journal of the Royal Statistical Society B*, **31**, 195–234.

Ericson, W. A. (1988). Bayesian inference in finite populations. In P. R. Krishnaiah and C. R. Rao (eds.), *Handbook of Statistics*, Vol. 6 *Sampling*. Amsterdam: Elsevier Science Publishers, pp. 213–246.

Faulkenberry, G. D., and Garoui, A. (1991). Estimating a population total using an area frame. *Journal of the American Statistical Association*, **86**, 445–449.

Félix Medina, M. H. (2000a). Analytical expressions for Rao–Blackwell estimators in adaptive cluster sampling. *Journal of Statistical Planning and Inference*, **84**, 221–236.

Félix Medina, M. H. (2000b). Contributions to the theory of adaptive sampling. Ph.D. dissertation. University Park, PA: Pennsylvania State University.

Félix Medina, M. H. (2002). Asymptotics in adaptive cluster sampling. *Environmental and Ecological Statistics*, to appear.

Félix Medina, M. H., and Thompson S. K. (1999). Adaptive cluster double sampling. *Proceedings of the Survey Research Section, Americal Statistical Association*.

Ferebee, B. (1983). An unbiased estimator for the drift of a stopped wiener process. *Journal of Applied Probability*, **20**, 94–102.

Fitzpatrick, S., and Scott, A. (1987). Quick simultaneous confidence intervals for multinomial proportions. *Journal of the American Statistical Association*, **82**, 875–878.

Foreman, E. K. (1991). *Survey Sampling Principles*. New York: Marcel Dekker.

Francis, R. I. C. C. (1984). An adaptive strategy for stratified random trawl surveys. *New Zealand Journal of Marine and Freshwater Research*, **18**, 59–71.

Frank, O. (1971). *Statistical Inference in Graphs*. Stockholm: Försvarets Forskningsanstalt.

Frank, O. (1977a). Survey sampling in graphs. *Journal of Statistical Planning and Inference*, **1**, 235–264.

Frank, O. (1977b). Estimation of graph totals. *Scandinavian Journal of Statistics*, **4**, 81–89.

Frank, O. (1978a). Estimating the number of connected components in a graph by using a sampled subgraph. *Scandinavian Journal of Statistics*, **5**, 177–188.

Frank, O. (1978b). Sampling and estimation in large social networks. *Social Networks*, **1**, 91–101.

Frank, O. (1979a). Estimation of population totals by use of snowball samples. In P. W. Holland and S. Leinhardt (eds.), *Perspectives on social Network Research*. New York: Academic Press, pp. 319–347.

Frank, O. (1979b). Moment properties of subgraph counts in stochastic graphs. *Annals of the New York Academy of Sciences*, **319**, 207–218.

Frank, O. (1981). A survey of statistical methods for graph analysis. *Sociological Methodology*, **1981**, 110–155.

Frank, O. (1988). Random sampling and social networks: A survey of various approaches. *Mathmatiques, Informatique et Sciences Humaines*, **26**, 19–33.

Frank, O. (1997). Composition and structure of social networks. *Mathmatiques, Informatique et Sciences Humaines*, **35**, 11–23.

Frank, O., and Harary, F. (1982). Cluster inference by using transitivity indices in empirical graphs. *Journal of the American Statistical Association*, **77**, 835–840.

Frank, O., and Snijders, T. (1994). Estimating the size of hidden populations using snowball sampling. *Journal of Official Statistics*, **10**, 53–67.

Freedman, D. A. (1991). Adjusting the 1990 census. *Science*, **252**, 1233–1236.

Friedman, S. R., Neaigus, A., Jose, B., Curtis, R., Goldstein, M., Ildefonso, G., Rothenberg, R. B., and Des Jarlais, D. C., (1997). Sociometric risk networks risk of HIV infection. *American Journal of Public Health*, **87**, 1289–1296.

Gates, C. E. (1979). Line transect and related issues. In R. M. Cormack, G. P. Patil, and D. S. Robson (eds.), *Sampling Biological Populations*. Fairland, MD: International Co-operative Publishing House, pp. 71–154.

Ghosh, M., and Meeden, G. (1997). *Bayesian Methods for Finite Population Sampling*. London: Chapman & Hall.

Ghosh, M., and Rao, J. N. K. (1994). Small area estimation: An appraisal. *Statistical Science*, **9**, 55–93.

Gilbert, R. O. (1987). *Statistical Methods for Environmental Pollution Monitoring*. New York: Van Nostrand Reinhold.

Godambe, V. P. (1955). A unified theory of sampling from finite populations. *Journal of the Royal Statistical Society B*, **17**, 269–278.

Godambe, V. P. (1966). A new approach to sampling from finite populations: I. *Journal of the Royal Statistical Society B*, **28**, 310–319.

Godambe, V. P. (1982). Estimation in survey sampling: Robustness and optimality. *Journal of the American Statistical Association*, **77**, 393–403.

Godambe, V. P. (1995). Estimation of parameters in survey sampling: Optimality. *Canadian Journal of Statistics*, **23**, 227–243.

Godambe, V. P., and Thompson, M. E. (1986). Parameters of superpopulation and survey population: Their relationships and estimation. *International Statistical Review*, **54**, 127–138.

Goodman, L. A. (1961). Snowball sampling. *Annals of Mathematical Statistics*, **32**, 148–170.

Govindarajulu, Z. (1999). *Elements of Sampling Theory and Methods*. Upper Saddle River, NJ: Prentice Hall.

Granovetter, M., and Cooper, M. P. (1976). Network sampling: Some first steps. *American Journal of Sociology*, **81**, 1287–1303.

Groves, R. M., and Couper, M. P. (1998). *Nonresponse in Household Interview Surveys*. New York: Wiley.

Hájek, J. (1960). Limiting distributions in simple random sampling from a finite population. *Publications of the Mathematical Institute of the Hungarian Academy*, **5**, 361–374.

Hájek, J. (1961). Some extensions of the Wald-Wolfowitz-Noether theorem. *Annals of Mathematical Statistics*, **32**, 506–523.

Hájek, J. (1971). Discussion of "An essay on the logical foundations of survey sampling, part one," by D. Basu. In V. P. Godambe and D. A. Sprott (eds.), *Foundations of Statistical Inference*. Toronto, Ontario, Canada: Holt, Rinehart and Winston, p. 236.

Hájek, J. (1981). *Sampling from a Finite Population*. New York: Marcel Dekker.

Hancock, H. (1960). *Theory of Maxima and Minima*. New York: Dover.

Hansen, M. M., and Hurwitz, W. N. (1943). On the theory of sampling from finite populations. *Annals of Mathematical Statistics*, **14**, 333–362.

Hansen, M. M., Hurwitz, W. N., and Madow, W. G. (1953). *Sample Survey Methods and Theory* (2 vol.). New York: Wiley.

Hansen, M. M., Madow, W. G., and Tepping, B. J. (1983). An evaluation of model-dependent and probability-sampling inferences in sample surveys. *Journal of the American Statistical Association*, **78**, 776–793.

Hansen, M. M., Dalenius, T., and Tepping, B. J. (1985). The development of sample surveys of finite populations. In A. C. Atkinson and S. E. Fienberg (eds.), *A Celebration of Statistics: The ISI Centenary Volume*. New York: Springer-Verlag, pp. 327–354.

Hartley, H. O., and Ross, A. (1954). Unbiased ratio estimates. *Nature*, **174**, 841–851.

Hayne, D. W. (1949). An examination of the strip census method for estimating animal populations. *Journal of Wildlife Management*, **13**, 145–157.

Hedayat, A. S., and Sinha, B. K. (1991). *Design and Inference in Finite Population Sampling*. New York: Wiley.

Henzinger, M. R., Heydon, A., Mitzenmacher, M., and Najork, M. (2000). On near-uniform URL sampling. *Proceedings of the 9th International World Wide Web Conference*. Amsterdam: Elsevier, pp. 295–308.

Hohn, M. E. (1988). *Geostatistics and Petroleum Geology*. New York: Van Nostrand Reinhold.

Holt, D., and Scott, A. J. (1981). Regression analysis using survey data. *The Statistician*, **30**, 169–178.

Horvitz, D. G., and Thompson, D. J. (1952). A generalization of sampling without replacement from a finite universe. *Journal of the American Statistical Association*, **47**, 663–685.

Isaki, C. T., and Fuller, W. A. (1982). Survey design under the regression superpopulation model. *Journal of the American Statistical Association*, **77**, 89–96.

Jansson, I. (1997). On statistical modeling of social networks. Ph.D. dissertation. Stockholm University.

Jensen, A. L. (1989). Confidence intervals for nearly unbiased estimators in single-mark and single-recapture experiments. *Biometrics*, **45**, 1233–1237.

Jessen, R. J. (1978). *Statistical Survey Techniques*. New York: Wiley.

Johnson, E. G., and Routledge, R. D. (1985). The line transect method: A nonparametric estimator based on shape restrictions. *Biometrics*, **41**, 669–679.

Jolly, G. M. (1965). Explicit estimates from capture–recapture with both death and immigration—stochastic model. *Biometrika*, **52**, 225–247.

Jolly, G. M. (1979). Sampling of large objects. In R. M. Cormack, G. P. Patil, and D. S. Robson (eds.), *Sampling Biological Populations*. Fairland, MD: International Co-operative Publishing House, pp. 193–201.

Journel, A. G. (1987). Geostatistics for the environmental sciences: An introduction. *Project CR 811893*, Las Vegas, NV: U.S. Environmental Protection Agency, Environmental Monitoring Systems Laboratory.

Journel, A. G. (1988). Non-parametric geostatistics for risk and additional sampling assessment. In L. Keith (ed.), *Principles of Environmental Sampling*. Washington, DC: American Chemical Society, pp. 45–72.

Journel, A. G. (1989). *Short Course in Geology*, Vol. 8, *Fundamentals of Geostatistics in Five Lessons*. Washington, DC: American Geophysical Union, 40 pp.

Kaiser, L. (1983). Unbiased estimation in line-intercept sampling. *Biometrics*, **39**, 965–976.

Kalton, G., and Anderson, D. W. (1986). Sampling rare populations. *Journal of the Royal Statistical Society A*, **149**, 65–82.

Karlberg, M. (1997). Triad count estimation and transitivity testing in graphs and digraphs. Ph.D. dissertation. Stockholm University.

Kendall, M. G., and Moran, P. A. P. (1963). *Geometrical Probability*. London: Griffin.

Khaemba, W. M., and Stein, A. (2001). Improved sampling of wildlife populations using airborn surveys. *Wildlife Research*, to appear.

Kimura, D. K., and Lemberg, N. A. (1981). Variability of line intercept density estimates: A simulation study of the variance of hydroacoustic biomass estimates. *Canadian Journal of Fisheries and Aquatic Sciences*, **38**, 1141–1152.

Kish, L. (1965). *Survey Sampling*, New York: Wiley.

Klovdahl, A. S. (1989). Urban social networks: Some methodological problems and possibilities. In M. Kochen (ed.), *The Small World*. Norwood, NJ: Ablex Publishing, pp. 176–210.

Kott, P. S. (1988). Model-based finite population correction for the Horvitz–Thompson estimator. *Biometrika*, **74**, 797–799.

Kremers, W. K. (1986). Completeness and unbiased estimation for sum–quota sampling. *Journal of the American Statistical Association*, **81**, 1070–1073.

Kremers, W. K. (1987). Adaptive sampling to account for unknown variability among strata. *Preprint 128*. Augsburg, Germany: Institut für Mathematik, Universität Augsburg.

Kremers, W. K., and Robson, D. S. (1987). Unbiased estimation when sampling from renewal processes: The single sample and $k$-sample random means cases. *Biometrika*, **74**, 329–336.

Kruskal, W., and Mosteller, F. (1980). Representative sampling: IV. The history of the concept in statistics, 1895–1939. *International Statistical Review*, **48**, 169–195.

Lauritzen, S. L. (1974). Sufficiency, prediction and extreme models. *Scandinavian Journal of Statistics*, **1**, 128–134.

Lawrence, S., and Giles, C. L. (1998). Searching the World Wide Web. *Science*, **280**, 98.

Lehmann, E. L. (1975). *Nonparametrics: Statistical Methods Based on Ranks*. Oakland, CA: Holden-Day.

Lessler, J. T., and Kalsbeek, W. D. (1992). *Nonsampling Error in Surveys*. New York: Wiley.

Levy, P. S. (1977). Optimum allocation in stratified random network sampling for estimating the prevalence of attributes in rare populations. *Journal of the American Statistical Association*, **72**, 758–763.

Levy, P. S., and Lemeshow, S. (1991). *Sampling of Populations: Methods and Applications*. New York: Wiley.

Little, R. J. A. (1982). Models for nonresponse in sample surveys. *Journal of the American Statistical Association*, **77**, 237–250.

Little, R. J. A. (1983). Estimating a finite population mean from unequal probability samples. *Journal of the American Statistical Association*, **78**, 596–604.

Little, R. J. A., and Rubin, D. B. (1987). *Statistical Analysis with Missing Data*. New York: Wiley.

Lohr, S. L. (1999). *Sampling: Design and Analysis*. Pacific Grove, CA: Duxbury.

Lucas, H. A., and Seber, G. A. F. (1977). Estimating coverage and particle density using the line intercept method. *Biometrika*, **64**, 618–622.

Madow, W. G. (1948). On the limiting distributions of estimates based on samples from finite universes. *Annals of Mathematical Statistics*, **19**, 535–545.

Manly, B. F. J. (1984). Obtaining confidence limits on parameters of the Jolly–Seber model for capture–recapture data. *Biometrics*, **40**, 749–758.

Matérn, B. (1960). Spatial variation. *Meddelanden från Statens Skogsforskningsinstitut*, **49**(5).

Matérn, B. (1986). *Spatial Variation*, 2nd ed. Berlin: Springer-Verlag.

McArthur, R. D. (1987). An evaluation of sample designs for estimating a locally concentrated pollutant. *Communications in Statistics—Simulation and Computation*, **16**, 735–759.

McBratney, R. W., Webster, R., and Burgess, T. M. (1981a). The design of optimal sampling schemes for local estimation and mapping of regionalized variables: I. *Computers and Geosciences*, **7**, 331–334.

McBratney, R. W., Webster, R., and Burgess, T. M. (1981b). The design of optimal sampling schemes for local estimation and mapping of regionalized variables: II. *Computers and Geosciences*, **7**, 335–336.

McDonald, J. F., and Palanacki, D. (1989). Interval estimation of the size of a small population from a mark–recapture experiment. *Biometrics*, **45**, 1223–1231.

McDonald, L. L. (1980). Line-intercept sampling for attributes other than coverage and density. *Journal of Wildlife Management*, **44**, 530–533.

Morgan, D. L., and Rytina, S. (1977). Comment on "Network sampling: Some first steps" by Mark Granovetter. *American Journal of Sociology*, **83**, 722–727.

Muller, W. G. (2001). *Collecting Spatial Data: Optimum Design of Experiments for Random Fields*, (2nd ed.) New York: Physica-Verlag.

Munholland, P. L., and Borkowski, J. J. (1993). Adaptive Latin square sampling + 1 designs. Technical Report No. 3-23-93, Department of Mathematical Sciences, Montana State University, Bozeman.

Murthy, M. N. (1957). Ordered and unordered estimators in sampling without replacement. *Sankhyā*, **18**, 379–390.

Murthy, M. N., and Rao, T. J. (1988). Systematic sampling with illustrative examples. In P. R. Krishnaiah and C. R. Rao (eds.), *Handbook of Statistics*, Vol. 6, *Sampling*. Amsterdam: Elsevier Science Publishers, pp. 147–185.

Nathan, G. (1976). An empirical study of response and sampling errors for multiplicity estimates with different counting rules. *Journal of the American Statistical Association*, **71**, 808–815.

Nathan, G., and Holt, D. (1980). The effect of survey design on regression analysis. *Journal of the Royal Statistical Society B*, **42**, 377–386.

Neaigus, A., Friedman, S. R., Goldstein, M. F., Ildefonso, G., Curtis, R., and Jose, B. (1995). Using dyadic data for a network analysis of HIV infection and risk behaviors among injection drug users. In R. H. Needle, S. G. Genser, and R. T. Trotter, II (eds.), *Social Networks, Drug Abuse, and HIV Transmission*. NIDA Research Monograph 151. Rockville, MD: National Institute of Drug Abuse. pp. 20–37.

Neaigus, A., Friedman, S. R., Jose, B., Goldstein, M. F., Curtis, R., Ildefonso, G., and Des Jarlais, D.C. (1996). High-risk personal networks and syringe sharing as risk factors for HIV infection among new drug injectors. *Journal of Acquired Immune Deficiency Syndromes and Human Retrovirology*, **11**, 499–509.

Neyman, J. (1934). On the two different aspects of the representative method: The method of stratified sampling and the method of purposive selection. *Journal of the Royal Statistical Society A*, **97**, 558–606.

Olea, R. A. (1982). Optimization of the high plains aquifer observation network. *Groundwater Series*, No. 7 Lawrence, KS: Kansas Geological Survey, University of Kansas, 73 pp.

Olea, R. A. (1984a). Sampling design optimization for spatial functions. *Mathematical Geology*, **16**, 369–392.

Olea, R. A. (1984b). Systematic sampling of spatial functions. *Series on Spatial Analysis*, No. 7. Lawrence, KS: Kansas Geological Survey, University of Kansas, 57 pp.

Orton, C. (2000). *Sampling in Archaeology*. Cambridge: Cambridge University Press.

Otis, D. L., Burnham, K. P., White, G. C., and Anderson, D. R. (1978). Statistical inference for capture data from closed populations. *Wildlife Monographs*, No. 62.

Otto, M. C., and Pollock, K. H. (1990). Size bias in line transect sampling: A field test. *Biometrics*, **46**, 239–245.

Overton, W. S. (1969). Estimating the number of animals in wildlife populations. In R. H. Giles (ed.), *Wildlife Management Techniques*, 3rd ed. Washington, DC: Wildlife Society, pp. 403–455.

Pathak, P. K. (1962). On simple random sampling with replacement. *Sankhyā A*, **24**, 287–302.

Petrucci, A.(1998). Adaptive sampling for environmental pollution data: Some simulation results. *Statistica Applicata*, **103**.

Pfeffermann, D. (1988). The effect of sampling design and response mechanism on multivariate regression-based predictors. *Journal of the American Statistical Association*, **83**, 824–833.

Pfeffermann, D., and Nathan, G. (1981). Regression analysis of data from a cluster sample. *Journal of the American Statistical Association*, **76**, 681–689.

Pollock, K. H. (1978). A family of density estimators for line transect sampling. *Biometrics*, **34**, 475–478.

Pollock, K. H. (1981). Capture–recapture models: A review of current methods, assumptions, and experimental design. In C. J. Ralph and J. M. Scott (eds.), *Estimating Numbers of Terrestrial Birds*. Studies in Avian Biology, No. 6 Oxford: Pergamon Press. pp. 426–435.

Pollock, K. H. (1991). Modeling capture, recapture, and removal statistics for estimation of demographic parameters for fish and wildlife populations: Past, present, and future. *Journal of the American Statistical Association*, **86**, 225–238.

Pollock, K. H., and Kendall, W. L. (1987). Visibility bias in aerial surveys: A review of estimation procedures. *Journal of Wildlife Management*, **51**, 502–510.

Pollock, K. H., and Otto, M. C. (1983). Robust estimation of population size in closed animal populations from capture–recapture experiments. *Biometrics*, **39**, 1035–1049.

Pollock, K. H., Hines, J. E., and Nichols, J. D. (1984). The use of auxiliary variables in capture–recapture and removal experiments. *Biometrics*, **40**, 329–340.

Pollock, K. H., Nichols, J. D., Hines, J. E., and Brownie, C. (1990). Statistical inference for capture–recapture experiments. *Wildlife Monographs*, **107**, 1–97.

Pontius, J. S. (1997). Strip adaptive cluster sampling: Probability proportional to size selection of primary units. *Biometrics*, **53**, 1092–1096.

Quang, P. X. (1990). Confidence intervals for densities in line transect sampling. *Biometrics*, **46**, 459–472.

Quang, P. X. (1991). A nonparametric approach to size-biased line transect sampling. *Biometrics*, **47**, 269–279.

Quang, P. X. (1993). Nonparametric estimators for variable circular plot surveys. *Biometrics*, **49**, 837–852.

Quang, P. X., and Lanctot, R. B. (1991). A line transect model for aerial surveys. *Biometrics*, **47**, 1089–1102.

Quenouille, M. H. (1949). Problems in plane sampling. *Annals of Mathematical Statistics*, **20**, 335–375.

Quinn, T. J., II (1981). The effect of group size on line transect estimators of abundance. In C. J. Ralph and J. M. Scott (eds.), *Estimating the Numbers of Terrestrial Birds, Studies in Avian Biology*, No. 6. Oxford: Pergamon Press, pp. 502–508.

Quinn, T. J., II and Gallucci, V. F. (1980). Parametric models for line transect estimators of abundance. *Ecology*, **61**, 293–302.

Raj, D. (1956). Some estimators in sampling with varying probabilities without replacement. *Journal of the American Statistical Association*, **51**, 269–284.

Raj, D. (1968). *Sampling Theory*. New York: McGraw-Hill.

Ramsey, F. L. (1979). Parametric models for line transect surveys. *Biometrika*, **66**, 505–512.

Ramsey, F. L., and Scott, J. M. (1979). Estimating population densities from variable circular plot surveys. In R. M. Cormack, G. P. Patil, and D. S. Robson (eds.), *Sampling Biological Populations*. Fairland, MD: International Co-operative Publishing House, pp. 155–181.

Ramsey, F. L., and Sjamsoe'oed, R. (1994). Habitat association in conjunction with adaptive cluster samples. *Environmental and Ecological Statistics*, **2**, 121–132.

Ramsey, F. L., Wildman, V., and Engbring, J. (1987). Covariate adjustments to effective area in variable-area wildlife surveys. *Biometrics*, **43**, 1–11.

Ramsey, F. L., Gates, C. E., Patil, G. P., and Taillie, C. (1988). On transect sampling to assess wildlife populations and marine resources. In P. R. Krishnaiah and C. R. Rao (eds.), *Handbook of Statistics*, Vol. 6 *Sampling*. Amsterdam: Elsevier Science Publishers, pp. 515–532.

Rao, C. R. (1965). *Linear Statistical Inference and Its Applications*, New York: Wiley.

Rao, C. R. (1973). *Linear Statistical Inference and Its Applications*, 2nd ed. New York: Wiley.

Rao, J. N. K. (1973). On double sampling for stratification and analytical surveys. *Biometrika*, **60**, 125–133.

Rao, J. N. K. (1975). Unbiased variance estimation for multistage designs. *Sankhyā C*, **37**, 133–139.

Rao, J. N. K. (1988). Variance estimation in sample surveys. In P. R. Krishnaiah and C. R. Rao (eds.), *Handbook of Statistics*, Vol. 6 *Sampling*. Amsterdam: Elsevier Science Publishers, pp. 427–447.

Rao, P. S. R.S. (1988). Ratio and regression estimators. In P. R. Krishnaiah and C. R. Rao (eds.), *Handbook of Statistics*, Vol. 6, *Sampling*. Amsterdam: Elsevier Science Publishers, pp. 449–468.

Reed, D. J. (1990). Personal communication. Fairbanks, AK: Alaska Department of Fish and Game.

Ripley, B. D. (1981). *Spatial Statistics*. New York: Wiley.

Rivest, L. P., and Crépeau, H. (1990). A two-phase sampling plan for the estimation of the size of a moose population. *Biometrics*, **46**, 163–176.

Robbins, H. (1952). Some aspects of the sequential design of experiments. *Bulletin of the American Mathematical Society*, **58**, 527–535.

Robins, G. L. (1998). Personal attributes in inter-personal contexts: Statistical models for individual characteristics and social relationships. Ph.D. dissertation. University of Melbourne.

Robinson, J. (1987). Conditioning ratio estimates under simple random sampling. *Journal of the American Statistical Association*, **82**, 826–831.

Roesch, F. A., Jr. (1993). Adaptive cluster sampling for forest inventories. *Forest Science*, **39**, 655–669.

Rosen, B. (1972a). Asymptotic theory for successive sampling with varying probabilities without replacement: I. *Annals of Mathematical Statistics*, **43**, 373–397.

Rosen, B. (1972b). Asymptotic theory for successive sampling with varying probabilities without replacement: II. *Annals of Mathematical Statistics*, **43**, 748–776.

Rosenbaum, P. R., and Rubin, D. B. (1983). The central role of the propensity score in observational studies for causal effects. *Biometrika*, **70**, 41–55.

Rothenberg, R. B., Woodhouse, D. E., Potterat, J. J., Muth, S. Q., Darrow, W. W., and Klovdahl, A. S. (1995). Social networks in disease transmission: The Colorado Springs study. In Needle, R. H., Genser, S. G., and Trotter, R. T., II, eds., *Social Networks, Drug Abuse, and HIV Transmission*. NIDA Research Monograph 151. Rockville, MD: National Institute of Drug Abuse, pp. 3–19.

Royall, R. M. (1970). On finite population sampling theory under certain linear regression models. *Biometrika*, **57**, 377–387.

Royall, R. M. (1976a). Likelihood functions in finite population sampling theory. *Biometrika*, **63**, 605–614.

Royall, R. M. (1976b). The linear least-squares prediction approach to two-stage sampling. *Journal of the American Statistical Association*, **71**, 657–664.

Royall, R. M. (1988). The prediction approach to sampling theory. In P. R. Krishnaiah and C. R. Rao (eds.), *Handbook of Statistics*, Vol. 6 *Sampling*. Amsterdam: Elsevier Science Publishers, pp. 399–413.

Royall, R. M., and Cumberland, W. G. (1978). Variance estimation in finite population sampling. *Journal of the American Statistical Association*, **73**, 351–358.

Royall, R. M., and Cumberland, W. G. (1981a). An empirical study of the ratio estimator and estimators of its variance. *Journal of the American Statistical Association*, **76**, 66–77.

Royall, R. M., and Cumberland, W. G. (1981b). The finite population linear regression estimator and estimators of its variance: An empirical study. *Journal of the American Statistical Association*, **76**, 924–930.

Royall, R. M., and Cumberland, W. G. (1985). Conditional coverage properties of finite population confidence intervals. *Journal of the American Statistical Association*, **80**, 355–359.

Royall, R. M., and Eberhardt, K. R. (1975). Variance estimates for the ratio estimator. *Sankhyā C*, **37**, 43–52.

Royall, R. M., and Herson, J. (1973a). Robust estimation in finite populations: I. *Journal of the American Statistical Association*, **68**, 880–889.

Royall, R. M., and Herson, J. (1973b). Robust estimation in finite populations: II. *Journal of the American Statistical Association*, **68**, 890–893.

**356**  REFERENCES

Royall, R. M., and Pfeffermann, D. (1982). Balanced samples and robust Bayesian inference in finite population sampling. *Biometrika*, **69**, 401–409.

Rubin, D. B. (1976), Inference and missing data. *Biometrika*, **63**, 581–592.

Rubin, D. B. (1987). *Multiple Imputation for Nonresponse in Surveys.* New York: Wiley.

Salehi, M. M. (1999). Rao–Blackwell versions of the Hansen–Hurwitz and Horvitz–Thompson estimators in adaptive cluster sampling. *Ecological and Environmental Statistics*, **6**, 183–195.

Salehi, M. M., and Seber, G. A. F. (1997a). Adaptive cluster sampling with networks selected without replacement. *Biometrika*, **84**, 209–219.

Salehi, M, M., and Seber G. A. F. (1997b). Two-stage adaptive cluster sampling. *Biometrics*, **53**, 959–970.

Salehi, M, M., and Seber, G. A. F. (2001). Unbiased estimators for restricted adaptive cluster sampling. *Australian and New Zealand Journal of Statistics*, to appear.

Sampath, S. (2001). *Sampling Theory and Methods.* New Delhi: Narosa Publishing House.

Särndal, C. E. (1978). Design-based and model-based inference in survey sampling. *Scandinavian Journal of Statistics*, **5**, 27–52.

Särndal, C. E. (1980a). A two-way classification of regression estimation strategies in probability sampling. *Canadian Journal of Statistics*, **8**, 165–177.

Särndal, C. E. (1980b). On $\pi$-inverse weighting versus best linear unbiased weighting in probability sampling. *Biometrika*, **67**, 639–650.

Särndal, C. E., and Swensson, B. (1987). A general view of estimation for two phases of selection with applications to two-phase sampling and nonresponse. *International Statistical Review*, **55**, 279–294.

Särndal, C. E., and Wright, R. L. (1984). Cosmetic form of estimators in survey sampling. *Scandinavian Journal of Statistics*, **11**, 146–156.

Särndal, C.-E., Swensson, B., and Wretman, J. (1992). *Model Assisted Survey Sampling.* New York: Springer-Verlag.

Satterthwaite, F. E. (1946). An approximate distribution of estimates of variance components. *Biometrics Bulletin*, **2**, 110–114.

Schafer, J. L. (1997). *Analysis of Incomplete Multivariate Data.* London: Chapman & Hall.

Schnabel, Z. E. (1938). The estimation of the total fish population of a lake. *American Mathematical Monthly*, **45**, 348–352.

Schreuder, H. T., Gregoire, T. G., and Wood, G. B. (1993). *Sampling Methods for Multiresource Forest Inventory.* New York: Wiley.

Schweder, T. (1977). Point process models for line transect experiments. In J. R. Barra, F. Brodeau, G. Romier, and B. Van Cutsem (eds.), *Recent Developments in Statistics.* Amsterdam: North-Holland, pp. 221–242.

Schweder, T. (1989). Independent observer experiments to estimate the detection capacity in line transect sampling of whales. *Technical Report.* Oslo: Norwegian Computing Center, University of Oslo.

Schweder, T., Øien, N., and Høst, G. (1991). Estimates of the detection probability for shipboard surveys of northeastern Atlantic minke whales, based on a parallel ship experiment. *Report of the International Whaling Commission*, **41**, 417–432.

Scott, A. J. (1977). On the problem of randomization in survey sampling. *Sankhyā C*, **39**, 1–9.

Scott, A. J., and Holt, D. (1982). The effect of two-stage sampling on ordinary least squares methods. *Journal of the American Statistical Association*, **77**, 848–854.

Scott, A. J., and Smith, T. M. F. (1973). Survey design, symmetry, and posterior distributions. *Journal of the Royal Statistical Society B*, **35**, 570–60.

Scott, A. J., and Wu, C. (1981). On the asymptotic distribution of ratio and regression estimators. *Journal of the American Statistical Association*, **76**, 98–102.

Seber, G. A. F. (1965). A note on the multiple-recapture census. *Biometrika*, **49**, 339–349.

Seber, G. A. F. (1970). The effects of trap response on tag–recapture estimates. *Biometrika*, **26**, 13–22.

Seber, G. A. F. (1973). *The Estimation of Animal Abundance*. London: Griffin.

Seber, G. A. F. (1977). *Linear Regression Analysis*. New York: Wiley.

Seber, G. A. F. (1979). Transects of random length. In R. M. Cormack, G. P. Patil, and D. S. Robson (eds.), *Sampling Biological Populations*. Fairland, MD: International Co-operative Publishing House, 183–192.

Seber, G. A. F. (1982). *The Estimation of Animal Abundance*, 2nd ed. London: Griffin.

Seber, G. A. F. (1986). A review of estimating animal abundance. *Biometrics*, **42**, 267–292.

Seber, G. A. F. (1992). A review of estimating animal abundance: II. *International Statistical Review*, **60**, 129–166.

Seber, G. A. F., and Manley, B. F. J. (1985). Approximately unbiased variance estimation for the Jolly–Seber mark–recapture census. In B. J. T. Morgan and P.M. North (eds.), *Statistics in Ornithology*. Lecture Notes in Statistics, No. 29. New York: Springer-Verlag, pp. 355–352.

Seber, G. A. F., and Thompson, S. K. (1993). Environmental adaptive sampling. In G. P. Patil and C. R. Rao, (eds.) *Handbook of Statistics*, Vol. 12, *Environmental Statistics*. New York: North-Holland/Elsevier Science Publishers, pp. 201–220.

Sedransk, J., and Smith, P. J. (1988). Inference for finite population quantiles. In P. R. Krishnaiah and C.R. Rao (eds.), *Handbook of Statistics*, Vol. 6, *Sampling*. Amsterdam: Elsevier Science Publishers, pp. 267–289.

Sekar, C. C., and Deming, W. E. (1949). On a method of estimating birth and death rates and the extent of registration. *Journal of the American Statistical Association*, **44**, 101–115.

Sen, A. R. (1953). On the estimate of variance in sampling with varying probabilities. *Journal of the Indian Society of Agricultural Statistics*, **5**, 119–127.

Sen, P. K. (1988). Asymptotics in finite population sampling. In P. R. Krishnaiah and C. R. Rao (eds.), *Handbook of Statistics*, Vol. 6, *Sampling*. Amsterdam: Elsevier Science Publishers, pp. 291–331.

Siegmund, D. (1985). *Sequential Analysis: Tests and Confidence Intervals*. New York: Springer-Verlag.

Silverman, B. W. (1986). *Density Estimation for Statistics and Data Analysis*. London: Chapman & Hall.

Singh, D., and Chaudhary, F. S. (1986). *Theory and Analysis of Sample Survey Designs*. New Delhi: Wiley Eastern.

Sirken, M. G. (1970). Household surveys with multiplicity. *Journal of the American Statistical Association*, **63**, 257–266.

Sirken, M. G. (1972a). Stratified sample surveys with multiplicity. *Journal of the American Statistical Association*, **67**, 224–227.

Sirken, M. G. (1972b). Variance components of multiplicity estimators. *Biometrics*, **28**, 869–873.

Sirken, M. G., and Levy, P. S. (1974). Multiplicity estimation of proportions based on ratios of random variables. *Journal of the American Statistical Association*, **69**, 68–73.

Skinner, C. J. (1983). Multivariate prediction from selected samples. *Biometrika*, **70**, 189–191.

Skinner, C. J., Holt, D., and Smith, T. M. F. (eds.) (1989). *Analysis of Complex Surveys*. New York: Wiley.

Smith, D. R., Conroy, M. J., and Brakhage, D. H. (1995). Efficiency of adaptive cluster sampling for estimating density of wintering waterfowl. *Biometrics*, **51**, 777–788.

Smith, T. M. F. (1976). The foundations of survey sampling: A review (with discussion). *Journal of the Royal Statistical Society A*, **139**, 183–204.

Smith, T. M. F. (1984). Sample surveys, present position and potential developments: Some personal views (with discussion). *Journal of the Royal Statistical Society A*, **147**, 208–221.

Snijders, T. A. B. (1992). Estimation on the basis of snowball samples: How to weight. *Bulletin de Methodologie Sociologique*, **36**, 59–70.

Solomon, H. (1978). *Geometric Probability*. Philadelphia: Society for Industrial and Applied Mathematics.

Solomon, H., and Zacks, S. (1970). Optimal design of sampling from finite populations: A critical review and indication of new research areas. *Journal of the American Statistical Association*, **65**, 653–677.

Spreen, M. (1992). Rare populations, hidden populations, and link-tracing designs: What and why? *Bulletin de Methodologie Sociologique*, **36**, 34–58.

Spreen, M. (1998). Sampling personal network structures: Statistical inference in ego-graphs. Ph.D. dissertation. University of Groningen.

Spreen, M., and Zwaagstra, R. (1994). Personal network sampling, outdegree analysis and multilevel analysis: Introducing the network concept in studies of hidden populations. *International Sociology*, **9**, 475–491.

Starks, T. H. (1986). Determination of support in soil sampling. *Mathematical Geology*, **18**, 529–537.

Steinhorst, R. K., and Samuel, M. D. (1989). Sightability adjustment methods for aerial surveys of wildlife populations. *Biometrics*, **45**, 415–425.

Sudman, S., Sirken, M. G., and Cowan, C. D. (1988). Sampling rare and elusive populations. *Science*, **240**, 991–996.

Sugden, R. A., and Smith, T. M. F. (1984). Ignorable and informative designs in survey sampling inference. *Biometrika*, **71**, 495–506.

Sukhatme, P. V., and Sukhatme, B. V. (1970). *Sampling Theory of Surveys with Applications*. Ames, IA: Iowa State University Press.

Tam, S. M. (1988). Some results on robust estimation in finite population sampling. *Journal of the American Statistical Association*, **83**, 242–248.

Thompson, M. E. (1997). *Theory of Sample Surveys*. London: Chapman & Hall.

Thompson, S. K. (1987). Sampling size for estimating multinomial proportions. *The American Statistician*, **41**, 42–46.

Thompson, S. K. (1988). Adaptive sampling. *Proceedings of the Section on Survey Research Methods of the American Statistical Association*, pp. 784–786.

Thompson, S. K. (1990). Adaptive cluster sampling. *Journal of the American Statistical Association*, **85**, 1050–1059.

Thompson, S. K. (1991a). Adaptive cluster sampling: Designs with primary and secondary units. *Biometrics*, **47**, 1103–1115.

Thompson, S. K. (1991b). Stratified adaptive cluster sampling. *Biometrika*, **78**, 389–397.

Thompson, S. K. (1993). Multivariate aspects of adaptive cluster sampling. In G. P. Patil and C. R. Rao, (eds.), *Multivariate Environmental Statistics*. New York: North-Holland/ Elsevier Science Publishers, pp. 561–572.

Thompson, S. K. (1996). Adaptive cluster sampling based on order statistics. *Environmetrics*, **7**, 123–133.

Thompson, S. K. (1997a). Effective sampling strategies for spatial studies. *Metron*, **55**, 1–21.

Thompson, S. K. (1997b). Spatial sampling (with discussion). In J. Goode (ed.), *Precision Agriculture: Spatial and Temporal Variability of Environmental Quality.* Ciba Foundation Symposium 210. New York: Wiley, pp. 161–172.

Thompson, S. K. (1997c). Adaptive sampling in behavioral surveys. In L. Harrison and A. Hughes (eds.), *The Validity of Self-Reported Drug Use: Improving the Accuracy of Survey Estimates.* NIDA Research Monograph 167. Rockville, MD: National Institute of Drug Abuse, pp. 296–319.

Thompson, S., and Frank, O. (2000). Model-based estimation with link-tracing sampling designs. *Survey Methodology*, **26**, 87–98.

Thompson, S. K., and Ramsey, F. L. (1983). Adaptive sampling of animal populations. *Technical Report 82.* Corvallis, OR: Department. of Statistics, Oregon State University.

Thompson, S. K., and Ramsey, F. L. (1987). Detectability functions in observing spatial point processes. *Biometrics*, **43**, 355–362.

Thompson, S. K., and Seber, G. A. F. (1994). Detectability in conventional and adaptive sampling. *Biometrics*, **50**, 712–724.

Thompson, S. K., and Seber, G. A. F. (1996). *Adaptive Sampling*. New York: Wiley.

Thompson, S. K., Ramsey, F. L., and Seber, G. A. F. (1992). An adaptive procedure for sampling animal populations. *Biometrics*, **48**, 1195–1199.

Tryfos, P. (1996). *Sampling Methods for Applied Research: Text and Cases.* New York: Wiley.

Valkenburg, P. (1990). Personal communication. Fairbanks, AK: Alaska Department of Fish and Game Wildlife Conservation Division.

van Meter, K. M. (1990). Methodological and design issues: techniques for assessing the representatives of snowball samples. In E. Y. Lambert (ed.), *The Collection and Interpretation of Data from Hidden Populations.* NIDA Monograph 98. Rockville, MD: National Institute on Drug Abuse, pp. 31–43.

Wald, A. (1947). *Sequential Analysis*. New York: Wiley.

Wasserman, S., and Faust, K. (1994). *Social Network Analysis: Methods and Applications.* New York: Cambridge University Press.

Watters, J. K., and Biernacki, P. (1989). Targeted sampling: Options for the study of hidden populations. *Social Problems*, **36**, 416–430.

Werner, C., Brantley, S. L., and Boomer, K. (2000). $CO_2$ emissions related to the Yellowstone volcanic system: 2. Statistical sampling, total degassing, and transport mechanisms. *Journal of Geophysical Research*, **105**, 831–846.

Wolter, K. M. (1984). An investigation of some estimators of variance for systematic sampling. *Journal of the American Statistical Association*, **79**, 781–790.

Wolter, K. M. (1985). *Introduction to Variance Estimation*. New York: Springer-Verlag.

Wolter, K. M. (1986). Some coverage error models for census data. *Journal of the American Statistical Association*, **81**, 338–346.

Wolter, K. M. (1991). Accounting for America's uncounted and miscounted. *Science*, **253**, 12–15.

Woodroofe, M. (1982). *Nonlinear Renewal Theory in Sequential Analysis*. Philadelphia: Society for Industrial and Applied Mathematics.

Wright, R. L. (1983). Finite population sampling with multivariate auxiliary information. *Journal of the American Statistical Association*, **78**, 879–884.

Yates, F. (1981). *Sampling Methods for Censuses and Surveys*. New York: Macmillan.

Yates, F., and Grundy, P. M. (1953). Selection without replacement from within strata with probability proportional to size. *Journal of the Royal Statistical Society B*, **15**, 253–261.

Yfantis, E. A., Flatman, G. T., and Behar, J. V. (1987). Efficiency of kriging estimation for square, triangular, and hexagonal grids. *Mathematical Geology*, **19**, 183–205.

Zacks, S. (1969). Bayes sequential designs of fixed size samples from finite populations. *Journal of the American Statistical Association*, **64**, 1342–1349.

Zahl, S. (1989). Line transect sampling with unknown probability of detection along the transect. *Biometrics*, **45**, 453–470.

Zhang, R., Warrick, A. W., and Myers, D. E. (1990). Variance as a function of sample support size. *Mathematical Geology*, **22**, 107–121.

# Author Index

**364**

Sedransk, J., 31, 357
Sekar, C.C., 235, 357
Sen, A.R., 55, 357
Sen, P.K., 61, 235, 357
Siegmund, D., 286, 357
Silverman, B.W., 207, 357
Simmons, M.A., 188, 347
Sinclair, R., 345
Singh, D., 125, 151, 164, 357
Sinha, B.K., 8, 111, 125, 151, 350
Sirken, M.G., 173, 176, 328, 344, 357, 358
Sjamsoe'oed, R., 288, 354
Skinner, C.J., 31, 105, 358
Smith, D.R., 287, 288, 358
Smith, P.J., 32, 357
Smith, T.M.F., 94, 111, 113, 356, 358
Snijders, T.A.B., 183, 184, 349, 358
Snowdon, C.B., 347
Solomon, H., 251, 253, 286, 358
Spreen, M., 184, 358
Starks, T.H., 276, 358
Stein, A., 288, 343, 351
Steinhorst, R.K., 194, 196, 197, 358
Stenger, H., 8, 346
Stevenson, D.C. 8
Sudman, S., 176, 234, 344, 358
Sugden, R.A., 111, 113, 358
Sukhatme, B.V., 8, 76, 125, 347, 358
Sukhatme, P.V., 8, 76, 125, 358
Swensson, B., 164, 169, 356

Taillie, C., 354
Tam, S.M., 93, 345, 358
Tepping, B.J., 350
Thompson, D.J., 53, 350
Thompson, M.E., 8, 31, 33, 62, 93, 97, 111, 288, 349, 358

Thompson, S.K., 8, 44, 105, 111, 113, 184, 197, 273, 275, 276, 279, 287, 288, 289, 290, 299, 300, 302, 310,311, 319, 320, 323,323,324, 333, 334, 335, 336, 337, 346, 348, 357, 358, 359
Tryfos, P., 8, 359

Valkenburg, P., 347, 359
van Meter, K.M., 184, 359
Vos, J.W.E., 55, 104, 286, 346

Wald, A., 286, 359
Warrick, A.W., 360
Wasserman, S., 184, 359
Watters, J.K., 184, 359
Webster, J.T., 121, 343
Webster, R., 352
Welsh, A.H., 210, 343
Werner, C., 288, 344, 359
Wildman, V., 354
Wolter, K.M., 31, 135, 234, 237, 240, 359, 360
Wood, G.B., 356
Woodhouse, D.E., 355
Woodroofe, M., 286, 360
Wretman, J.H., 345, 356
Wright, R.L., 93, 356, 360
Wu, C.F.J., 31, 76, 91, 347, 357

Yates, F., 8, 55, 360
Yfantis, E.A., 272, 360

Zacks, S., 8, 105, 286, 344, 358, 360
Zahl, 226, 360
Zhang, R., 276, 360
Zwaagstra, R., 184, 358

# Subject Index

# WILEY SERIES IN PROBABILITY AND STATISTICS
ESTABLISHED BY WALTER A. SHEWHART AND SAMUEL S. WILKS

Editors
*David J. Balding, Peter Bloomfield, Noel A. C. Cressie, Nicholas I. Fisher,
Iain M. Johnstone, J. B. Kadane, Louise M. Ryan, David W. Scott,
Adrian F. M. Smith, Jozef L. Teugels*
Editors Emeriti: *Vic Barnett, J. Stuart Hunter, David G. Kendall*

The **Wiley Series in Probability and Statistics** is well established and authoritative. It covers many topics of current research interest in both pure and applied statistics and probability theory. Written by leading statisticians and institutions, the titles span both state-of-the-art developments in the field and classical methods.

Reflecting the wide range of current research in statistics, the series encompasses applied, methodological and theoretical statistics, ranging from applications and new techniques made possible by advances in computerized practice to rigorous treatment of theoretical approaches.

This series provides essential and invaluable reading for all statisticians, whether in academia, industry, government, or research.

ABRAHAM and LEDOLTER · Statistical Methods for Forecasting
AGRESTI · Analysis of Ordinal Categorical Data
AGRESTI · An Introduction to Categorical Data Analysis
AGRESTI · Categorical Data Analysis
ANDĚL · Mathematics of Chance
ANDERSON · An Introduction to Multivariate Statistical Analysis, *Second Edition*
*ANDERSON · The Statistical Analysis of Time Series
ANDERSON, AUQUIER, HAUCK, OAKES, VANDAELE, and WEISBERG ·
    Statistical Methods for Comparative Studies
ANDERSON and LOYNES · The Teaching of Practical Statistics
ARMITAGE and DAVID (editors) · Advances in Biometry
ARNOLD, BALAKRISHNAN, and NAGARAJA · Records
*ARTHANARI and DODGE · Mathematical Programming in Statistics
*BAILEY · The Elements of Stochastic Processes with Applications to the Natural
    Sciences
BALAKRISHNAN and KOUTRAS · Runs and Scans with Applications
BARNETT · Comparative Statistical Inference, *Third Edition*
BARNETT and LEWIS · Outliers in Statistical Data, *Third Edition*
BARTOSZYNSKI and NIEWIADOMSKA-BUGAJ · Probability and Statistical Inference
BASILEVSKY · Statistical Factor Analysis and Related Methods: Theory and
    Applications
BASU and RIGDON · Statistical Methods for the Reliability of Repairable Systems
BATES and WATTS · Nonlinear Regression Analysis and Its Applications
BECHHOFER, SANTNER, and GOLDSMAN · Design and Analysis of Experiments for
    Statistical Selection, Screening, and Multiple Comparisons
BELSLEY · Conditioning Diagnostics: Collinearity and Weak Data in Regression
BELSLEY, KUH, and WELSCH · Regression Diagnostics: Identifying Influential
    Data and Sources of Collinearity
BENDAT and PIERSOL · Random Data: Analysis and Measurement Procedures,
    *Third Edition*

*Now available in a lower priced paperback edition in the Wiley Classics Library.

BERRY, CHALONER, and GEWEKE · Bayesian Analysis in Statistics and
    Econometrics: Essays in Honor of Arnold Zellner
BERNARDO and SMITH · Bayesian Theory
BHAT · Elements of Applied Stochastic Processes, *Second Edition*
BHATTACHARYA and JOHNSON · Statistical Concepts and Methods
BHATTACHARYA and WAYMIRE · Stochastic Processes with Applications
BILLINGSLEY · Convergence of Probability Measures, *Second Edition*
BILLINGSLEY · Probability and Measure, *Third Edition*
BIRKES and DODGE · Alternative Methods of Regression
BLISCHKE AND MURTHY · Reliability: Modeling, Prediction, and Optimization
BLOOMFIELD · Fourier Analysis of Time Series: An Introduction, *Second Edition*
BOLLEN · Structural Equations with Latent Variables
BOROVKOV · Ergodicity and Stability of Stochastic Processes
BOULEAU · Numerical Methods for Stochastic Processes
BOX · Bayesian Inference in Statistical Analysis
BOX · R. A. Fisher, the Life of a Scientist
BOX and DRAPER · Empirical Model-Building and Response Surfaces
*BOX and DRAPER · Evolutionary Operation: A Statistical Method for Process
    Improvement
BOX, HUNTER, and HUNTER · Statistics for Experimenters: An Introduction to
    Design, Data Analysis, and Model Building
BOX and LUCEÑO · Statistical Control by Monitoring and Feedback Adjustment
BRANDIMARTE · Numerical Methods in Finance: A MATLAB-Based Introduction
BROWN and HOLLANDER · Statistics: A Biomedical Introduction
BRUNNER, DOMHOF, and LANGER · Nonparametric Analysis of Longitudinal Data in
    Factorial Experiments
BUCKLEW · Large Deviation Techniques in Decision, Simulation, and Estimation
CAIROLI and DALANG · Sequential Stochastic Optimization
CHATTERJEE and HADI · Sensitivity Analysis in Linear Regression
CHATTERJEE and PRICE · Regression Analysis by Example, *Third Edition*
CHERNICK · Bootstrap Methods: A Practitioner's Guide
CHILÈS and DELFINER · Geostatistics: Modeling Spatial Uncertainty
CHOW and LIU · Design and Analysis of Clinical Trials: Concepts and Methodologies
CLARKE and DISNEY · Probability and Random Processes: A First Course with
    Applications, *Second Edition*
*COCHRAN and COX · Experimental Designs, *Second Edition*
CONGDON · Bayesian Statistical Modelling
CONOVER · Practical Nonparametric Statistics, *Second Edition*
COOK · Regression Graphics
COOK and WEISBERG · Applied Regression Including Computing and Graphics
COOK and WEISBERG · An Introduction to Regression Graphics
CORNELL · Experiments with Mixtures, Designs, Models, and the Analysis of Mixture
    Data, *Third Edition*
COVER and THOMAS · Elements of Information Theory
COX · A Handbook of Introductory Statistical Methods
*COX · Planning of Experiments
CRESSIE · Statistics for Spatial Data, *Revised Edition*
CSÖRGÖ and HORVÁTH · Limit Theorems in Change Point Analysis
DANIEL · Applications of Statistics to Industrial Experimentation
DANIEL · Biostatistics: A Foundation for Analysis in the Health Sciences, *Sixth Edition*
*DANIEL · Fitting Equations to Data: Computer Analysis of Multifactor Data,
    *Second Edition*
DAVID · Order Statistics, *Second Edition*

*Now available in a lower priced paperback edition in the Wiley Classics Library.

*DEGROOT, FIENBERG, and KADANE · Statistics and the Law
 DEL CASTILLO · Statistical Process Adjustment for Quality Control
 DETTE and STUDDEN · The Theory of Canonical Moments with Applications in
    Statistics, Probability, and Analysis
 DEY and MUKERJEE · Fractional Factorial Plans
 DILLON and GOLDSTEIN · Multivariate Analysis: Methods and  Applications
 DODGE · Alternative Methods of Regression
*DODGE and ROMIG · Sampling Inspection Tables, *Second Edition*
*DOOB · Stochastic Processes
 DOWDY and WEARDEN · Statistics for Research, *Second Edition*
 DRAPER and SMITH · Applied Regression Analysis, *Third Edition*
 DRYDEN and MARDIA · Statistical Shape Analysis
 DUDEWICZ and MISHRA · Modern Mathematical Statistics
 DUNN and CLARK · Applied Statistics: Analysis of Variance and Regression, *Second
    Edition*
 DUNN and CLARK · Basic Statistics: A Primer for the Biomedical Sciences,
    *Third Edition*
 DUPUIS and ELLIS · A Weak Convergence Approach to the Theory of Large Deviations
*ELANDT-JOHNSON and JOHNSON · Survival Models and Data Analysis
 ETHIER and KURTZ · Markov Processes: Characterization and Convergence
 EVANS, HASTINGS, and PEACOCK · Statistical Distributions, *Third Edition*
 FELLER · An Introduction to Probability Theory and Its Applications, Volume I,
    *Third Edition*, Revised; Volume II, *Second Edition*
 FISHER and VAN BELLE · Biostatistics: A Methodology for the Health Sciences
*FLEISS · The Design and Analysis of Clinical Experiments
 FLEISS · Statistical Methods for Rates and Proportions, *Second Edition*
 FLEMING and HARRINGTON · Counting Processes and Survival Analysis
 FULLER · Introduction to Statistical Time Series, *Second Edition*
 FULLER · Measurement Error Models
 GALLANT · Nonlinear Statistical Models
 GHOSH, MUKHOPADHYAY, and SEN · Sequential Estimation
 GIFI · Nonlinear Multivariate Analysis
 GLASSERMAN and YAO · Monotone Structure in Discrete-Event Systems
 GNANADESIKAN · Methods for Statistical Data Analysis of Multivariate Observations,
    *Second Edition*
 GOLDSTEIN and LEWIS · Assessment: Problems, Development, and Statistical Issues
 GREENWOOD and NIKULIN · A Guide to Chi-Squared Testing
 GROSS and HARRIS · Fundamentals of Queueing Theory, *Third Edition*
*HAHN · Statistical Models in Engineering
 HAHN and MEEKER · Statistical Intervals: A Guide for Practitioners
 HALD · A History of Probability and Statistics and their Applications Before 1750
 HALD · A History of Mathematical Statistics from 1750 to 1930
 HAMPEL · Robust Statistics: The Approach Based on Influence Functions
 HANNAN and DEISTLER · The Statistical Theory of Linear Systems
 HEIBERGER · Computation for the Analysis of Designed Experiments
 HEDAYAT and SINHA · Design and Inference in Finite Population Sampling
 HELLER · MACSYMA for Statisticians
 HINKELMAN and KEMPTHORNE: · Design and Analysis of Experiments, Volume 1:
    Introduction to Experimental Design
 HOAGLIN, MOSTELLER, and TUKEY · Exploratory Approach to Analysis
    of Variance
 HOAGLIN, MOSTELLER, and TUKEY · Exploring Data Tables, Trends and Shapes
*HOAGLIN, MOSTELLER, and TUKEY · Understanding Robust and Exploratory
    Data Analysis

*Now available in a lower priced paperback edition in the Wiley Classics Library.

*Now available in a lower priced paperback edition in the Wiley Classics Library.

*Now available in a lower priced paperback edition in the Wiley Classics Library.

NELSON · Applied Life Data Analysis

NEWMAN · Biostatistical Methods in Epidemiology

OCHI · Applied Probability and Stochastic Processes in Engineering and Physical Sciences

OKABE, BOOTS, SUGIHARA, and CHIU · Spatial Tesselations: Concepts and Applications of Voronoi Diagrams, *Second Edition*

OLIVER and SMITH · Influence Diagrams, Belief Nets and Decision Analysis

PANKRATZ · Forecasting with Dynamic Regression Models

PANKRATZ · Forecasting with Univariate Box-Jenkins Models: Concepts and Cases

*PARZEN · Modern Probability Theory and Its Applications

PEÑA, TIAO, and TSAY · A Course in Time Series Analysis

PIANTADOSI · Clinical Trials: A Methodologic Perspective

PORT · Theoretical Probability for Applications

POURAHMADI · Foundations of Time Series Analysis and Prediction Theory

PRESS · Bayesian Statistics: Principles, Models, and Applications

PRESS and TANUR · The Subjectivity of Scientists and the Bayesian Approach

PUKELSHEIM · Optimal Experimental Design

PURI, VILAPLANA, and WERTZ · New Perspectives in Theoretical and Applied Statistics

PUTERMAN · Markov Decision Processes: Discrete Stochastic Dynamic Programming

*RAO · Linear Statistical Inference and Its Applications, *Second Edition*

RENCHER · Linear Models in Statistics

RENCHER · Methods of Multivariate Analysis, *Second Edition*

RENCHER · Multivariate Statistical Inference with Applications

RIPLEY · Spatial Statistics

RIPLEY · Stochastic Simulation

ROBINSON · Practical Strategies for Experimenting

ROHATGI and SALEH · An Introduction to Probability and Statistics, *Second Edition*

ROLSKI, SCHMIDLI, SCHMIDT, and TEUGELS · Stochastic Processes for Insurance and Finance

ROSS · Introduction to Probability and Statistics for Engineers and Scientists

ROUSSEEUW and LEROY · Robust Regression and Outlier Detection

RUBIN · Multiple Imputation for Nonresponse in Surveys

RUBINSTEIN · Simulation and the Monte Carlo Method

RUBINSTEIN and MELAMED · Modern Simulation and Modeling

RYAN · Modern Regression Methods

RYAN · Statistical Methods for Quality Improvement, *Second Edition*

SALTELLI, CHAN, and SCOTT (editors) · Sensitivity Analysis

*SCHEFFE · The Analysis of Variance

SCHIMEK · Smoothing and Regression: Approaches, Computation, and Application

SCHOTT · Matrix Analysis for Statistics

SCHUSS · Theory and Applications of Stochastic Differential Equations

SCOTT · Multivariate Density Estimation: Theory, Practice, and Visualization

*SEARLE · Linear Models

SEARLE · Linear Models for Unbalanced Data

SEARLE · Matrix Algebra Useful for Statistics

SEARLE, CASELLA, and McCULLOCH · Variance Components

SEARLE and WILLETT · Matrix Algebra for Applied Economics

SEBER · Linear Regression Analysis

SEBER · Multivariate Observations

SEBER and WILD · Nonlinear Regression

SENNOTT · Stochastic Dynamic Programming and the Control of Queueing Systems

*SERFLING · Approximation Theorems of Mathematical Statistics

SHAFER and VOVK · Probability and Finance: It's Only a Game!

SMALL and McLEISH · Hilbert Space Methods in Probability and Statistical Inference

*Now available in a lower priced paperback edition in the Wiley Classics Library.

*Now available in a lower priced paperback edition in the Wiley Classics Library.